Manufacturing Engineering

MANUFACTURING ENGINEERING AND MATERIALS PROCESSING

A Series of Reference Books and Textbooks

SERIES EDITORS

Geoffrey Boothroyd
Chairman, Department of Industrial and Manufacturing Engineering
University of Rhode Island
Kingston, Rhode Island

George E. Dieter
Dean, College of Engineering
University of Maryland
College Park, Maryland

1. Computers in Manufacturing, *U. Rembold, M. Seth and J. S. Weinstein*
2. Cold Rolling of Steel, *William L. Roberts*
3. Strengthening of Ceramics: Treatments, Tests and Design Applications, *Henry P. Kirchner*
4. Metal Forming: The Application of Limit Analysis, *Betzalel Avitzur*
5. Improving Productivity by Classification, Coding, and Data Base Standardization: The Key to Maximizing CAD/CAM and Group Technology, *William F. Hyde*
6. Automatic Assembly, *Geoffrey Boothroyd, Corrado Poli, and Laurence E. Murch*
7. Manufacturing Engineering Processes, *Leo Alting*
8. Modern Ceramic Engineering: Properties, Processing, and Use in Design, *David W. Richerson*
9. Interface Technology for Computer-Controlled Manufacturing Processes, *Ulrich Rembold, Karl Armbruster, and Wolfgang Ülzmann*
10. Hot Rolling of Steel, *William L. Roberts*
11. Adhesives in Manufacturing, *edited by Gerald L. Schneberger*
12. Understanding the Manufacturing Process: Key to Successful CAD/CAM Implementation, *Joseph Harrington, Jr.*
13. Industrial Materials Science and Engineering, *edited by Lawrence E. Murr*
14. Lubricants and Lubrication in Metalworking Operations, *Elliot S. Nachtman and Serope Kalpakjian*
15. Manufacturing Engineering: An Introduction to the Basic Functions, *John P. Tanner*
16. Computer-Integrated Manufacturing Technology and Systems, *Ulrich Rembold, Christian Blume, and Ruediger Dillmann*
17. Connections in Electronic Assemblies, *Anthony J. Bilotta*

18. Automation for Press Feed Operations: Applications and Economics, *Edward Walker*
19. Nontraditional Manufacturing Processes, *Gary F. Benedict*
20. Programmable Controllers for Factory Automation, *David G. Johnson*
21. Printed Circuit Assembly Manufacturing, *Fred W. Kear*
22. Manufacturing High Technology Handbook, *edited by Donatas Tijunelis and Keith E. McKee*
23. Factory Information Systems: Design and Implementation for CIM Management and Control, *John Gaylord*
24. Flat Processing of Steel, *William L. Roberts*
25. Soldering for Electronic Assemblies, *Leo P. Lambert*
26. Flexible Manufacturing Systems in Practice: Applications, Design, and Simulation, *Joseph Talavage and Roger G. Hannam*
27. Flexible Manufacturing Systems: Benefits for the Low Inventory Factory, *John E. Lenz*
28. Fundamentals of Machining and Machine Tools, Second Edition, *Geoffrey Boothroyd and Winston A. Knight*
29. Computer-Automated Process Planning for World-Class Manufacturing, *James Nolen*
30. Steel-Rolling Technology: Theory and Practice, *Vladimir B. Ginzburg*
31. Computer Integrated Electronics Manufacturing and Testing, *Jack Arabian*
32. In-Process Measurement and Control, *Stephan D. Murphy*
33. Assembly Line Design: Methodology and Applications, *We-Min Chow*
34. Robot Technology and Applications, *edited by Ulrich Rembold*
35. Mechanical Deburring and Surface Finishing Technology, *Alfred F. Scheider*
36. Manufacturing Engineering: An Introduction to the Basic Functions, Second Edition, Revised and Expanded, *John P. Tanner*

OTHER VOLUMES IN PREPARATION

Manufacturing Engineering

An Introduction to the Basic Functions

Second Edition, Revised and Expanded

John P. Tanner

John P. Tanner & Associates
Orlando, Florida

CRC Press
Taylor & Francis Group
Boca Raton London New York

CRC Press is an imprint of the
Taylor & Francis Group, an **informa** business

First published 1991 by Marcel Dekker

Published 2020 by CRC Press
Taylor & Francis Group
6000 Broken Sound Parkway NW, Suite 300
Boca Raton, FL 33487-2742

ISBN 13: 978-0-8247-8402-7 (hbk)

Library of Congress Cataloging-in-Publication Data

Tanner, John.
Manufacturing engineering : an introduction to the basic functions / John P. Tanner. -- 2nd ed., rev. and expanded.
p. cm. -- (Manufacturing engineering and materials processing : 36)
Includes bibliographical references and index.
ISBN 0-8247-8402-2
1. Production engineering. I. Title. II. Series.
TS176.T366 1991
670.42--dc20 90-20937
CIP

Preface

The continuing growth and expansion of manufacturing technology worldwide, and the renaissance of American manufacturing in particular in the past several years, emphasize the importance of good manufacturing engineering as the driving force behind this phenomenon. The manufacturing engineers are the key players in this growth. It is they who must design the processes, the methods, and the tooling, then provide the planning, implementation, and follow-up that ensure success. The widespread acceptance of this book as an introduction to manufacturing engineering has led to the need to update, improve, and expand the original edition to keep pace with technology and continue as the primary introductory text and reference source on manufacturing engineering.

The purpose of the book remains the same, to provide the important, practical essentials of manufacturing engineering in a single, convenient reference source for use by all engineers and managers who need to know the basics of this important branch of engineering, as well as by the practicing manufacturing engineer who must review these essentials from time to time. Greater emphasis is now being given to the use of the book as a text for those studying manufacturing engineering in technical colleges and universities. To that end, questions have been added at the end of each chapter. An answer book will be provided for teachers.

A new chapter on manufacturability has been added because of recognition of the favorable impact that sound product design has on manufacturing cost and quality. This chapter covers the essential ingredients for a producible product design and highlights pitfalls to be avoided. Also new is an appendix of basic machining data including feeds and speeds, drill selection tables, sheet metal bend deductions, and similar data.

The chapter on cost estimating has been rewritten and expanded to fully treat all aspects of product cost estimating. New or updated material has been added to the chapters on preproduction planning, capital equipment and facilities, tooling, special processes, methods and detailed work planning, mechanization and automatic assembly and manufacturing engineering systems. The chapter on manufacturing research and development has been completely revised to emphasize the functions, facilities required, and goals and objective of this important aspect of manufacturing engineering. Finally, the appendix on manufacturing engineering terminology has been expanded with many new terms and definitions.

As before, the reader is assumed to have a working knowledge of engineering fundamentals, including some basic knowledge in manufacturing processes and engineering materials. The new material covering statistical process control requires a basic knowledge and understanding of statistics.

John P. Tanner

Contents

Introduction

HISTORY AND DEVELOPMENT OF MANUFACTURING ENGINEERING

Manufacturing engineering is just now beginning to emerge as a separate and recognized engineering discipline. Much of what manufacturing engineering is today and hopes to be in the future stems from the past. The Industrial Revolution, which took place in the eighteenth and nineteenth centuries, also gave birth to an engineering discipline devoted exclusively to the engineering of manufacturing. With the Industrial Revolution came the beginning of the factory system as we know it today, the invention of the steam engine to replace water as the principal source of power, and advances in metal cutting and production of machine tools.

Early pioneers in manufacturing engineering include Matthew Boulton and James Watt, who fabricated and assembled Watt's steam engine and then, as partners, produced the steam engine commercially. Eli Whitney, with his invention of the cotton gin, made it possible to turn out 50 lb of cleaned cotton a day. This revolutionized the textile industry and made cotton the leading industry in the South. Of greater significance was his invention of interchangeable parts manufactured for muskets at his factory near New Haven, Conn. in 1798.

Joseph Whitworth, between 1840 and 1850, developed a measuring machine and a standard system of uniform measures and gauges, and, perhaps of greater importance, a uniform system of screw threads. Whitworth in 1857 perfected a hexagonal-barreled rifle of great range, accuracy, and penetrative power which was far superior to the Enfield, which was then largely in use. In 1865 Whitworth invented a rifled cannon that would fire a 250-lb shell some 6 miles. His invention of compressed cast steel for ordnance manufacture was another outstanding achievement.

Another pioneer in manufacturing engineering was Alfred Krupp, a German inventor and metallurgist, who discovered a method for casting steel into large ingots of up to 2 tons with no flaws. He also developed a new Bessemer steel for the production of weapons and railway wheels, and a method for hardening armor plate.

Frederic W. Taylor was an outstanding early pioneer in both industrial and manufacturing engineering. His achievements in scientific management are perhaps better known than his invention of the Taylor-White process of treating modern high-speed cutting tools, for which he received a gold medal at the Paris Exposition in 1900. He also wrote *The Art of Cutting Metals* in 1905. Taylor's studies on the use of fluids to aid the cutting of metals proved to machinists of his era that a heavy stream of water, flooding the tool-chip-workpiece contact area during cutting, would allow cutting speeds to be increased 30 to 40%.

The primary reason for Taylor's successes was that he had identified all of the variables involved in metal cutting by machine, and then applied the scientific rule of varying only one factor at a time. He identified some 12 variable factors in metal cutting, and from this devised a slide rule for the use of manufacturing engineers which substituted scientific control of the machine shop for the rule-of-thumb methods of the machinists of his day.

The most spectacular results of all were achieved when Taylor began to vary the composition of his steel tools and the degree of heat applied to them before letting them harden in air. From the experience of 14 years of experimental work he concluded that the best comparative basis of tool performance was the exact cutting speed in feet per minute which would cause the tool to be ruined after a 20-min run, all the other factors being held constant. These tool tests were the climax of Taylor's long years of experimentation. Taylor had succeeded in proving that in steel that contained 7% tungsten, 2% carbon, and 2.5% manganese, it was the manganese content which gave the steel its self-hardening characteristic.

Working in conjunction with Maunsel White in the Bethlehem Steel Works in 1892, Taylor made a discovery that astonished the engineering world of the time. He discovered that tungsten chrome tool steel performance would continue to improve with heating until the fusion point was reached. When such a tool was brought to cherry-red heat before cooling in air, a cutting speed of 30 ft/min was enough to cause failure in the prescribed period of 20 min. When a similar tool was heated to 2000°F, this being just below the fusion point, cutting speed could be increased to 80 or 90 ft/min before failure occurred. Up until this time all the recongized authorities on steels had insisted that the new alloy tool steels must not be heated above 1550°F.*

Other early pioneers in manufacturing engineering include Charles Norton, who worked with grinding wheels and grinding machines, Henry Ford, with the mass production of the automobile, and many more

---

*Portions of the previous historical material adapted from L.T.C. Holt, *A Short History of Machine Tools*, MIT Press, Cambridge, 1967.

who have helped evolve the manufacturing engineering discipline to the stature and recognition that it has attained today.

In the early shops and factories of the eighteenth and ninteenth centuries, and in the early twentieth century prior to the World War II, the manufacturing engineering duties and responsibilities were part of the shop supervisor's job. There were no specialists in manufacturing engineering, production control, quality control, industrial engineering, and so on. Methods, processes, and tools were primitive for the most part, and labor was cheap.

As manufacturing organizations grew larger and manufacturing became more complex and specialized, the foreman could no longer perform as a supervisor of the production work force and have the time or knowledge to handle the planning, tooling, material control, or product inspection requirements necessary to keep the shop or factory operating. The need for the technical training and expertise of engineers to plan production, design processes and tools, and select and specify production equipment became more apparent as the industrialization of the Western world continued.

Manufacturing engineering evolved from the earlier traditional disciplines of civil, mechanical, and electrical engineering. Also, many of the technologies of traditional industrial engineering such as plant layout, methods, work measurement, cost analysis, and material handling are part of the manufacturing engineering discipline as we know it today.

Manufacturing engineering is the specialized knowledge and skill in engineering science and analysis applied to the design, operation, and control of manufacturing processes and systems. Included in this broad definition are the following primary activities:

1. Selection and design of manufacturing processes
2. Determination of sequences and methods for product fabrication, assembly, and test
3. Selection and design of production equipment
4. Selection and design of tools and test equipment
5. Layout of factory buildings, machines, equipment, materials, and storage facilities
6. Determination of standard times for manufacturing operations
7. Selection and design of manufacturing systems and computer-aided manufacturing techniques
8. Manufacturing cost estimating, cost analysis, and cost trade studies
9. Manufacturing research and development
10. Review of product designs and specifications to ensure manufacturing producibility
11. Management, coordination, and control of manufacturing operations

Manufacturing engineering is defined by the Society of Manufacturing Engineers (SME) as "that specialty of professional engineering which requires such education and experience as is necessary to understand, apply, and control engineering procedures in manufacturing processes and methods of production of industrial commodities and products, and requires the ability to plan the practices of manufacturing, to research and develop the tools, processes, machines and equipment, and to integrate the facilities and systems for producing quality with optimal expenditure."

The manufacturing engineer provides the link between engineering design and product build. He has the responsibility for the translation of highly technical engineering information into simple, easy-to-understand work instructions enabling production operators to easily understand the tasks they must perform to build the product. Not only must the manufacturing engineer determine the manufacturing process, but it must be the most efficient and most economical process. He must determine how the manufacturing process and materials must be configured to ensure maximum productivity. The manufacturing engineer must consider all manufacturing aspects: what is to be manufactured, in what quantity, what the schedule requirements are, how it will be built, and what it takes to build it.

It has only been in the past few years that there has been college-level, professional engineering training leading to a degree in manufacturing engineering. Now fully accredited programs in manufacturing engineering are offered at the associate, bachelor's, and master's levels. Associate and bachelor's degrees are also offered in manufacturing technology. Most manufacturing engineers received their college-level training in either industrial, mechanical, or electrical engineering. Many of the professional engineering societies such as the Society of Manufacturing Engineers offer specialized training and certification to practicing manufacturing engineers through seminars, short courses, and home study programs.

Certification indicates an engineer's ability to meet a given set of standards related to the many aspects of manufacturing engineering. In addition to academic requirements, certification pertains to the actual experience required of the manufacturing engineer. To meet the SME requirements for certification as a manufacturing engineer, an applicant must have a good record of 10 years or more in responsible charge of important manufacturing engineering projects, which have included:

—Planning and selecting methods for manufacture
—Designing equipment for manufacturing
—Research and development leading to the creation of new manufacturing methods or the improvement of existing processes
—Administration of manufacturing operations covering long-range planning, materials, production control, product configuration, manufacturing facilities, and production management systems

The Institute of Industrial Engineers has established separate divisions for industrial engineers working in manufacturing systems, production and inventory control, facilities planning and design, work measurement, and methods engineering. The American Society of Mechanical Engineers and the Institute of Electrical and Electronic Engineers have sections specializing in manufacturing engineering.

A manufacturing engineer is, by definition of the Society of Manufacturing Engineers, an engineer who has acquired certain standards of professional accomplishment. Many states now offer professional registration to manufacturing engineers. Manufacturing engineering is a branch of engineering which a candidate for registration can be examined and then licensed.

In our high-technology society, the professional manufacturing engineer is a key participant. His knowledge, expertise, and experience, along with those of professional engineers in other branches of engineering, make our standard of living what it is today. From the high-volume mass production of consumer goods to the manufacture of a highly complex military weapons system, the manufacturing engineer plays an important role. In the years to come the continued advance of the technology of manufacturing will depend upon the contribution of competent professional manufacturing engineers.

SCOPE OF MANUFACTURING ENGINEERING

All manufacturing is conversion: the conversion of raw materials into finished goods, the conversion of coal and iron ore into steel, and then the conversion of the steel into a modern automobile. The mining of bauxite and its conversion into aluminum bar stock, which is ultimately converted into screw machine parts, is among the almost limitless examples.

The manufacturing engineer must be there, planning and tooling the process of manufacture. Manufacturing engineers are found in almost every industry where manufacturing takes place. If the industry is highly specialized and has a technology all its own, then the manufacturing engineers in that industry must be knowledgeable and experienced in that technology. This results in a high degree of specialization in the manufacturing engineering field, yet there is a common, fundamental discipline of manufacturing engineering used by all practitioners, and that is what this book is about.

Manufacturing engineering grew and developed as a genuine engineering discipline in the metalworking industry, because in the early years of this century that was where the frontier of industrial technology existed. This frontier later moved to electronics, and now today into computers. Manufacturing engineers are found in mass production and assembly, such as in the automotive and consumer electronics industries.

The vast majority of industry today, in the modern industrial world, consists of job shop or job order manufacturing. Job order manufacturing offers a special challenge to the manufacturing engineer, in that production runs are short, and the costs of job planning and setup can easily eliminate job profitability if they are not controlled.

The majority of manufacturing engineers today work in private industry. Some few work for government agencies, educational institutions, or management and engineering consulting firms. They are found in the manufacturing of primary metal products, machinery, electrical and electronic equipment, transportation equipment, and miscellaneous manufactured products. Job responsibilities include developing methods of manufacture, designing tools and equipment for manufacturing, as well as administrative and supervisory responsibilities.

The manufacturing engineering function is subdivided into various areas of specialization. How the function is subdivided depends largely on the needs of the corporate structure. The subdivisions include process engineering, tool engineering, standards, plant engineering, administration and control, and research. The next section, on the organization of the manufacturing engineering department, will show how this should be done for best results.

ORGANIZING FOR MANUFACTURING ENGINEERING

As in organizing for any other company function, the same basic principles of organization must be followed. These principles include as a minimum:

1. Authority, responsibility, and acountability must be clearly defined.
2. The organization function must be integrated into the company plan of organization.
3. The new department must be able to perform its assigned tasks effectively.

The table of contents of this book shows a wide variety of activities that a manufacturing engineering department can perform. In many instances these activities may not all be included in the tasks assigned to a given company manufacturing engineering department.

Place in Company Organization

The scope of activities assigned in a given company is a prime factor in deciding where to fit the manufacturing engineering department in the plan of organization and how to organize it internally. In the majority of cases the manufacturing engineering department is part of the manufacturing or operations division, and reports to the chief operations or manufacturing executive. This is the recommended reporting level. Manufacturing engineering should be on the same organizational level as manufacturing shops, production control, material control, and plant engineering. This is the most effective organization placement for most companies. Figure 1 shows such an organization plan.

Often we find manufacturing engineering called production engineering, process engineering, or industrial and production engineering. The function is still the same: to provide manufacturing engineering services. Today, as never before, industry depends upon the manufacturing engineer to start the wheels of production turning. The proper placement of manufacturing engineering in the organization ensures that this will take place.

Internal Department Organization

The manufacturing engineering department may have an overall organization paralleling that of the shop or production operations department, or it may be organized into subdivisions by related functions. There are advantages and disadvantages associated with both forms of organization. First, the advantages of such an organization structure are as follows:

1. Engineers working on all aspects of an operating department's problems receive very broad training and a good understanding

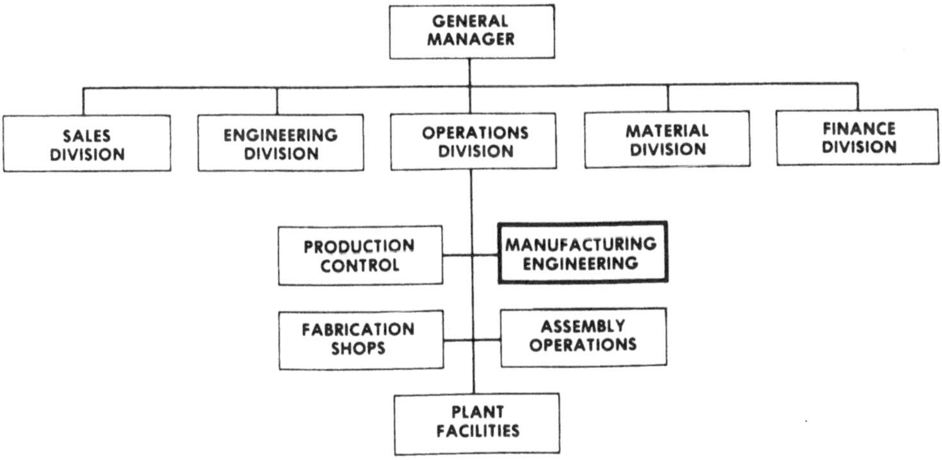

FIG. 1 Manufacturing engineering in the overall company organization.

of how manufacturing engineering services sustain and complement the operating department they serve.
2. This form of organization provides excellent specialized training for engineers for subsequent transfer as line supervisors of the department served.
3. This arrangement ensures that the highest priority problem in the shop department gets first attention.
4. Good rapport and a good working relationship usually develop between manufacturing engineers and shop supervision.
5. Since manufacturing engineering is highly technical and specialized, this arrangement allows assignment of the most qualified engineers to the appropriate shop operating department.

Figure 2 shows such an organization plan. The disadvantages of this kind of organization include the following:

1. Individual engineering proficiency in any one discipline is not likely to be obtained, since total support is required to cover the entire spectrum of services provided by manufacturing engineering.
2. Manufacturing engineering must be staffed with the degree and depth of experience needed to successfully support such an organizational arrangement.

In the department organization that groups related functions, the primary advantages are as follows:

1. Manufacturing engineers specialize within a narrow field, thus rapidly attaining a high degree of technical proficiency.
2. This organization form allows use of engineers with more limited experience because of the limited range of assigned duties.
3. Manufacturing engineers can easily be assigned to tasks that best suit their talents, interest, and experience.
4. This specialization makes it easier to recruit and staff the manufacturing engineering group because fewer engineers with heavy, diversified background experience are required.
5. A department organization that groups related functions is easier to manage and control.

Figure 3 shows a typical manufacturing engineering department organization that groups related functions. Disadvantages of this form of organization include:

1. The engineer specialist is less likely to be able to relate the effect of his work to the overall operation of the shop department he is supporting.

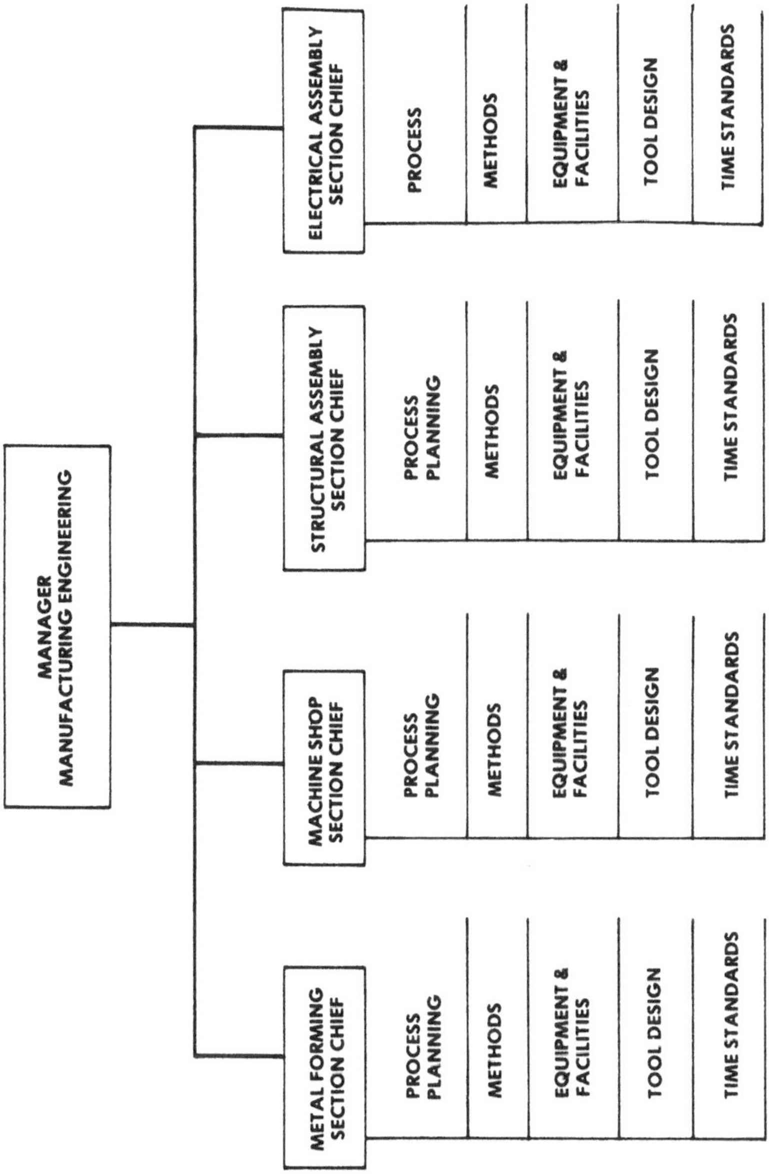

FIG. 2 Manufacturing engineering organization paralleling shop organization.

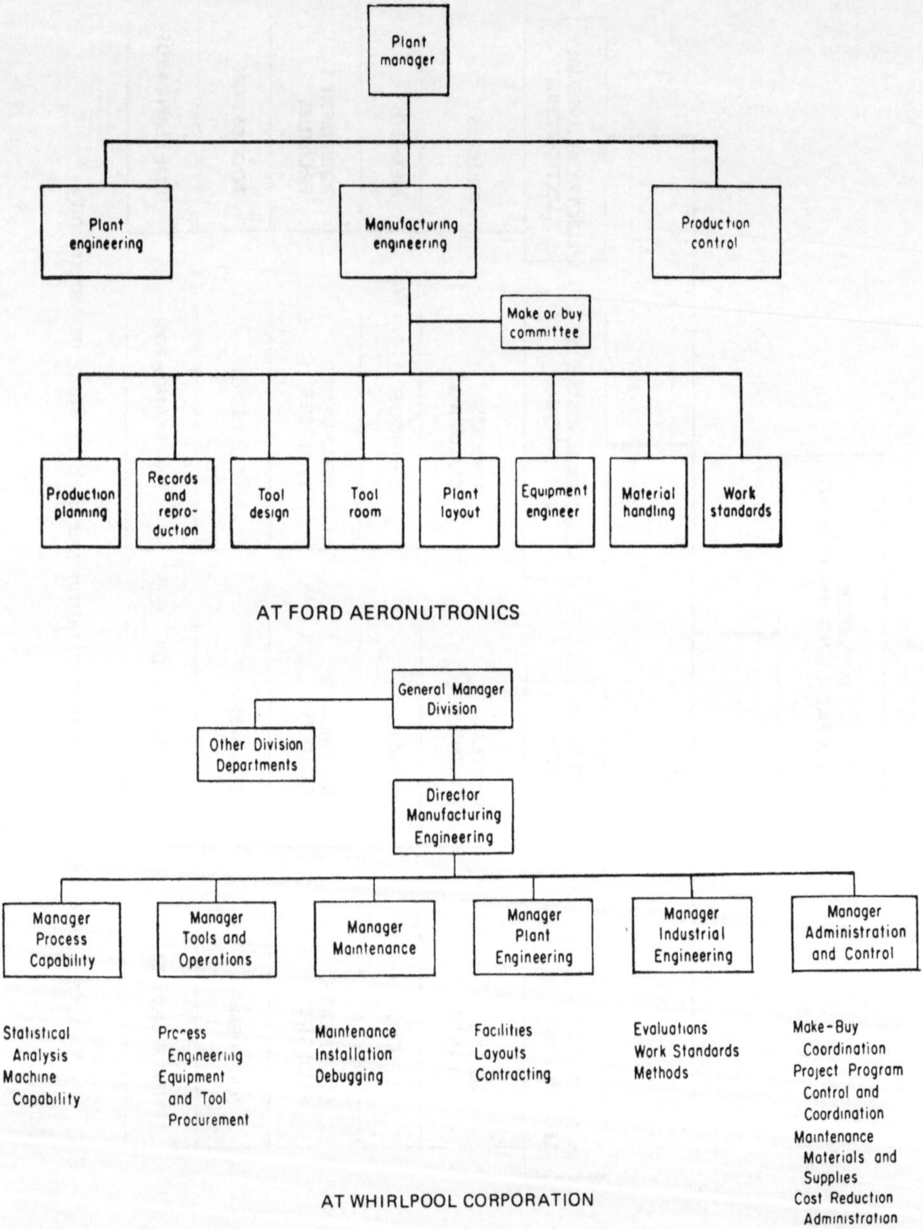

FIG. 3 Two related-function manufacturing engineering organizations. (Courtesy of the Ford Motor Company, Dearborn, Mich., and the Whirlpool Corporation, Benton Harbor, Mich.)

2. Because the manufacturing engineers are specialists and work in all shop departments, close working relationships are seldom developed with line management in any one department.
3. Cross-training of manufacturing engineers is more difficult, and the ability to shift personnel to areas of need from other specialized assignments is limited in the manufacturing engineering group.

Whichever organization form is selected, it should be pointed out that the degree of complexity of the manufacturing operations, not the organization form, is the determining factor in the amount of manufacturing engineering support required and, ultimately, the size of the manufacturing engineering department required.

There is no one best department organization. The department organization that parallels the shop organization is recommended for plants which primarily work to job order or short-run production, and the organization that groups related functions is best for plants where production runs are long and the number of jobs or product lines is small. The number and variety of internal organization forms of a manufacturing engineering department are almost infinite. Company size, however, plays a key role in determining department size and organization. The small company will typically have a senior manufacturing engineer reporting directly to the manufacturing manager or superintendent, with duties consisting of planning product manufacture and designing and ordering the tooling and production machines and equipment. Reporting to the senior manufacturing engineer may be a tool designer and a junior engineer. The senior manufacturing engineer in this situation must have heavy experience and be a qualified professional in every sense of the word.

In the medium-sized company, the organization of the manufacturing engineering department should take the form shown in Fig. 4. The various sections parallel the shop organization because of the need for a close working relationship with shop supervision. The manufacturing engineering department would be headed by a manager who reports to a director of manufacturing or operations. Sections within the department should be headed by section chiefs. Grade levels ranging from associate engineer, through engineer and senior engineer, up to lead engineer would be appropriate. Technician grades would include analyst, tool designer, and numerical control (N/C) programmer. Even though the organization is more formal, the emphasis is the same as for the small company. The need for the key positions to be filled by qualified professionals with heavy experience is extremely important.

The large-company organization should take the form shown in Fig. 5. A large company would have 900 or more employees, and the medium-sized company described above would have from 300 to 900 employees. The small company would have fewer than 300 employees. The large company with greater resources would employ more specialists

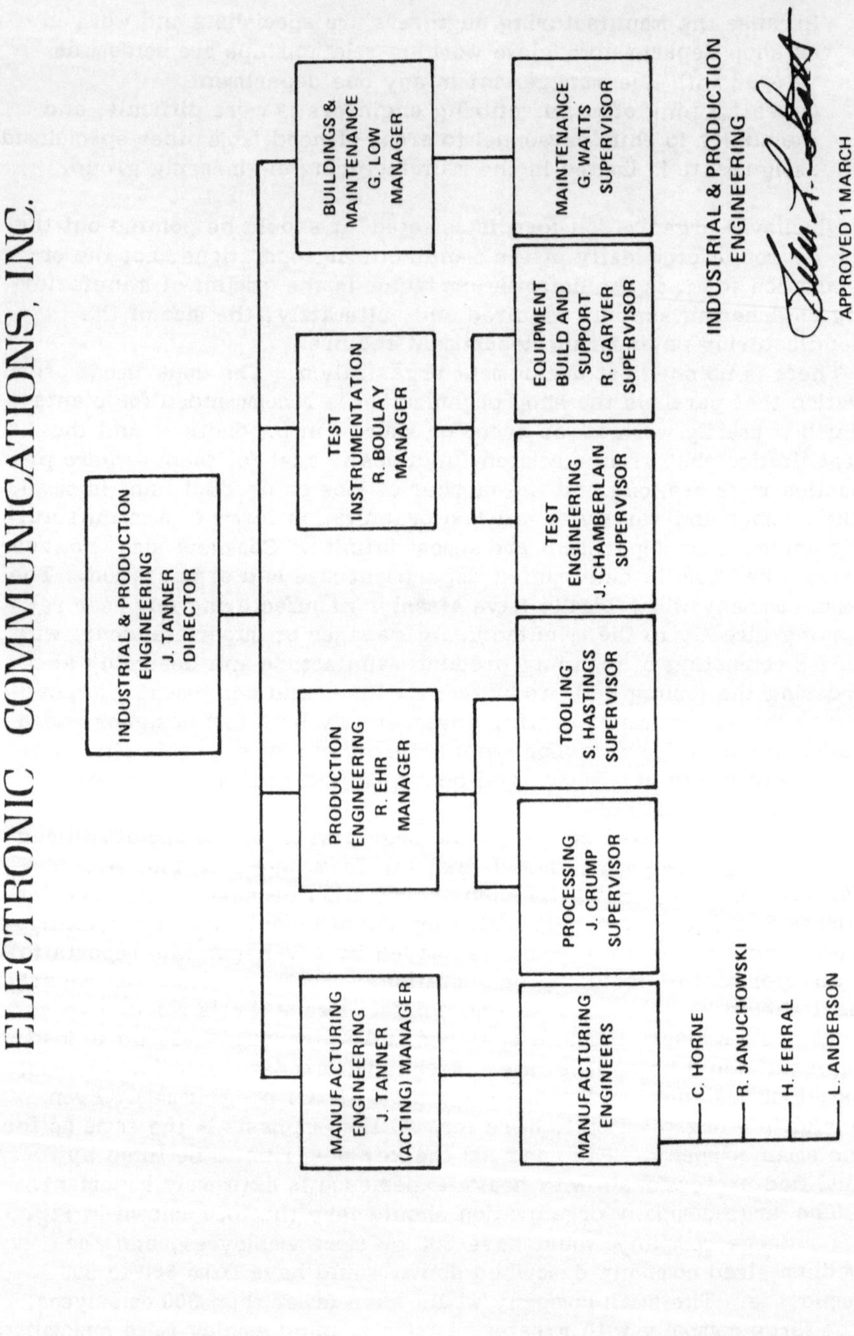
ELECTRONIC COMMUNICATIONS, INC.
INDUSTRIAL & PRODUCTION ENGINEERING J. TANNER DIRECTOR
MANUFACTURING ENGINEERING J. TANNER (ACTING) MANAGER
PRODUCTION ENGINEERING R. EHR MANAGER
TEST INSTRUMENTATION R. BOULAY MANAGER
BUILDINGS & MAINTENANCE G. LOW MANAGER
MANUFACTURING ENGINEERS
PROCESSING J. CRUMP SUPERVISOR
TOOLING S. HASTINGS SUPERVISOR
TEST ENGINEERING J. CHAMBERLAIN SUPERVISOR
EQUIPMENT BUILD AND SUPPORT R. GARVER SUPERVISOR
MAINTENANCE G. WATTS SUPERVISOR
C. HORNE
R. JANUCHOWSKI
H. FERRAL
L. ANDERSON
INDUSTRIAL & PRODUCTION ENGINEERING
APPROVED 1 MARCH

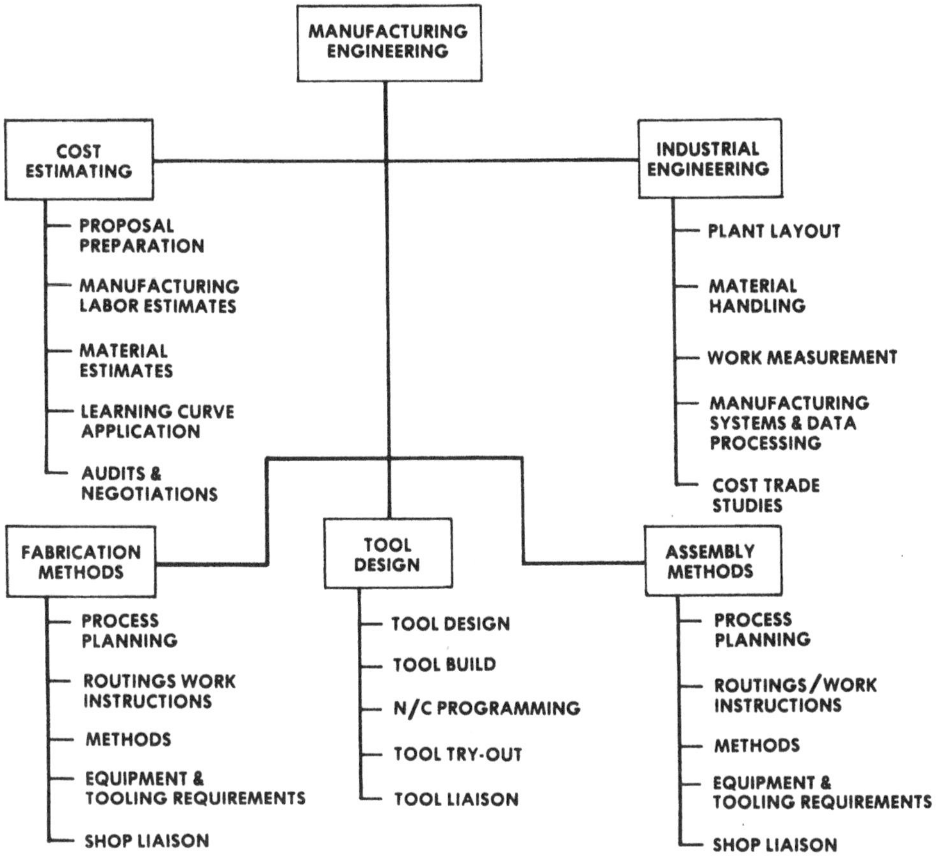

FIG. 5 Typical large-company manufacturing engineering organization.

FIG. 4 Related-function manufacturing—engineering organization in a medium-sized company. (Courtesy of the Electronic Communications Division of E Systems Corporation, St. Petersburg, Fl.)

in manufacturing engineering on its staff. In the larger organization more time and emphasis are placed upon such functions as manufacturing research and development, producibility review of new product design, and cost estimating for new work. Usually the larger company's manufacturing engineering organization is structured in the grouping of related functions pattern. The greater depth and diversity in projects and services performed, increased professional staff, and more specialized staff lead to potential problems in control of various activities, which makes the grouping by related function the preferable organizational structure. The large company organization may be headed by a director of manufacturing engineering who reports to the manufacturing or operations vice president. Department sections would be headed by managers, and subsections by supervisors or chiefs. Other job titles and levels by grade would be the same as in the medium-sized company.

Interface with Other Functional Departments

For the manufacturing engineering department to be truly effective, it must work effectively, and with the least amount of friction possible, with other key manufacturing and support departments. The most critical working interface is with the production control department. Production control schedules and monitors the progress of all manufacturing operations. Its kits and issues material to production, based upon the operation sequences and steps called out by manufacturing engineering in routings, operation work instructions, and operation parts lists. It issues and controls tools according to the call-outs by manufacturing engineering. Most important to manufacturing engineering, it reproduces, issues, updates status, controls, and acts as the central file for all operation routings, operators work instructions, operation parts lists, work orders, and other shop paperwork generated by the manufacturing engineering department. Production control and manufacturing engineering cannot perform without each other's help and support. They must work together effectively because they are the primary support and service organizations for the manufacturing shops and operating departments.

The manufacturing engineering department is a service organization. Its primary function is to support the manufacturing operating departments with production planning, processes, tooling, work instructions, work methods, labor standards, production equipment, and help in resolving manufacturing problems. The working interface with the manufacturing operating departments must ensure rapid and accurate exchange of information, a feeling of trust and mutual support, and respect for the problems and responsibilities of the other. There must be confidence that manufacturing engineering will be there when needed to help with problems, and that the total service provided by manufacturing engineering is and will continue to be of the highest quality.

Other critical functional interfaces for manufacturing engineering include quality control and design engineering. Quality control determines inspection and test sequences which are incorporated into manufacturing routings. Quality control also participates in tool try-out and first article build prove-out. There must be free and open flow of information and support between the two departments. New products must be designed before they can be manufactured. The released design configuration is the primary determinant of manufacturing cost. That is why the manufacturing engineering interface with design engineering is so important. Information on how to make the new design producible must be given to design engineering in sufficient time to be acted upon, before the new design is frozen and the drawings are released.

Other important functional interfaces for manufacturing engineering include purchasing and accounting. Purchasing provides feedback on outside material and tooling costs, orders production equipment as determined and specified by manufacturing engineering, and may require the services of manufacturing engineering to survey the production capabilities and facilities of prospective suppliers and subcontractors. Accounting provides historical cost data, correct labor rates and burden rates, and equipment depreciation information needed for cost estimating and cost trade studies.

Manufacturing Engineering in the Matrix Organization

A matrix organization attempts to combine the advantages of a functional organization structure with those of a project or program organization structure. In this kind of organization each program manager reports directly to the company president or plant general manager. Each program manager has total responsibility and accountability for his program's success. The functional departments headed by managers also report directly to the company president or plant general manager. Each functional manager has full responsibility and accountability for the performance of his functional department. Figure 6 shows how a matrix organization structure works. There are certain ground rules that must be followed for the matrix to work:

1. The program managers must spend full time on their programs.
2. There must exist both horizontal and vertical control and communications.
3. All managers must have an input into the planning process.
4. All managers in the matrix must be willing to negotiate for resources.
5. The horizontal line of control and communications must be permitted to operate as a separate entity, except for administrative purposes.

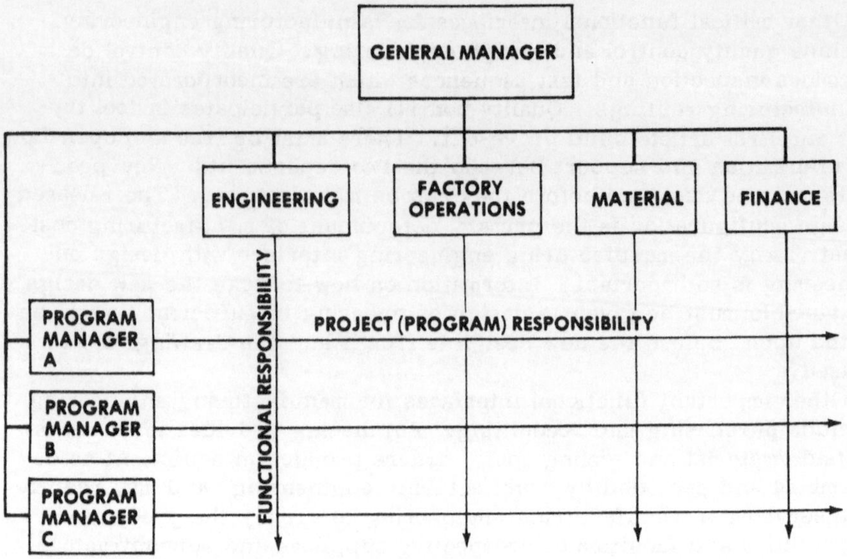

FIG. 6 Matrix organization structure.

In a number of large companies where the matrix organization is used extensively, the manufacturing engineering department will also be organized as a matrix. Program manufacturing engineers will be assigned to head up the manufacturing engineering effort on each program. They become in effect program managers for manufacturing engineering, reporting directly to the director of manufacturing engineering. They are assigned resources from the various functional sections of manufacturing engineering. At the same time, the program manufacturing engineer would report directly to his overall program manager for his program direction and his budget and, in effect, would represent manufacturing engineering on the program.

The manufacturing engineering program team would include electrical manufacturing engineers, structural manufacturing engineers, tool designers, industrial engineers, and production test engineers, all reporting to a program manufacturing engineer.

SOURCE, TRAINING, AND DEVELOPMENT OF MANUFACTURING ENGINEERS

Until recent years, most manufacturing engineers came up through the ranks, with many working first as skilled machinsts or semiskilled production operators and then advancing to group leader, assistant foreman, and foreman before starting in the manufacturing engineering department as a trainee. Formal training as an engineer, while highly desirable, was the exception rather than the rule. The years spent in the shops and the operating departments provided an experience background in key manufacturing technologies which could not be obtained in any other way.

As manufacturing processes became more complex, requiring formal training in engineering and technology as well as in computer science, and with the rapid growth of computer-controlled processes, the manufacturing engineering staff must of necessity also include personnel with formal training in engineering science and technology. The ideal manufacturing engineer should be a graduate professional engineer with three to five years of hands-on shop work experience as a skilled or semiskilled operator or setup man. He should also have in-depth training in computer science and programming and a working knowledge of the more common computer-controlled processes. Finding manufacturing engineering professionals with all of these qualifications is very difficult.

The opportunity should continue to exist for skilled and semiskilled shop personnel to move into manufacturing engineering, but they should be given formal training in engineering fundamentals and computer science after regular working hours, on their own time but at company expense. Their advancement in the manufacturing engineering organization above the associate engineer level would be contingent upon completing this training. The recent engineering graduate interested in a manufacturing engineering career should, as part of his training, be assigned to work in the shop as an apprentice for a short period of time to become familiar with shop operations, processes, and equipment. This kind of shop familiarization training should also be given to newly hired, experienced engineers before they undertake any assignment in the organization.

There is also a continuing need for training to enhance the skills of experienced manufacturing engineers and to keep them abreast of the latest advances in their particular manufacturing technology. This training can be accomplished by attendance at short courses and seminars, by attending trade shows and exhibits, and by attending in-house training courses specifically designed for this purpose.

EFFECTIVE MANAGEMENT OF THE MANUFACTURING ENGINEERING FUNCTION

The manufacturing engineering manager and his supervisors and chiefs provide the leadership and direction needed to make the department function as a team and provide the quality and level of service required so that the manufacturing shops and operating departments can effectively build and ship the manufactured product. The manager must establish department goals and objectives and must organize and staff the department effectively. He must plan the work, establish targets and deadlines, and must provide for necessary controls and coordination.

An important part of the manufacturing engineering manager's job is the establishment of policies and operating procedures for his department, within the framework of company policies and procedures. These policies and procedures establish how the department does business. If policies and procedures are cumbersome and poorly thought out and coordinated, the results will be poor performance and the inability to react quickly and effectively to shop needs or problems. In many manufacturing organizations, product configuration is controlled by the manufacturing engineering documentation on the shop floor, and the temptation is strong to create a paperwork system that guarantees no possibility for error, either by the operator or by the manufacturing engineer. Not only is this impossible to do, but the cost of maintaining and servicing such a system is prohibitive. Many otherwise capable manufacturing engineering departments are so involved in "the paper chase" that there is little time left to do the important manufacturing engineering work. Even the computer cannot solve the problem if the system is too cumbersome, and in fact the introduction of the system on the computer may even create more problems.

A well-organized and well-managed manufacturing engineering department will have a system of checks and balances which ensure that necessary work is accomplished on schedule, that it is done correctly, and that the cost of doing the work is within budget. This can be done quite well where the work can be planned and scheduled and can be assigned to available engineering staff, whose availability for the assignment was planned for. Many times, especially in high-technology manufacturing firms, unplanned engineering changes occur in rapid succession after the drawings have been formally released, the routings, tools, and work instructions are on the shop floor, material has been purchased, and production build of the product has already begun. The manufacturing engineering manager must respond quickly to these changes by assessing the impact on documentation, tooling, material, and work in process, and deciding on means of effecting the changes which minimize the impact on production.

He must then assign the work of accomplishing the change to staff already doing scheduled work so as to minimize the impact on his planned and scheduled work. In some companies that are primarily engaged in the design and development of new products which have never been in production, this unplanned change activity can account for more than half of the work in the manufacturing engineering department. The manager should insist that design engineering provide him with projected change curves showing anticipated numbers of engineering changes by month over the life of the program. With this limited visibility, he can justify the budget and additional staff needed for this effort.

The manufacturing engineering department exists primarily to service and support the manufacturing operating departments. A key measure of its performance is how well the plant or company does in building quality products, on schedule, and at optimum cost. The sum total of production planning, process design, equipment selection, tooling, and support to manufacturing is reflected in the efficient output of production. This needs to be continuously kept in mind by the manager of manufacturing engineering as he plans, directs, and controls the services performed by his department.

An excellent checklist used to evaluate how well the manufacturing engineering function is being managed has been developed by the Naval Surface Weapons Center in Dahlgren, Va., for determining whether a prospective contractor is ready to go into production on a newly designed weapon system. Parts of this checklist are shown in Table 1. The list focuses on performance and tangible results.

TABLE 1 Manufacturing Engineering Production Readiness Assessment

1. Determine whether the product as designed and developed is producible within the current state of the art and contractor's capability.
2. Determine whether design status is frozen. Identify risk and impact of undefined portions.
3. Determine extent and application of risk analysis. Were production capability problems resolved during the design stages? Will production methods or techniques be required?
4. Is manufacturing engineering "in the loop" for assessing the impact of proposed design changes prior to change implementation?
5. Are design changes effectively controlled and efficiently incorporated into the production system?
6. Have potential production bottlenecks been identified?
7. Evaluate the system of tool tryout to determine whether adequate controls exist before tools are put into production.
8. Review the system used to develop methods and processes in manufacturing engineering. Is producibility engineering brought into the design phase early enough to ensure producible hardware?
9. Determine whether the company has a recognized manufacturing methods program.

TABLE 1 (Continued)

10.	Determine the levels of approval required to initiate a study of manufacturing methods.
11.	Determine utilization of the latest manufacturing methods and techniques.
12.	Identify and evaluate manufacturing engineering computer techniques.
13.	Determine whether specified methods or processes have been given trial runs prior to initiation of production and assess effectiveness.
14.	Determine that planned production equipment will produce quality parts and support the planned production rate.
15.	Evaluate the tooling philosophy for the planning, design, fabrication, and control of special tools and test equipment.
16.	Evaluate planned use of numerically controlled machines for effectiveness and efficiency of operations.
17.	Evaluate the amount of tool design and fabrication which is subcontracted.

Courtesy of Naval Surface Weapons Center, Dahlgren, Va.

THE FUTURE OF MANUFACTURING ENGINEERING

According to a study funded by the Society of Manufacturing Engineers, in future years rapid and massive changes will take place in the factory and in manufacturing as we know it today. These include:

1. The continued expansion of applied computer-oriented technology to manufacturing
2. The automation of groups of operations in manufacturing, together with automated assembly, automatic inspection, and the application of adaptive controls and diagnostics, leading ultimately to the computer-integrated factory
3. Group technology
4. Expanded use of lightweight materials in products
5. Development of energy-efficient manufacturing methods and materials-conservative product designs
6. Compliance with regulatory or legal requirements related to product safety, worker hazards, and manufacturing wastes, effluents, and emissions
7. Job enrichment in the manufacturing industry

These changes will result in changes in job responsibilities of manufacturing engineers and in a stronger orientation toward formal, professional training in engineering and computer sciences, as opposed to practical training and experience.

Again, according to the SME study, the major recent influence on the manufacturing engineer has been the development of computer-based manufacturing techniques. Computers are being applied to transfer line operations, flexible manufacturing systems, and stand-alone N/C machines. The increased acceptance, utilization, and refinement of these techniques have prompted industry to consider manufacturing increasingly from the systems viewpoint. Recent concepts that have emerged from the application of the systems approach include automated factories, industrial robots, flexible manufacturing systems, and group technology.

Also, wider utilization of nontraditional machining techniques and the development of new techniques are forecast. Application of lasers and electron beams for cutting, welding, and hardening is expected to increase in the future by a significant amount. New developments in adhesives for materials joining and new techniques for forming and shaping are anticipated. Major advances in machine tool technology are also expected.

The manufacturing engineer will be confronted with issues related to factory operations that have developed over the past decade. These issues concern job safety, environmental protection, energy conservation, and worker job satisifaction. Changing job roles for the manufacturing engineer will include stronger emphasis on cost justification and the development of external constraints on manufacturing operations. Manufacturing engineers are being assigned increased responsibility for planning, designing, and justifying the overall system for design and production of products. Use of the team approach in industry is growing and is breaking down the distinction between engineering disciplines. Closer collaboration between technical personnel performing related functions is taking place.

QUESTIONS

1. Who were some of the early pioneers in manufacturing engineering, and what were they noted for?
2. What is the definition of manufacturing engineering adopted by the Society of Manufacturing Engineers (SME)?
3. How does the small manufacturing company obtain occasional specialized manufacturing engineering services?
4. In a small manufacturing company there may be as few as one, or as many as five manufacturing engineers on the staff. Would you expect these manufacturing engineers to be specialists or to work across the entire spectrum of manufacturing engineering, and why?
5. How would you expect the manufacturing engineering function to be organized in an aerospace company that uses a matrix or program type organization structure? Why?
6. What is the primary mission of the manufacturing engineering department, and what are some key measures of its performance?
7. Since all manufacturing is conversion, what are the conversion steps from iron ore and coal to a finished automobile?
8. What would the primary job responsibilities of a manufacturing engineer be in a job-order, precision machine shop?
9. The latest trend in industry today is to place the manufacturing engineering organization on the same reporting level as design engineering and manufacturing operations, rather than as part of the manufacturing operations organization. What are some of the advantages and disadvantages of such an arrangement?
10. Why should today's manufacturing engineer be a graduate engineer, and what engineering disciplines would provide the best formal education background if an engineering degree program is not available in manufacturing engineering?
11. What roles would you expect manufacturing engineering and quality control to play in setting up a statistical process control system in a machine shop?

1
Cost Estimating

1.0 INTRODUCTION TO COST ESTIMATING

The cost estimate is one of the keys to the acquisition of new or follow-on business. Its preparation cannot be left to amateurs or accountants. If the potential new business acquisition involves the manufacture of a product, whether newly designed or one that has been in production before, the manufacturing engineer must play a key role and participate in preparing the estimate. Success in acquiring this and future new business may well depend upon how well the job of estimating is done.

This chapter addresses the essentials of developing a truly competitive cost estimate. Preparation of the manufacturing labor estimate is one of the most important jobs or duties of the manufacturing engineer. It requires experience, judgment, engineering know-how, and knowledge of the product, including what it takes to produce it in a cost-effective manner.

1.1 COST ESTIMATING OVERVIEW

It is common knowledge in engineering work that if you can fully define a problem, the solution usually becomes readily apparent. The same holds true for estimating manufacturing costs; if the cost estimate requirements are fully known and understood, preparing the estimate itself becomes very routine. Before beginning to prepare the cost estimate a few questions must first be answered.

1.1.1 Why Is The Cost Estimate Being Prepared?

The cost estimate is prepared to determine how much time and money will go into the preparation of the manufacturing cost estimate and set the stage for further analysis. Some reasons for cost estimates when there are no bids or price quotes to submit are: (1) to verify quotations from outside vendors, (2) to determine if a proposed product line will be profitable, (3) to provide the basis for a make or buy decision, or (4) to evaluate design alternatives. Another important part of this is to determine who the prospective customer and competition are.

1.1.2 What Is The Product To Be Manufactured?

Product design essentially determines manufacturing costs; therefore, a thorough and complete product definition is critical in developing cost estimates. Inadequate product specifications can lead to parts or materials that do not meet or greatly exceed the required quality levels. Incomplete specifications may result in estimates that do not meet the costs of carrying out the final engineering design or reliability requirements.

Product definition normally involves a set of drawings that describe the product or parts to be made. This documentation should be provided by the prospective customer if it is part of a bid package. Sometimes, shop process or routing sheets are also available, and often the product or parts are currently in production. This makes it a relatively easy matter to determine the steps or sequences of manufacture and assembly, and to develop the manufacturing cost estimate.

If the drawing package describes something never before made in your plant, a cautious approach is needed to put together a sound cost estimate. The package must be carefully analyzed to determine exactly what is required and how it will be manufactured.

Often, it is necessary to estimate the cost of producing an item that is not yet fully designed. Such estimates, known as concept estimates, are made from sketches, notes, and design layouts. Extreme care should be taken to make ceratin the concept as envisioned by the design engineer is fully and completely understood by the estimator. It is not uncommon for such preliminary designs to become much more complex before the final manufacturing drawings are released. In some cases, the parts count required to make the product increases by as much as 40%.

1.1.3 What Are The Delivery Requirements?

Delivery requirements can greatly affect the final cost of the product. The time needed to make the cost estimate, prepare new facilities, obtain production equipment and raw materials, and actually manufacture the product must all be taken into account. The importance of knowing shop flow times, material lead times, the capacity of shop machines and other production equipment, as well as production loads in the time period the proposed work would be done can be pivotal in developing accurate cost estimates.

The answers to these questions will lead to the formulation of a bid strategy along with the ground rules and assumptions to be used as a basis for preparing the cost estimate. An understanding of cost estimate requirements, a well thought out bid strategy, and a good manufacturing plan almost guarantees the cost estimate will come close to actual production costs. How well the cost estimate requirements are understood and utilized can determine whether a company survives and grows or declines and goes out of business.

1.2 THE MAKE OR BUY ANALYSIS

The first task of the manufacturing engineer is to review the product documentation and to develop an indented parts list if one is not provided. From the indented parts list, the manufacturing engineer establishes a manufacturing bill of material and a preliminary make or buy plan. He identifies the obvious buy and obvious make items utilizing his knowledge of the company's capability to fabricate, assemble and test the various materials, parts, subassemblies, and the final assembly of the product. There will be many items that could be made in-house or purchased outside, depending on a number of factors which are primarily economic in nature. In the preliminary make or buy plan, he should assume these items will be made in-house.

Purchased parts and materials, as well as the subcontracted items, can account for as much as 70% of manufactured product cost. This is why it is of such importance to have a qualified manufacturing engineer, knowledgeable in company or plant capability to manufacture the items on the bill of material, make the initial or preliminary assessment of what should be made and what should be bought. For example, if your company is primarily an assembly house, then all fabricated parts would fall into the "must buy" category, as would all hardware and bulk material items such as paints, solvents and chemicals. Items that fall into the "can make/can buy" category are the ones requiring a detailed analysis from a cost and shop load standpoint.

1.2.1 Minimum Requirements

As a minimum, your company's make or buy planning should:

1. Maximize solicitation of subcontractors and other suppliers with proven track records in cost, schedule, and quality performance.

2. Send formal written requests for quotations with requirements spelled out in detail.
3. Hold bidder's conference with subcontractors on critical procurement items.
4. Show your evaluation of suppliers from a technical and performance standpoint.
5. Make provision for negotiation of statements of work, cost, schedule, and technical specifications.

In large companies, a formal make or buy committee with representation from all key functions is chartered and is chaired by a senior member of management (usually a vice president or director of either operations or manufacturing). This committee formally decides on make or buy based upon recommendations from all the disciplines concerned, including manufacturing engineering. Figure 1.1 shows the make or buy decision cycle in a large company.

1.2.2 Decision Cycle

Make or buy decisions are often influenced by factors other than cost and in-house shop loads. For example, does it make good sense to create an in-house capability to wind coils, when coil-winding companies that specialize in nothing but coil winding, and have years of experience in this specialized field, are available to provide this capability, even though at a somewhat higher price than is estimated for doing it in-house? A closer look reveals a requirement for a major capital expenditure for coil winders, ovens, and other specialized equipment such as Q-meters and capacitance bridges for testing. There would also be a requirement for trained engineers, setup men, and coil winders. Also, the current requirement for coils is for a relatively

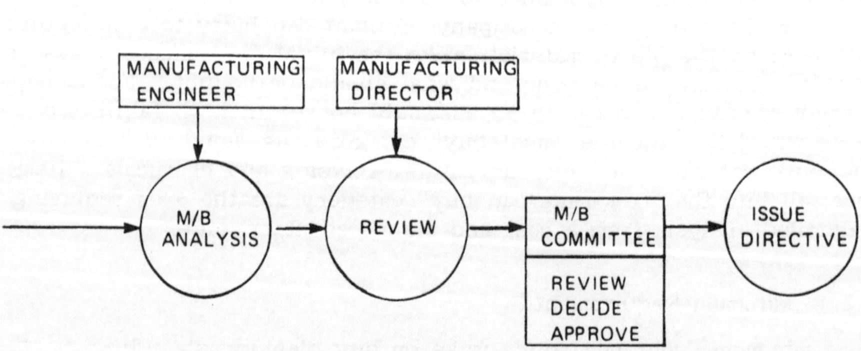

FIG. 1.1 Make or buy decision cycle. Factors: total cost, quality, reliability, schedule, subcontractors, manpower, facilities.

short period of time with little or no requirement beyond the current product line.

These are the kinds of tough questions that must be asked, when on the surface it appears attractive to do certain kinds of work in-house which would require creating a new manufacturing capability. At the same time, we must be careful not to subcontract the really critical work that controls product quality and delivery. The make or buy decisions made at the front end of a new job or program can have a major long-term impact over the life of the production run. The best skills and judgement in the company should participate in the make or buy decision cycle.

1.2.3 Buy Item Follow-up

Once the make or buy list has been firmed up, the manufacturing engineer must move out and firm up his manufacturing bill of material. This is basically a bill of material that breaks the in-house make items down to the lowest piece part and material level. It includes the buy items, but does not break them down to the level of the engineering parts list or bill of material. The manufacturing bill of material then becomes the baseline for developing the estimate by listing everything that must be priced to develop product cost.

1.3 ESTIMATING METHODOLOGY

Since manufacturing costs are made up of material costs as well as labor costs, it is important that the manufacturing engineer recognize this fact and thoroughly analyze material costs as well as labor costs. In some manufacturing companies, material and the cost subcontracts can be as much as 70% of product cost, while direct factory labor can be as little as 12%.

Any cost estimate is a prediction of what the product will cost to manufacture at some future point in time. Estimating is not an exact science, yet a thorough cost estimate prepared by professionals will come very close to actual costs incurred in many instances. If the estimating methodology outlined in this chapter is followed, the cost estimate will be very close to the cost actually incurred.

1.3.1 Cost Estimate Requirements

Essential requirements for any manufacturing cost estimate include a thorough pre-production plan, and a set of ground rules and assumptions for preparing the estimate. However, before this can happen, certain basic information must be available:

Product specifications
Delivery schedules and desired production rates
Bills of material
Detail manufacturing drawing, or at least, clear concept drawing and sketches (see Fig. 1.2)
Production routings, work instructions, or a conceptual plan to manufacture
Tooling and equipment requirements
Inspection and test requirements
Packaging and shipping instructions
A well thought out make or buy plan

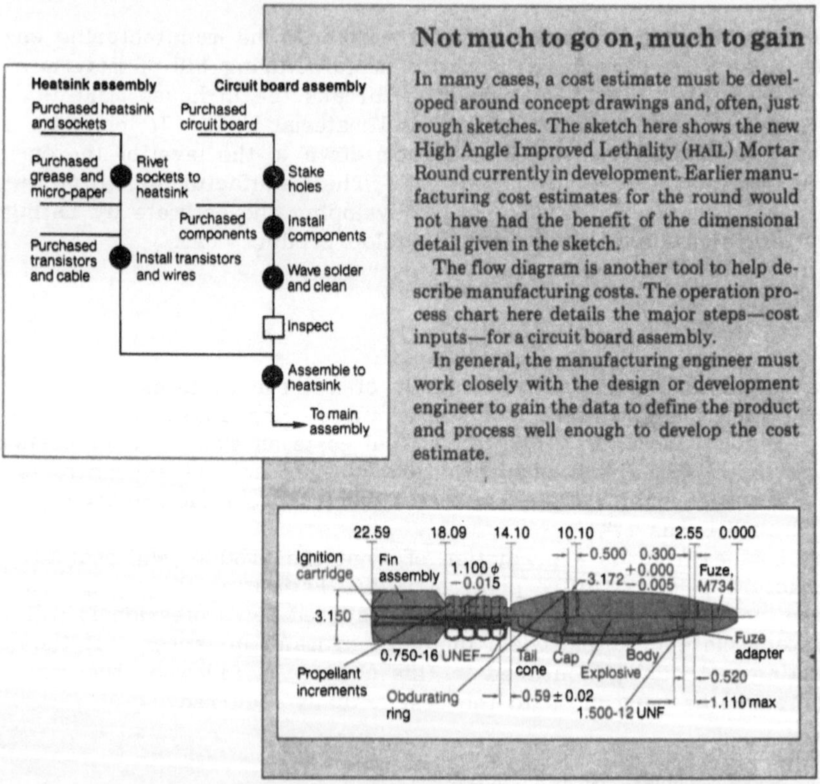

FIG. 1.2 The operation process chart is the baseline manufacturing plan. The sketch provides an example of a concept product definition. (Courtesy of Penton Publications and McDonnell Douglas Corporation.)

1.3.2 Basic Approaches

The two approaches to cost estimating in manufacturing are the cost projection approach and the simulation approach. The cost projection approach is the most commonly used and accepted. With it, past costs are collected and analyzed, then used as a basis for the projection of costs to be incurred on a future similar product. This approach is especially useful where the details of product definition are sketchy or ill-defined. It should be used only where past costs are properly collected and reflect accurately the costs a given product actually incurred.

The simulation approach is the predicting or modeling of costs to be incurred. It is most often used for a new product. Product definition may or may not be in detail, but the overall approach is essentially driven by the lack of real cost history.

Using the simulation approach each part or assembly to be made in-house is reviewed and analyzed to determine the sequence of operations required to fabricate, assemble, and test. At this point the manufacturing engineer should construct an operation process chart depicting the sequences of operations and how parts and materials flow into the process of manufacture (see Figure 1.2). The operation process chart becomes a production plan in outline form when the machines, equipment and major tools required are listed for each operation shown on the chart. From this the tabulation can be made of the machines, equipment, and tools required to manufacture the product. Figure 1.3 shows such a tabulation for an orbital rivet machine.

In material and subcontract cost estimating, purchasing is the primary support to the manufacturing engineer in determining these costs. For best results, direct quotes from vendors and subcontractors should be obtained but, where time to prepare the estimate is limited, purchasing should have price listings for the most commonly purchased parts and materials.

Many times, because time to prepare the cost estimate is limited, purchasing may use historical cost data for the same or similar parts and materials, and factor these cost data for inflation and other expected cost growth. Many possible pitfalls are inherent in such an approach, though this is a common occurrence in industry today. Because material pricing is so important and has such a large effect on product cost in manufacturing, the problems in material estimating should be considered. The chief dangers in estimating material costs include the following:

1. Inadequate product specifications or product definition may cause the use of prices for parts or materials that are below or above the needed quality level.

Orbital Riveter (Roll Swage)

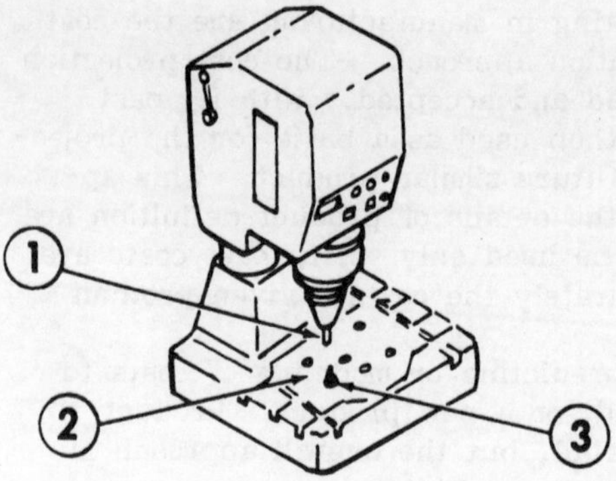

BK-100/NM Riveter

Form Punch		Parts Being Roll Swaged		
1	SPT-2-3230001	3235444	(7)	MS16535-66
		3235444	(10)	2541065-2
Fixture		3235432	(2)	2541065-2
		3235665	(7)	2557544-12
2	AF2540001TDS	3235665	(96)	SE12XCO4S
		3235671	(4)	2557544-12
Anvils		3235675	(48)	SE12XCO4S
3	SPTMS16535	3235915	(4)	2541065-3
3	SPT2541065	3235917	(4)	2541065-3
3	SPT2557544	3235919	(2)	2541065-3
3	SPTSE12XCO4S	3235920	(2)	2541065-3
		3235924	(4)	2541065-3

Time

$$190 \text{ parts} \times \frac{5 \text{ swages}}{\text{part}} = 950 \text{ swages}$$

$$950 \text{ swages} \times \frac{0.05 \text{ min}}{\text{swage}} = 47.5 \text{ min}$$

$$\frac{47.5 \text{ min}}{60 \text{ min/hr}} = 0.79 \text{ hr}$$

FIG. 1.3 Machine and parts it makes plus time. (Courtesy of McDonnell Douglas Corporation, St. Louis, Mo.)

2. Incorrect or changing delivery requirements or vendor delivery failures may demand more expensive substitutes or production changes that require more set-ups or production delays.
3. Incomplete product specifications may result in material estimates that do not cover the material costs involved in the final engineering design or product quality and reliability requirements.
4. Material price levels may change upward and exceed the estimated material costs. This is particularly serious when estimating material costs for products with long manufacturing lead times.
5. Material price breaks may be anticipated and planned on but not realized because of such factors as delivery schedule changes, or changes in inventory policy.

The manufacturing engineer should also consider detection of areas of material cost improvement part of his responsibility. Some problem areas may include:

Overspecification by engineering.
The possibility of obtaining lower-cost substitutions, or even less expensive sources by Purchasing.
The potential for scrap reduction by manufacturing.

A first class manufacturing estimate cannot be developed by simply reflecting experience in material estimates. These problems can and should be addressed by the buyer and the manufacturing engineer working as a team. The buyer can develop contingency factors for price increases and material cost growth with time. Estimating the cost of manufacturing labor varies in importance with each manufacturing situation. In most companies, it is a vital element of cost. The accuracy of the labor cost estimate depends upon:

1. The accuracy of the product configuration baseline and associated manufacturing planning
2. The accuracy of the labor standards available to the manufacturing engineer
3. The time allowed for preparation of the cost estimate

The detail and precision of the standards varies with the situation. Usually, the higher the ratio of labor cost to total cost, the greater the need for more precise standards. In modern day manufacturing, standards have many uses other than just for estimating. The primary use is for labor cost control

Good labor standards are vital to so many aspects of manufacturing that it is startling how often they are either totally undeveloped or grossly inaccurate. The main deterrent seems to be the cost of developing and maintaining them. The effort more than pays for itself, if the standards are fully utilized. It is a true statement that, if you need standards, you are already paying for them. If standards are unavailable, the manufacturing engineer must use historical cost data from accounting, or personal knowledge and experience, to develop the labor estimate.

A manufacturing labor estimate can be developed as either a bottom-up or a top-down estimate. A bottom-up estimate is developed from standards and is built up from detailed estimates of manufacturing labor for individual piece parts, subassemblies and assemblies to the final product cost. The top-down estimate is just the opposite; it is developed downward from a management or customer assessment of what the product should cost. Obviously, the top-down estimate is a high risk situation and should not be used except under very unusual situations or circumstances.

Let us assume that you are the manufacturing engineer on the staff of a small company that manufactures background music systems. The company is requested to quote on an order for some (200) amplifier assemblies. The assemblies are a standard product line and, as such, have been produced many times over in the past. Standard times exist for all operation sequences, which are shown in Figs. 1.4 to 1.6 in minutes per operation for the two circuit card subassemblies and the main amplifier assembly. These three operation process charts provide all of the information needed to prepare the basic labor estimate. If we total the standard minutes and then divide by 60, we have the total standard time in decimal hours for the two subassemblies and the main amplifier assembly:

(2000–973) Circuit card assembly	0.65 hr
(2000–973) Circuit card assembly	0.47 hr
(S/A–100) Amplifier assembly	1.27 hr
Total	2.39 hr

If we assume that the units will be produced in two lots of 100 each, then there will be two setups in addition to the run times shown above. Setup time is estimated to be 4.50 hr, and with two setups we would have a total of 9.00 hr for setup and 478.00 hr of total run time for 200 units. This is a total standard time of 2.44 hr per unit for 200 units.

Previous production of the amplifier units followed an 85% learning curve, with standard being attained at unit 1000. Since these units have been previously manufactured, some learning would transfer to these units, allowing a point of entry on the learning curve at unit number 12, giving us an average time per unit of 5.15 hr. This

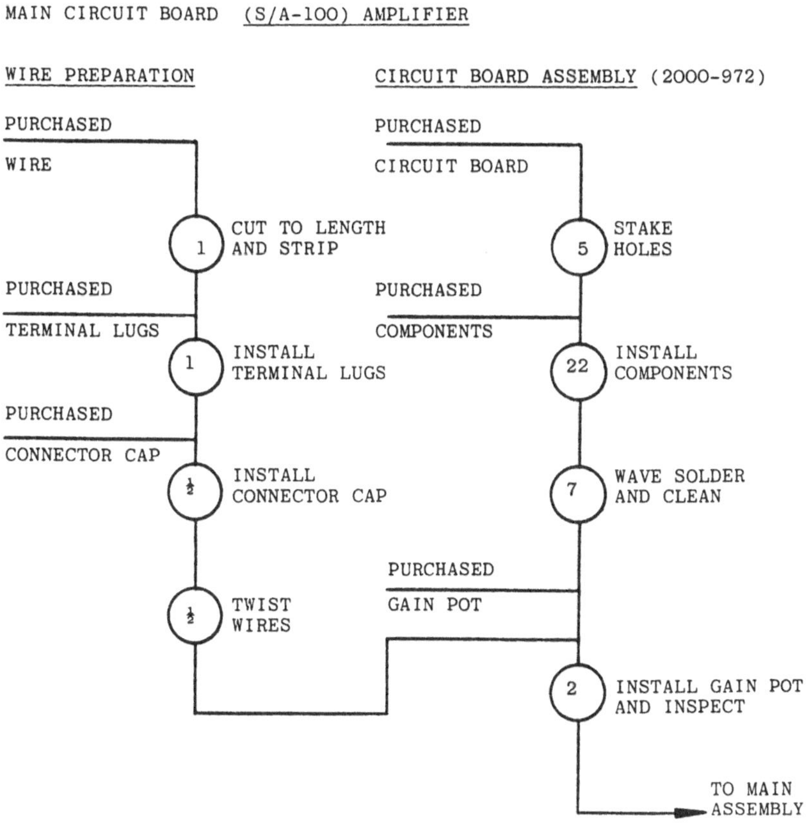

FIG. 1.4 Main circuit board subassembly.

represents what the run of 200 units will actually cost to produce. When multiplied by the hourly labor rate, and overhead added, we have the labor cost per unit to manufacture the 200 amplifier assemblies. To this is added the cost of material and material burden to give the total unit cost. When profit is added, we have the unit selling price.

1.3.3 Recurring versus Nonrecurring Costs

Nonrecurring costs are all of the one-time costs, such as the initial cost of tooling, the planning and manufacturing engineering documentation costs, and the cost of engineering development, equipment, and facilities. Recurring costs include all costs directly associated

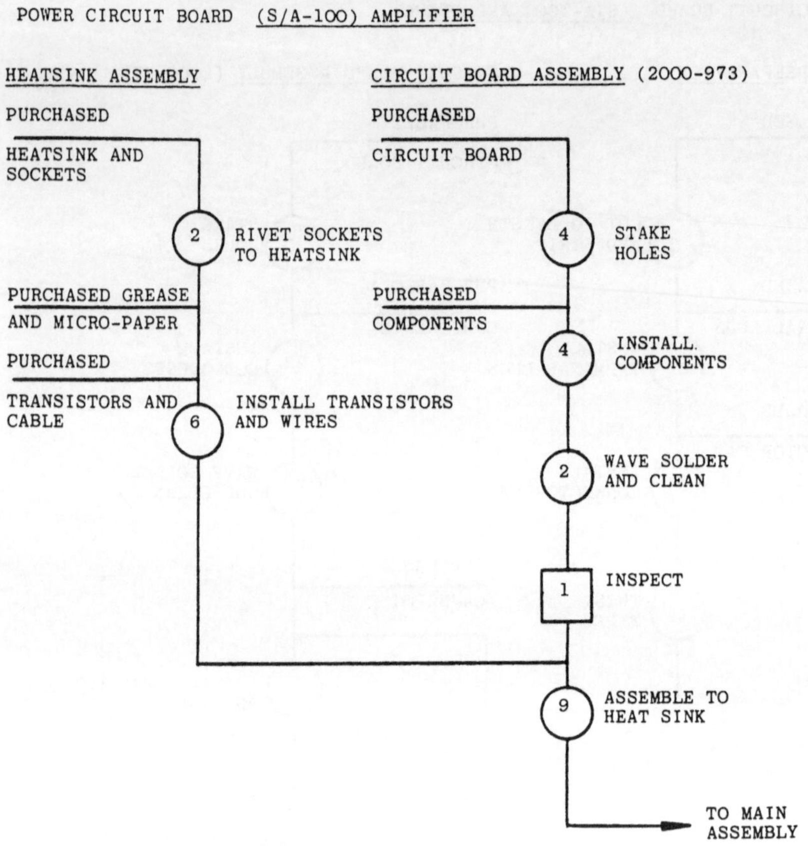

FIG. 1.5 Power circuit board subassembly.

with product manufacture, including factory labor to fabricate, assemble, inspect, and test the product. Also included would be labor for manufacturing engineering support to the factory on a day-to-day basis, the cost of supervision, and production control labor associated with keeping production operations supplied with material, expediting, and tracking the movement of semifinished and finished product.

It is important in developing the manufacturing labor estimate to always keep these costs separated. Nonrecurring costs are start-up costs, which for the most part are independent of the number of units of product that will be manufactured. Recurring costs are the unit costs for the manufacture of the product and vary as number of units produced.

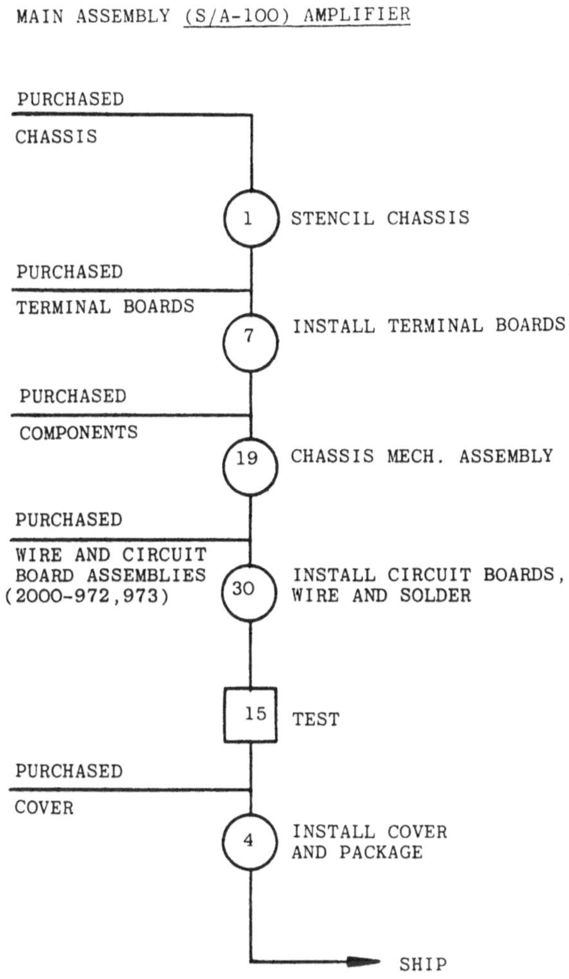

FIG. 1.6 Main amplifier assembly.

1.3.4 The Tooling Estimate

Tooling is required in the manufacture of product hardware. The most obvious example is in metal fabrication, where cutting tools, punches, dies, and fixturing are important cost elements. The two types of tooling are durable and perishable. Durable tooling can be further categorized as standard and special design.

Perishable tooling may be considered as part of factory overhead. However, where tooling costs are appreciably higher for a given machining center or centers, such perishable tooling costs should be included either as a direct tooling cost or in the burden for that machining center and not in the general factory overhead rate.

Also, if a given job or product requires exceptionally high consumption of perishable tools such as from tools, cutters for high speed turning, or broaching tools, then such tooling cost must be charged to the job or product rather than included in a burden rate. Generally, costs of perishable tools are estimated from vendor's prices by current pricing data obtained from purchasing and estimated tool usage over the life of the job.

Durable tools are used in repeated production runs and usually last the life of the job or product. Because they are used to manufacture specific parts or assemblies, they can be charged directly to product cost. Manufacturing engineers must estimate such tooling costs based upon the tooling concepts in the pre-production planning for the product or job. This ensures that tooling design and build costs will reflect actual costs incurred.

It must be remembered that tooling costs are major elements of total manufacturing cost and must be thoroughly analyzed and identified.

1.3.5 Facilities

Facility and capital equipment costs to support the production of a new product or product line often can be quite high, especially when no comparable equipment or facilities exist within the company. A good example of this would be a metal finishing or plating line, which would also require waste treatment and disposal capability. Many times, the cost of needed facilities or equipment to manufacture a new product, or to continue to manufacturing an existing product line, and remain competitive with others is such that it cannot be written off over the life of a single or even several product lines. In cases such as this, management may choose to invest in the new facilities with the knowledge that some risk may be involved, or decide to subcontract the work to outside sources.

1.3.6 Potential Cost Estimating Problems

A cost estimate is a forecast of what is involved in manufacturing a product. Its ultimate goal is prediction of costs. Cost estimating in manufacturing is a vital function. The company's success—even its survival—depends on how well the estimating job is done. The more complex the product, the greater the need for good estimating. Consider the problems in developing a sound manufacturing cost estimate:

1. Often, the data upon which to develop the cost estimate are inadequate.
2. Inadequate staff and time to prepare the estimate.
3. Lack of coordination and communication between key organizational elements in the company, such as accounting, purchasing, and manufacturing.
4. Careless estimating.
5. Optimistic estimating.
6. Poor estimator selection.
7. Management inertia.

Many, if not all, of these problems can be resolved, or at least considerably minimized by management. One approach is to set up a dedicated proposal team for a major manufacturing cost estimate. It should consist of the best professionals from every discipline in the company organization, including top talent manufacturing engineers to lead the effort.

The manufacturing labor estimate should be completely documented with all notes, backup data, and calculations carefully filed in a folder. The documentation should be clearly labeled and dated and should incude even very rough handwritten notes and sketches. This file should enable the manufacturing engineer and others to go to the file long after the estimate was prepared, and easily determine how the estimate was prepared, what backup data are available to support the estimate, and answer any questions about the estimate that may arise. Such data are also very useful in preparing estimates for similar products or in updating an earlier estimate.

In summary, for the manufacturing cost estimate to be valid and as accurate as possible, the following data are required:

1. Product specifications
2. Product quantity and delivery schedules
3. Manufacturing bill of material
4. Engineering drawings, or concept sketches of the product
5. Operation process charts, routings, and work instructions
6. A make or buy plan
7. Tooling and equipment requirements

8. Inspection and testing requirements
9. Product packaging and shipping instructions

A cost estimate is a prediction or forecast of what it will take to manufacture a product. Its ultimate goal is a prediction of costs. The cost estimate is an essential ingredient in the acquisition of new business and growth of the company.

1.4 LEARNING CURVES

Once the basic costs of material, labor, and tooling have been developed, consideration must be given to learning or improvement curves, which determine labor realization factors and material cost improvement trends, and provide a measure of productivity for the company. The following are some important points to remember about the learning curve and its concepts:

1. The phenomenon of the learning curve is primarily the result of management action and management learning.
2. The learning curve effect itself is beyond any question.
3. The learning curve is so universal that its absence is almost a warning of mismanagement.
4. Understanding of the underlying causes of why learning occurs is still far from perfect.
5. It is known that costs are more certain to decline if it is generally expected that they should and will.

The cost of value added declines approximately 25% in real terms each time accumulated experience is doubled. These cost declines continue indefinitely as long as the growth rate is continuous. Such cost declines do not occur automatically; they require management and often capital investment. The learning curve is an observable fact. Learning curve effects can be observed and measured in any business, in any industry, and in any element of cost, almost anywhere.

1.4.1 Definition

The mathematical development of the learning curve is relatively simple. The relationship between average cost and increasing quantity has been observed to approximate a power function. It has been shown by correlation and other statistical methods that a plot of actual performance data may be described by the following equation with a high degree of significance:

$$Y = kx^{-n}$$

where Y is the cumulative average recurring labor cost stated in terms of labor hours for all units produced through a quantity of x units, and k is the theoretical first unit cost. The exponent (−n), which is a negative number with an absolute value less than 1, defines the rate at which the average cost decreases as quantity increases. This exponential curve plots on arithmetic coordinates as a concave curve which flattens out as quantity increases (see Fig. 1.7). This same curve will plot as a straight line on logarithmic coordinates (see Fig. 1.8).

The rate of improvement is called the slope of the learning curve. One of the primary characteristics of the learning curve is that each time the production quantity is doubled, the average cost is reduced by some constant percentage determined by the slope. In the case of a 90% slope, the cumulative average cost through two lots of equal size is predicted to be 90% of the cost for the first lot. For this to occur, the second lot must cost 80% as much as the first lot. Doubling the quantity again reduces the average cost to 90% of the average through two lots, or to 85% (90% × 90%) of the initial observation, and so forth.

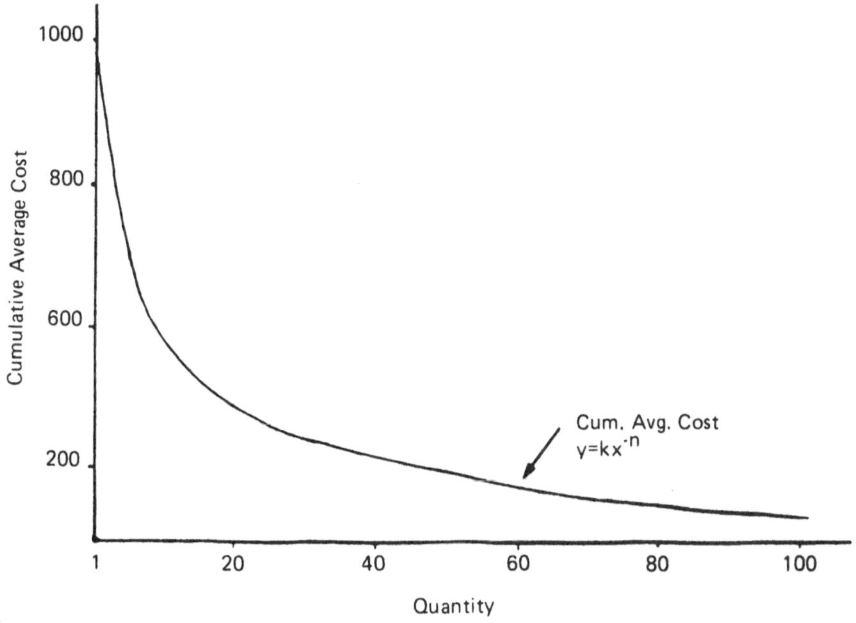

FIG. 1.7 Learning curve plotted on arithmetic coordinates. The relationship between average cost and increasing quantity approximates a power function. (Courtesy of McDonnell Douglas Corporation, St. Louis, Mo.)

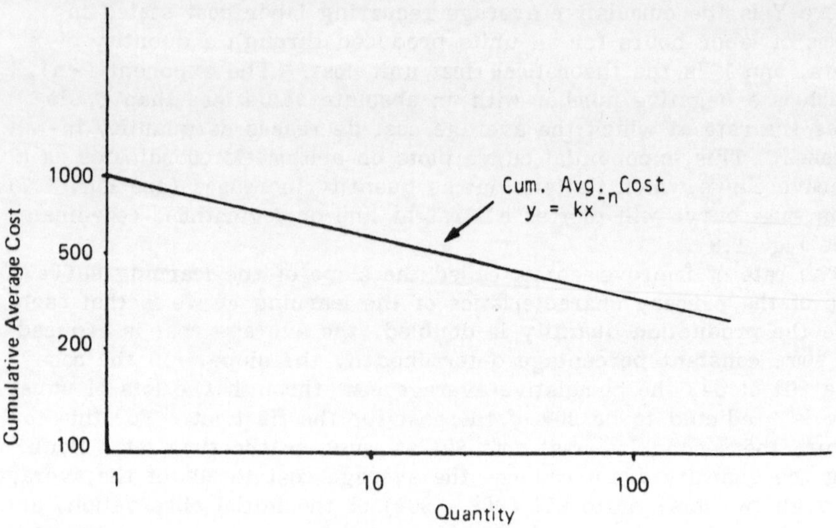

FIG. 1.8 Learning curve plotted on logarithmic coordinates. The exponential cost/quantity relationship plots as a straight line on logarithmic coordinates. (Courtesy of McDonnell Douglas Corporation, St. Louis, Mo.)

The learning curve slope is a function of a number of variables. These include characteristics of the type of work that is being performed, production variations, and uncontrolled external factors. Different slopes can be expected for different manufacturing processes. In general, the more labor-intensive the operations, the steeper the slope that can be expected. Processes and operations that are machine-controlled, or where a high degree of mechanization or automation exists, exhibit a much flatter slope. The reason for this is that the mechanized or automated process starts out at or near optimum performance, and a much lesser degree of learning or improvement can be expected over a given production run. Appendix A contains a more complete and detailed explanation of learning curve theory and development.

1.4.2 Some Practical Approaches

Products do not all have the same learning curve. The use of learning curve concepts as an estimating tool involves more than the insertion of variables into an equation. The actual behavior of manufacturing cost trends is influenced by a number of key factors which must be considered, in terms of both their effect on the actual data base, and of what will then influence the cost of new or follow-on work.

Depending upon planned production quantity, constraints in lead time, limitations in availability of funds for investment, and product design maturity, tooling may range from nonexistent (build on setup) to a fully engineered tooling approach which is designed to reduce product costs over a large production run. In the case of the minimum tooling approach, the starting performance can be expected to be higher than if the tooling was designed to achieve reduced product production costs.

Cost improvement in the case of minimum tooling depends in large measure on improvements in productivity by shop personnel. Should tooling for efficient rate production be employed, start-up costs can be expected to reflect problems associated with tool shakedown, and the learning curve should reflect the rapid resolution of any problems. The successful resolution of start-up tooling problems will result in a steeper curve in the early production phase than a minimal tooling approach would allow.

Automatic test equipment can often enable dramatic cost savings in production. If the equipment has been completely debugged prior to its use in production, the effect on cost will show on the learning curve as a step reduction at the quantity where the equipment was brought on line.

1.4.3 Impacts

The primary impacts that affect the learning curve can be summarized as follows:

1. The change from development, to pilot production, to full-rate production.
2. Production rate constraints and dictated schedule changes. Higher rates allow more units in work at a given time, which allows better manpower utilization. Lower rates mean fewer units in work and less efficient manpower utilization. Lower production rates result in smaller lot sizes, increasing the number of setups required for the same number of units.
3. Number of setups affects average fabrication costs. Detail parts fabrication incurs both setup and run costs. Setup costs occur each time a lot of parts is fabricated, but are not dependent on number of parts in the lot. Run costs are incurred each time a part is fabricated. The average cost of parts in a lot (total cost divided by number of parts fabricated) is directly dependent on the number of parts that are fabricated with each setup.
4. Interruptions in production for any reason will result in a loss of learning.

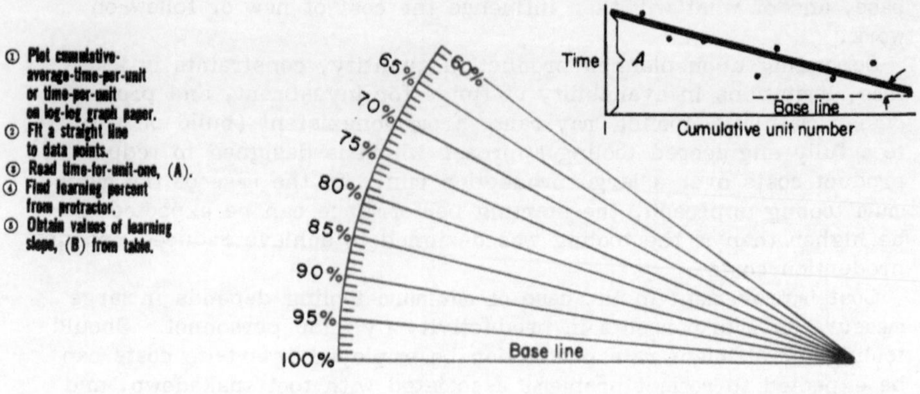

%	B	B+1	%	B	B+1	%	B	B+1	%	B	B+1
60	−0.737	0.263	70	−0.514	0.485	80	−0.322	0.678	90	−0.152	0.847
61	−0.713	0.287	71	−0.494	0.506	81	−0.304	0.696	91	−0.136	0.864
62	−0.690	0.310	72	−0.474	0.526	82	−0.286	0.714	92	−0.120	0.880
63	−0.667	0.333	73	−0.454	0.546	83	−0.269	0.731	93	−0.105	0.895
64	−0.644	0.356	74	−0.434	0.566	84	−0.255	0.748	94	−0.089	0.911
65	−0.621	0.379	75	−0.415	0.585	85	−0.234	0.764	95	−0.074	0.926
66	−0.599	0.401	76	−0.369	0.604	86	−0.218	0.782	96	−0.059	0.941
67	−0.578	0.422	77	−0.377	0.623	87	−0.200	0.800	97	−0.044	0.956
68	−0.556	0.444	78	−0.358	0.642	88	−0.184	0.816			
69	−0.535	0.465	79	−0.340	0.660	89	−0.168	0.832			

⑥ Substitute values of A, B, and $B+1$ to find learning curves*

Learning curve	If unit times were plotted	If cumulative average times were plotted
⑦ Time for R-th Unit ($R = 1, 2, 3 \cdots$)	AR^B	$(B+1)\,AR^B$
⑧ Cumulative average time for R-th unit	$\frac{AR^B}{B+1}$	AR^B
⑨ Total Time for R Units	$\frac{AR^{B+1}}{B+1}$	AR^{B+1}

* For best accuracy R should be greater than 10

FIG. 1.9 Learning curve slope protractor.

5. Timely resolution of manufacturing problems is a major factor in the realization of the improvement process.
6. Continued improvement depends upon management learning and management action.

1.4.4 Pitfalls

When correctly applied, the learning curve is one of many valuable tools available to the manufacturing engineer. However some caution should be exercised in the selection and use of labor cost data in plotting the learning curve:

1. If a company receives a large portion of its raw material in a finished state, its labor input per unit of product should decline. This decline, however, is merely the result of shifting labor input from the company to the plant of a supplier. It would be erroneous to consider the decline as a real reduction in labor hours. The same work is being performed in total, and no net saving has been realized.
2. It may be possible to generate a direct labor saving of 1000 hr by spending 2000 hr on additional tooling and manufacturing engineering. It is obvious that there would be no real savings as a result. It is clear from this that direct-touch labor costs cannot be considered as separate from the changes in the other elements of cost.
3. There is always the danger of false labor savings resulting from the reshuffling of accounting records. The learning curve plots direct labor only. What happens when greater use is made of supervisory labor, which is classified as indirect? If the emphasis is placed on direct labor savings alone, without considering indirect labor, a distorted picture of direct labor productivity will result.

1.4.5 Applications

The applications of learning curves are almost infinite in number. If we begin with an application that has a given first-unit or T_1 value and a given or predetermined learning curve slope, we have a problem that is very common in cost estimating. This example is solved by using calculations and the methods discussed earlier in this chapter and in the theory discussion in Appendix A. It is also possible to solve this problem utilizing the learning curve template or protractor shown in Fig. 1.9.

PROBLEM

Given an estimated first-unit cost of 1200 hr and an anticipated 88% improvement curve slope, calculate projected costs for the first 50 units and for 100 units following the first 50 units.

Solution

Approach: This problem can be solved by solving the cumulative average equation $y = kx^{-n}$ after determining $-n$ by using equation

$$-n = \frac{\log(p \div 100)}{\log 2}$$

Procedure:

1. Determine −n given that p = 88%

$$-n = \frac{\log(0.88)}{\log 2} = \frac{-0.0555}{0.3010} = -0.1844 \qquad (1)$$

2. Substitute 1200 for k and −0.1844 for −n in the cumulative average cost equation:

$$y = 1200 \times x^{-0.1844} \qquad (2)$$

3. Solve for y at x = 50 and at x = 150:

$$\text{Cum. avg. for } 50 = 1200 \times 50^{-0.1844} = 583.2 \qquad (3)$$

$$\text{Cum. avg. for } 150 = 1200 \times 150^{-0.1844} = 496.3 \qquad (4)$$

4. Determine cost for 50 units:

$$\text{Total for } 50 = 50 \times 583.2 = \underline{29{,}160} \qquad (5)$$

5. Determine cost for first 150 units;

$$\text{Total for } 150 = 150 \times 476.3 = 71{,}445 \qquad (6)$$

6. Determine cost of 100 units following 50 units by subtracting Eq. (5) from Eq. (6):

$$71{,}445 - 29{,}160 = \underline{42{,}285}$$

Another common application of the learning curve is in determining the slope of the curve, and the indicated first-unit (T_1) value, when the hours for the first two blocks or lots of production units are known.

PROBLEM

Given that the first 3 units cost a total of 2539 hr and the next block of 4 units cost 2669 hr, determine theoretical cost of the first unit and the improvement curve slope indicated. Assume that the difference in block size did not affect the trend.

Solution

Approach: The cumulative average cost through 3 and through 7 units can be readily calculated and the cumulative average cost equation $y = kx^{-n}$ can be solved to determine the first-unit cost (k) and the curve exponent (−n). The slope percentage equation $p \div 100 = 2^{-n}$ can then be used to determine the curve slope percentage.

Procedure:

1. Determine cumulative average costs at 3 and at 7 units;

   At $x = 3$, $y = 2539 \div 3 = 846.3$

   At $x = 7$, $y = (2539 + 2669) \div 7 = 5208 \div 7 = 744.0$

2. Substitute these values into the equation $y = kx^{-n}$:

$$846.3 = k \times 3^{-n} \tag{1}$$

   and

$$744.0 = k \times 7^{-n} \tag{2}$$

3. To solve these two simultaneous equations, first divide Eq. (1) by Eq. (2) to eliminate:

$$(846.3 \div 744.0) = k \times 3^{-n} \div k \times 7^{-n} \tag{3}$$

   or

$$1.1375 = (3 \div 7)^{-n} = 0{,}4286^{-n}$$

4. To solve for −n, take the logarithm of both sides of the equation:

$$\log(1.1375) = -n \log(0.4286)$$

   Using a scientific calculator, this becomes:

$$0.05595 = -n(-0.36795)$$

5. Solve for −n:

$$-n = \frac{0.05595}{-0.36795} = -0.15206$$

6. Determine p:

$$p \div 100 = 2^{-n} = 2^{-0.15206} = \underline{90\%}$$

7. Substitute −n into Eq. (1):

$$846.3 = k \times 3^{-0.15206} = k \times 0.8462$$

8. Solve for T_1 cost (k):

$$k \frac{846.3}{0.8462} = \underline{1000\ \text{hr}}$$

PROBLEM

Assuming that the improvement trend in the previous problem continues, calculate the cost of an additional 12 units.

Solution

Approach: The T_1 cost and n factor were determined in the preceding problem. The cumulative average method can be used to determine the cost of the additional units.

Procedure:

1. Given that T_1 = 1000 and −n = −0.15206, calculate the cumulative average through 3 + 4 + 12 or 19 units:

$$y = 1000 \times 19^{-0.15206} = 639.1$$

2. Compute the total cost through 19 units:

$$\text{Total cost } yx = 639.1 \times 19 = 12{,}142.9$$

3. Calculate the cost of the last 12 of these units, given that the first 7 cost 5208 hr, as determined in the preceding problem:

$$12{,}143 - 5208 = \underline{6{,}935\ \text{hr}}$$

(Note that the block average cost for these additional units is: 6935 ÷ 12 = 577.9)

As in the previous problem, both of these problems can be worked without calculations by using the learning curve slope protractor and plotting the data and learning curve on a piece of log-log paper. It is recommended that both techniques be used as a method of checking or verifying an answer.

1.5 COMPUTER-AIDED COST ESTIMATING

Recently computer assistance techniques have significantly increased the repeatability and reliability of manufacturing cost estimating to the extent that those companies without this aid may find themselves at a competitive disadvantage. One such system utilized in the estimating of contract machining work utilizes both a keyboard and stylus input to develop highly reliable cost estimates in a fraction of the time required using traditional methods.

The system hardware and software consists of a professional microcomputer, which in addition to memory, utilizes a keyboard, one diskette, a hard disk for information storage, a printer, a sonic digitizer, and a stylus to trace and gather input data. Wherever possible the stylus is used. The workpiece drawing is placed on any flat surface. Data about materials, cutters, and machines are keyed into the system. Other data such as dimensions of the raw workpiece are also entered. The stylus is then used to simulate the cutter paths. From this, new data is generated for developing and computing the estimate. The system has the capability for both two- and three-dimensional work including pocketing and sculpturing routines. The system works by storing standard data such as machining methods, speeds, feeds, tooling, and material in the program. Distance of cuts, such as depth of a bored hole, the cutter path of a milled part, or the length of a bar to be turned are quickly and accurately found by tracing over the piecepart drawing with the stylus. As the manufacturing engineer emulates the cutter path with the stylus, the sonic digitizer traps the X and Y coordinates. This information is automatically transferred to the microcomputer, where it is converted to machine cutting time and machining cost.

Declining costs and the increasing power of microcomputers are now placing computer-aided cost estimating within reach of virtually all firms doing contract machining work. Computer-aided cost estimating systems for machining are available on the market, but the following cautions should be kept in mind:

1. A system should not be selected on the basis of cost alone, or on the assumption that all systems are equally capable of handling estimating-related problems.
2. There are a number of software systems on the market that are designed for a particular machine, or type of machines. If your firm has a mix of machines, look for software that will easily adapt to more than one use.
3. Some systems have editing capabilities, some do not. Look for a system that will enable you to add, modify, or delete existing processes on the software.
4. If you need the benefit hard copy offers, find out if the system has printout capability.

5. Does the system come with training, consultation support and enhancements? Look into the supplier's service record and capability.
6. Many systems can be run by someone with less training and experience than a manufacturing engineer. Despite computerized advantages, the work of cost estimating should never be assigned to those with little or no machining experience.
7. If a system cannot work according to its advertised features during a demonstration, then those features are probably not in the program.

1.6 CONCLUSIONS

Cost estimating for manufacturing requires experience and skill. It also requires engineering judgment and attention to detail. Availability of the source date to prepare an accurate cost estimate is also very important. These kinds of data needed and where they can usually be found are shown in Table 2.1. Information requirements for a sound cost estimate vary with the industry and the type of business.

A company producing machine tools or heavy equipment will have different data requirements than a small electronics company or an investment casting house. A high-volume producer of stereo equipment will have data requirements similar to those of the custom shop manufacturing space electronics, but one difference in emphasis will be substantial. Nevertheless, four important rules should be followed consistently in preparing any manufacturing cost estimate:

1. *Secure sufficient information to define the product.* Drawings, specifications, bills of material, weights, and shop hours are the most critical facts. This information must be supplied by the originator of the inquiry, and should be checked by him for errors before it is released for cost estimating. When the originator is outside the company and beyond its discipline, sales or engineering will have to perform this review. An important and often neglected point here is the need to set acceptable data stndards so that customers routinely provide sufficient information and, just as important, provide it in a way in which it can be used without costly conversion or rework.
2. *Match product requirements against available resources.* Determining what will be needed in floor space, machines, equipment, tools, and manpower is the next important phase of estimating. The manufacturing engineer should make this assessment. Company management must know the impact the proposed product is likely to have on current and near-future status of the factory.

TABLE 1.1 Cost Estimating Data Requirements and Sources

Data description	Source
Design specifications, drawings, technical data, design concepts, parts lists, engineering bills of material	Design engineering, sales, contracts, marketing
Quantities and rates of production	Request for estimate, sales, marketing
Manufacturing and tooling plans	Manufacturing engineering
Make or buy plan, manufacturing bill of material	Manufacturing engineering
Machine, equipment, floor space, manpower requirements	Manufacturing engineering
Material cost data, subcontractor costs, packaging and transportation cost data	Purchasing, procurement, subcontracts
Test and inspection requirements, tooling, procedures	Quality control, engineering, sales, marketing, contracts
Standard time data	Manufacturing engineering, industrial engineering
Historical cost data, previous cost estimates	Accounting, manufacturing engineering, industrial engineering

3. *Always make a comparative review and secure additional data only where necessary.* While estimates are still in a preliminary stage, they should be checked against an accumulation of past experience and data. Injecting the judgement of senior technical staff and managers at this point offers two advantages. First, gross errors, omissions, or inconsistencies will be detected early and resolved before a great deal of time and expense is incurred pursuing false starts. Second, the preliminary estimate may be entirely adequate for the purpose at hand, making additional work unnecessary.
4. *Review final cost estimates with accounting before release.* Cash requirements for the manufacture of the proposed item and a careful audit of such significant parameters as the overhead rates or inventory values used are aspects best left to accounting.

BIBLIOGRAPHY

Ostwald, Phillip F., *Cost Estimating*, Second Edition, Prentice-Hall, In., Englewood Cliffs, New Jersey, 1984.

Vernon, Ivan R., *Realistic Cost Estimating for Manufacturing*, Society of Manufacturing Engineers, Dearborn, Michigan, 1968.

Malstorm, Eric M., *What Every Engineer Should Know About Manufacturing Cost Estimating*, Marcel Dekker, New York, 1981.

Gibson, D. C., *Manufacturing Improvement Curves; Concept and Application*, McDonnell-Douglas Corporation, St. Louis, Missouri, 1981.

Taylor, *Handbook of Electronics Industry Cost Estimating Data*, John Wiley and Sons, New York, 1985.

QUESTIONS

1. Why should the manufacturing engineer play a key role in the preparation of a manufacturing cost estimate?
2. What are some of the reasons for preparing manufacturing cost estimates?
3. What is meant by the statement that "product design essentially determines manufacturing cost"?
4. What is a concept estimate, and what are come of the potential pitfalls in making such an estimate?
5. How can delivery requirements impact an otherwise sound manufacturing cost estimate?
6. Why is the manufacturing engineer the most qualified person to determine preliminary make or buy?
7. What are some of the factors other than cost and in-house capabilities that may influence a make or buy decision?
8. What minimum basic information is required to prepare a valid manufacturing cost estimate?
9. What is the most commonly used approach to cost estimating and how does it work?
10. What are some of the problem areas in material costs in which the manufacturing engineer can assist in resolving?
11. What three factors most influence the accuracy of the manufacturing labor estimate?
12. What is the difference between recurring and non-recurring costs in manufacturing estimating?
13. What is the underlying theory of the learning curve, and what causes it to happen?
14. What is meant by the term "slope" of a learning curve, and what determines the degree of slope for a given learning curve?
15. What are some of the factors that affect learning curves both favorably and adversely?

2
Product Manufacturability

2.0 DESIGNING FOR MANUFACTURABILITY

In the product design and development environment some frequently heard expressions are, "design for quality," "design for reliability," "design for automated assembly," and "design for manufacturability." The manufacturing engineer who has the opportunity to work with product design groups whose work reflects these expressions is indeed fortunate. Even more fortunate is the manufacturing engineer who has the opportunity to participate in the early stages of the product design to influence incorporation of design features favorable to good manufacturing practices.

Unfortunately, a close working relationship between design and manufacturing engineering is not a naturally occurring phenomena in all companies, but it is too important to be left to chance, since it always results in enhanced product designs and improved manufacturing processes. The design/manufacturing interface can be developed and maintained through the same good human relations practices that work in any other relationship, and must include courtesy, mutual trust, and respect. There is an excellent return in benefits to the manufacturing engineer who spends the time and effort to initiate and sustain this relationship, and to establish credibility as a contributor to the design/manufacturing engineering team effort. The overall goal of this team effort is to ensure that design functionality and manufacturability are compatible features of a good product design.

2.1 DESIGNING FOR MANUFACTURABILITY

Manufacturability can be thought of as the art and science of designing a product so that it is easy to manufacture. Manufacturability does not just happen. There must be a concerted effort by the design and manufacturing engineers at the earliest stages of the design concept phase. The manufacturing engineer must be there when the concepts are being discussed and he must actively pariticpate in the preliminary design reviews. He may provide cost estimates of the various design alternatives to help the design engineer narrow down his solution range. He must point out design approaches which have caused problems in the past. In short, the manufacturing engineer must be an active participant in the design and development process. The manufacturing engineer must be at the drawing board or computer terminal when preliminary and final design layouts are made. Problems or potential problems from a manufacturing standpoint must be found and corrected before the design is frozen and drawings are released.

As a minimum before the design is released, there should be a worst-case dimensional analysis to determine whether there are any tolerance stacks at assembly. If testing is required, the assembly should lend itself to stage-by-stage testing before incorporation into the next higher assembly. The product design should lend itself to mechanized and automated assembly techniques. Any test points or adjustments should be readily accessible. Cost trade-offs should be made between such design considerations as castings versus forgings, sheet metal structure versus machined hog outs, chemical process finish versus paint, and sheet metal versus composite structure, for example.

Manufacturability also includes materials. Standard materials, standard parts, and standard hardware means standard costs and procurement lead times. Special materials, nonstandard parts, and nonstandard hardware mean special manufacturing processes and special costs and lead times. In the final analysis, it is the product design that truly determines the cost of manufacture. Improved methods, processes and tooling have minimal impact when the product is not really a producible design.

2.2 THE NEED FOR MANUFACTURABILITY

Some companies in the high technology field, where new and untried product designs are frequently rushed into production before they are ready, may have their management invoke formal manufacturing design standards to ensure producible products. Such standards are prepared by manufacturing engineers and concurred in by design engineers, and define the key manufacturability requirements that

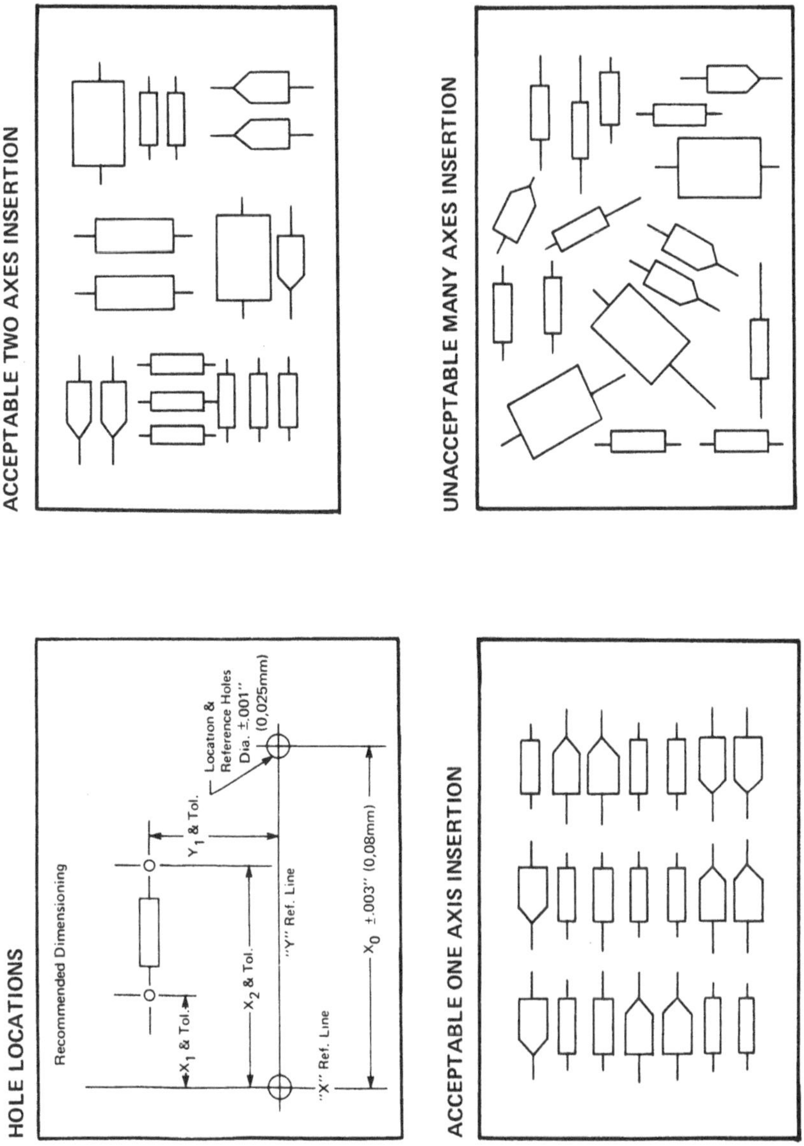

FIG. 2.1 Design guideline for printed circuit layout. (Courtesy of the Universal Instruments Corporation, Binghamton, N.Y.)

every new product must meet. Such standards cover castings, forgings, machined parts, formed parts, nonmetallic parts, structural assembly, printed circuit cards, assembly, paint, finish, and so on. Figure 2.1 shows an example of a manufacturing design standard for printed circuit card assemblies to ensure that they are designed and laid out properly for automatic assembly.

When the design is frozen, and before the drawings are released to manufacturing, the manufacturing engineer should be required to sign off on all the drawings in the technical documentation package before they can be released. This final review and sign off prior to release is just that: a last review to ensure that the product can be produced in manufacturing. Without the earlier working interface with the design engineer in the initial concept stages, through the design layout stage, having a final sign-off of the drawings prior to release to manufacturing may not be worth very much.

2.3 THE DESIGN/MANUFACTURING ENGINEERING INTERFACE

The product knowledge obtained from direct contact with the design engineer becomes the cornerstone of the information database that the manufacturing engineer must build to successfully perform the design function that is his primary responsibility—the design of the manufacturing process. At the same time that the manufacturing engineer is obtaining the product knowledge he will use to develop the manufacturing process, he must assist the design engineer by providing inputs to make the product compatible with the most economical and effective manufacturing techniques. This knowledge of, and experience with manufacturing processes and practices provides the basis for the manufacturing engineer's contribution to the team effort, complementing the design engineer's expertise in the areas of aesthetic and functional product design. The success of this team in creating product designs which are functional and adaptable to cost-effective production are critical to the long term success of the company.

2.4 AREAS OF OPPORTUNITY FOR REDUCING MANUFACTURED PRODUCT COST

There are many areas of opportunity where the manufacturing engineer can act to reduce the manufactured cost of the product and eliminate potential problems during the manufacturing phase. A planned methodical review of the design starting at early concepts and continuing throughout the total design cycle will be the most effective and productive way to accomplish this task.

Checklists are convenient tools when doing manufacturability analysis. They help prevent overlooking or forgetting items that could be opportunities for design improvements and, in turn, those missed opportunities to reduce manufacturing costs. The lists that are provided here are not all-inclusive and are intended as a starting point. The manufacturing engineer should develop a list which is less generic and more specific to the product line produced by his company, as he gains experience in manufacturability analysis and becomes more familiar with the product. As with most engineering skills, manufacturability knowledge and expertise improves with practice.

2.5 MATERIALS, PARTS AND COMPONENTS

The checklist for materials, parts and components address many of the major cost drivers in the design and manufacture of most lines of product:

1. Can the material be obtained in standard stock directly from supplier stock, or is a special order and special mill run required? (Bar stock, standard sheet, standard extrusions, etc.)
2. Is the material compatible with the most economical process of manufacture? (Ease of forming, casting, or machining.)
3. Is the material subject to wide price fluctuations over time?
4. Is the hardware of standard size and configuration?
5. Can the number of hardware sizes be reduced?
6. Are specified components (such as switches, locks, clamps, hinges, electronic components, etc.) standard catalog items?
7. Can the parts and component count be reduced, and the quantity of different sizes and values be reduced?
8. Are material callouts allowing for fabrication trimming and fixturing during the manufacturing cycle?
9. Are callouts for adhesives, potting compounds, paint, solder, wire, and similar items specific as to quantity required per unit of product, or is it left up to manufacturing to determine?

2.6 FABRICATED PARTS

1. Are specified tolerances reasonable for functional product requirements?
2. Are required tolerances attainable within the normal capability of the manufacturing process and equipment to be used?
3. Are data points/surfaces and tooling points clear, and are they accessible?

4. Does part configuration minimize the need for special processes, special tooling and special equipment?
5. Are flat or round surfaces specified instead of contoured surfaces?
6. Are sharp, reverse and closely spaced bends avoided on sheet metal parts?
7. Are corners properly radiused to avoid sharp bends?

2.7 PRODUCT ASSEMBLY

The following checklist for reviewing manufacturability of an electronic assembly is representative of the kind of questions the manufacturing engineer should be asking as he looks at the design concepts and design layouts for any manufactured product assembly:

1. Does the dimensioning facilitate manufacturing, and are the tolerances realistic?
2. Is all marking and stenciling defined and visible?
3. Are assembly notes complete and definitive?
4. Is internal wiring critial? If so, is the location clearly specified?
5. Are test points and adjustments accessible?
6. Is wire harness development required? If so, can the harness be fabricated outside the unit and installed later as a subassembly?
7. Does the design lend itself to mechanized or automated assembly techniques?
8. Does the design avoid the need for select-at-test component matching?
9. Are component parts accessible for assembly? Can all assembly and wiring be done without restriction?
10. Can required testing be performed without disassembling the unit?
11. If wire wrap is used, does the design facilitate automatic or machine wire wrap?
12. Has a prototype unit been built and tested?
13. Are standard connectors and assembly hardware used?
14. If circuit card assemblies are included as part of the top assembly, are they designed to plug in, or must they be wired in?
15. Are there mechanical loads such that printed circuit epoxy glass boards are in compression?
16. Is the maximum power rating of the assembly exceeded? Electrically? Thermally?
17. Has the assembly been analyzed to meet the vibration and shock requirements?
18. Has consideration been given to using printed circuit flex cable in place of hard wiring of the assembly?

19. Can plastic tie-wraps be used in place of lacing or spot ties on wire bundles?

One of the manufacturability techniques used by a manufacturer of power supplies is to design the wiring so that it can be grouped into wire bundles to form a harness with all wire ends soldered to connectors or terminal strips, or with crimped terminal lugs. This allows all the wiring in the form of a harness subassembly to be installed at final assembly with a power screwdriver.

No discussion of manufacturability would be complete without considering the requirements for automatic assembly. Boothroyd et al., in *Automatic Assembly*, (1982) have summarized the rules of product design for automatic or mechanized assembly as follows:

1. Minimize the number of parts.
2. Ensure that the product has a suitable base part on which to build the assembly.
3. Ensure that the base part has features that will enable it to be readily located in a stable position in the horizontal plane.
4. If possible, design the product so that it can be built up in layer fashion, each part being assembled from above and positively located so that there is no tendency for it to move under the action of horizontal forces during the machine index cycle.
5. Try to facilitate assembly by providing chamfers or tapers which will help to guide and position the parts correctly.
6. Avoid expensive and time-consuming fastening operations such as the use of threaded hardware and soldering.

The rules for individual parts design for automatic assembly were also included:

1. Avoid projections, holes, or slots that will cause tangling with identical parts when placed in bulk in the feeder. This may be achieved by arranging so that the holes or slots are smaller than the projections.
2. Attempt to make the parts symmetrical to avoid the need for extra orienting devices with a corresponding loss in feeder efficiency.
3. If symmetry cannot be achieved, exaggerate asymmetric features to facilitate orienting or, alternately, provide corresponding asymmetrical features that can be used to orient the parts.

As a minimum, an ongoing manufacturability effort will ensure that product drawings released for production:

1. Are dimensioned and have datum surfaces that are compatible with accepted machining and fabricating practices

2. Have sufficient stock allowances on castings, forgings, and stampings to provide for any mismatch or distortion which may result from heat treating
3. Have maximum allowable tolerances on nonfunctional features and characteristics
4. Have realistic tolerances on functional characteristics.
5. Have adequate provision for clamping and locating.
6. Provide sufficient clearance and access for assembly of all component parts
7. Call out standard parts and materials which can be processed and assembled with general purpose machines, equipment, and tools

2.8 DESIGN STANDARDS FOR MANUFACTURABILITY

The most critical areas in the design for manufacturability for most manufactured products include sheet metal, fabricated parts, machined surfaces and associated tolerances, printed circuit card layout, assembly involving the use of adhesives, and automatic assembly or automation. Accordingly, the remainder of this chapter will address these key areas, providing detailed design standards, which if followed will ensure that there will be few, if any, manufacturability problems. The manufacturing engineer should begin with these design standards, and with his added product knowledge gained as a member of the design team, should develop more specific design standards oriented around his company's line of products.

2.9 SHEET METAL FORMING AND FABRICATION

A critical area where manufacturing problems and cost drivers are likely to occur is in product manufacturability associated with metal forming and fabrication. Problems in metal forming and in fabrication are most likely to surface in the material selection and initial product design layout phases. This is also the best time to correct such problems, while the design is still fairly fluid. Material selection criteria include the following:

1. Use only standard sizes and tolerances. Nonstandard sizes may require a special mill run and special handling which escalate product cost and create manufacturing schedule problems.
2. Parts should be kept within standard sheet and plate widths and length, with allowance for between part trim and finish.
3. The preferred minimum sheet material thickness is .032 inch with .020 inch as the thinnest reasonable size.

Most sheet metal parts are made from aluminum which has a number of variations which the design team needs to consider:

1. Aluminum alloy 7075 is one of the strongest grades and is generally used in aircraft structural fabrication. When this material thickness exceeds 3/32 inch, it does not lend itself to shearing operations due to tearing and leaving the sheared edges ragged. This then requires that the parts be sheared oversize and then machined to drawing dimensions.
2. Condition '0' aluminum is soft and difficult to machine due to the galling effect it has on punches, dies, and cutters. Slower feeds and speeds are required to overcome this problem while excessively large burrs are evident after the machining of '0' material. Smaller bend radii are possible but distortion during handling and heat treatment leads to an expensive straightening or aligning operation.
3. Parts in the 'T' or 'H' condition can be formed satisfactorily provided the bend radius is large enough to prevent cracking. Lower temper materials such as T3 have a tendency to age harden at room temperature, so care should be taken in determining the proper bend radius to fit the design needs if this material is used.

The lowest level alloy which will do the job should be selected. This means lowest strength properties, highest ductility, and most commonly available material.

When considering sheet metal shearing or blanking operations, the following criteria should apply.

1. Design parts so that all edges may be sheared or blanked. This requires attention to forming procedures so that local polishing to avoid edge cracks is unnecessary. It should be specified that a sheared surface is permitted.
2. Allow punched holes for fasteners whenever possible.
3. Where blanks prior to forming are of a size or nature which require routing, be sure to permit corner and fillet radii which are compatible with standard cutters.

Numerous times at the drawing board stage, bend relief slots and notches are made so large that many extra hours of welding, cleaning and blending are required to make the fabricated assembly or part presentable and workable. In many instances the drawing tolerance only permits that power brake forming be within ±.010 inch, yet the slot for brake clearance is greater than .062 inch wide. This requires an additional .045 inch to .062 inch of slot to be filled with weld. This means that the accuracy of ±.010 inch in layout, notching

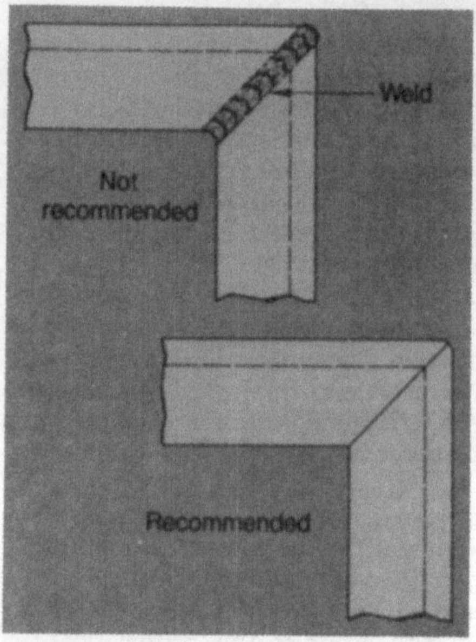

FIG. 2.2 Comparison of relief notching methods.

and forming is sacrificed for a later gaping welding operation. Figure 2.2, graphically illustrates this situation. To form a part and have the raw edges butt and join within .010 inch, it is only necessary to notch (vertex of 90 degree notch) to the inside of the bend dimension, plus one inside bend allowance. This disallows a bend relief hole and makes it very convenient for an inaccurate notching operation and off-dimension forming operation, but it adds hours to the job when welding is required to fill the gaps. Additional advantages are that the manufacturing operations close to welded areas can be performed while the part is in the flat, whereas these operations must be done after forming and welding when you have large areas to fill with weld.

Normally, deburring or any other handwork is the biggest cost driver in sheet metal parts fabrication. To avoid this, the drawings should never have a "break all edges" note. On sheet metal parts of any size this means extensive hand work and inspection. Use "remove burrs" note, but omit "and sharp edges." The latter may be construed to mean 90 degree edges with a small corner radius and result in extensive handwork.

As indicated earlier, it is very economical to fabricate sheet metal parts in the flat prior to forming. By punching holes and cutouts in the flat state, and keeping them far enough away from formed bends to ensure no distortion in forming, the need for drill jigs and fixtures after forming is eliminated.

Parts should be dimensioned and tolerances calculated to permit heat-treated condition forming. Heat treatment after forming creates part warp and requires subsequent reforming and straightening operations, but may well impact part-to-part uniformity with subsequent assembly difficulties.

Forming or bending is an easy way to convert flat sheet metal stampings into parts that are three dimensional, however, each forming operation is an added cost. Figures 2.3–2.5 illustrate ways to hold forming operations and costs to a minimum. Figure 2.3 shows a part in two different configurations, yet both do the same job of locating critical hole centerlines. The "not recommended" configuration requires two forming operations: one for each direction of bending, whereas the "recommended" configuration only one forming operation. The elongated Z bend is easy to include in the forming die, and is inexpensive. The "not recommended" configuration in Figure 2.4, requires a process known as channel forming. Severe angles,

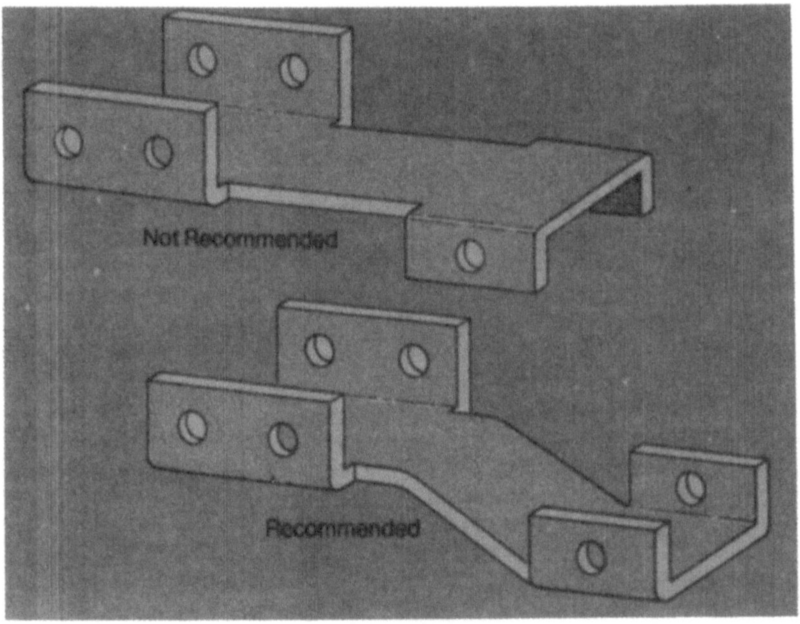

FIG. 2.3 Minimize bends.

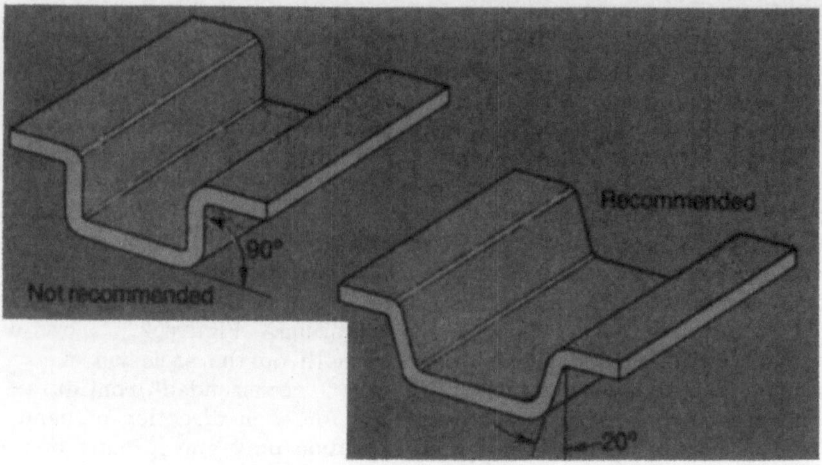

FIG. 2.4 When channel forming, avoid severe angles.

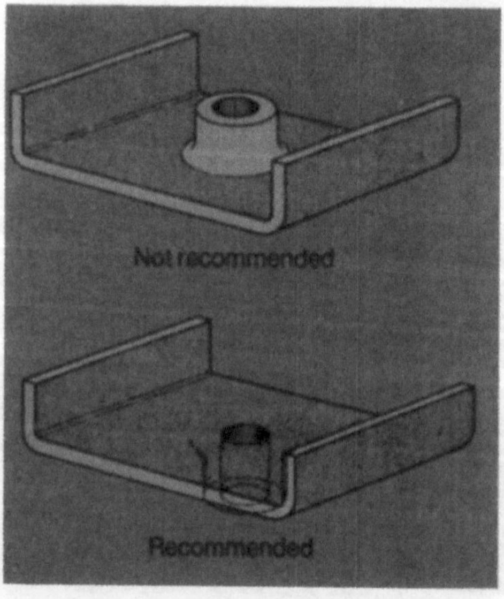

FIG. 2.5 Minimize extrusion operations.

near 90 degrees, make die removal difficult, create high stresses in the metal, and require high bending forces. The "recommended" configuration, also channel formed, has slight bending angles and is therefore easy and inexpensive to form.

In Fig. 2.5, the hole and boss in the center of this formed part must be extruded in both configurations. In the "not recommended" configuration, extruding is an additional operation, because the forming press must move in an opposite direction. In the "recommended" configuration, the part can be formed and the center hole and boss extruded in a single operation on the forming press.

One final comment on forming: avoid compound contour forming when simple curvatures will do the job. Compound contours require expensive draw or stretch dies that can cause localized thinning of the material. This may degrade the design of the product. Also, parts should be designed to be stretch-formed, not shrink-formed. Make full use of features such as raised reliefs and depressions in the part web, which will eliminate shrink-forming of flanges.

2.10 FINISHES AND TOLERANCES FOR MACHINED PARTS

A major cost driver on many manufactured products is the apparent requirement for microfinishes and precision tolerances on machined parts that go into the product. In many instances this may exceed the quality level needs of the finished product by an order of magnitude, and are specified by engineers unfamiliar with shop processes and machine capabilities. It is of the utmost importance that the design team make certain that economical surface finish and tolerance requirements for drilling, milling, turning, and grinding are selected for machined parts.

In the design of machined parts the goal should be maximum simplicity in functional and physical characteristics. The materials selected should be of the lowest cost that will meet design criteria and yet facilitate ease of machining and finishing. Inherent in the design should be consideration for ease of locating, setting up and holding the part. Parts should be designed for minimum number of separate operations in machining, finishing, forming, molding, casting, forging, and assembly. The bottom line is that machined parts should be designed for the most economical production methods taking into consideration their inherent producibility limitations.

Figure 2.6, specifies the range of acceptable economical tolerances for common machining processes. Tolerances on surface finish should be no greater than that required for acceptable functional operation of the part and the finished product. Surface finishes that are normally obtainable with selected manufacturing processes are shown in Figure 2.7. There is also a direct relationship between tolerances,

Range of sizes		Tolerances								
From	Through									
0.000	0.599	0.00015	0.0002	0.0003	0.0005	0.0008	0.0012	0.002	0.003	0.005
0.600	0.999	0.00015	0.00025	0.0004	0.0006	0.001	0.0015	0.0025	0.004	0.006
1.000	1.499	0.0002	0.0003	0.0005	0.0008	0.0012	0.002	0.003	0.005	0.008
1.500	2.799	0.00025	0.0004	0.0006	0.001	0.0015	0.0025	0.004	0.006	0.010
2.800	4.499	0.0003	0.0005	0.0008	0.0012	0.002	0.003	0.005	0.008	0.012
4.500	7.799	0.0004	0.0006	0.001	0.0015	0.0025	0.004	0.006	0.010	0.015
7.800	13.599	0.0005	0.0008	0.0012	0.002	0.003	0.005	0.008	0.012	0.020
13.600	20.999	0.0006	0.001	0.0015	0.0025	0.004	0.006	0.010	0.015	0.025

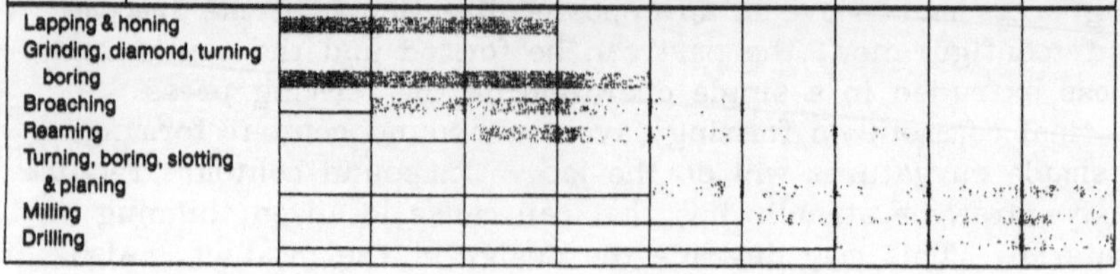

FIG. 2.6 Tolerance chart.

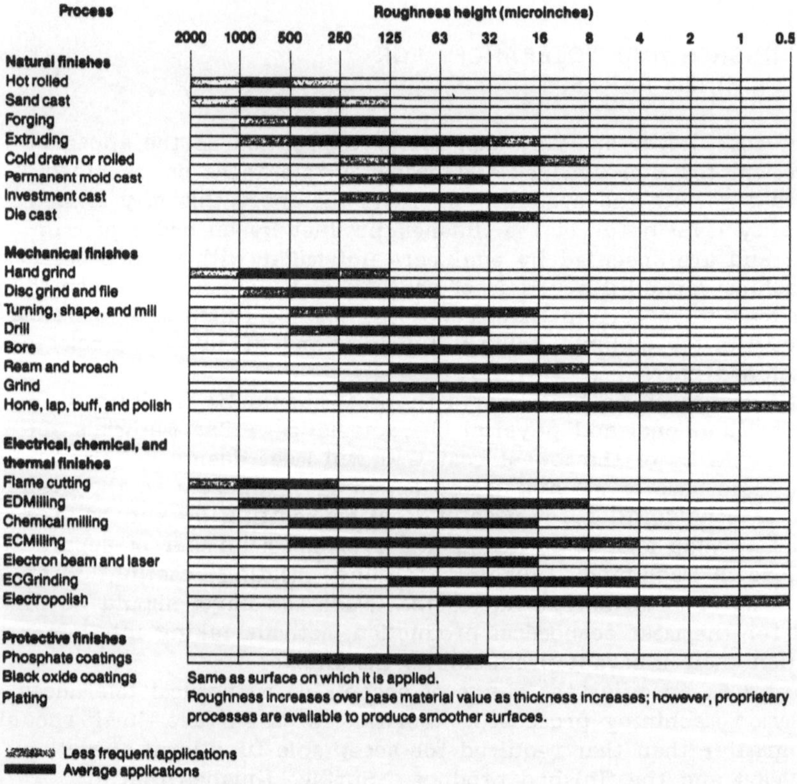

FIG. 2.7 Normal machine tolerances.

surface finishes, and cost. This is shown in Fig. 2.8. Tolerance and finish requirements must be established based upon the part's intended function. It should be kept in mind that as more stringent tolerance and finish requirements are placed upon a part, additional, more precise machining operations are required which drives up cost. Figure 2.9 shows how relative costs of machining increase as tolerance and surface finish requirements are tightened. Normal working tolerances on hole drilling range as follows:

Drill size	Tolerance
80	+.0023 −.0005
75	+.0027 −.0005
60	+.0034 −.0005
50	+.0041 −.0010
35	+.0047 −.0010
30	+.0050 −.0010
2	+.0060 −.0010

If closer tolerances are required, the cost will increase. Hole depth should be as shallow as functionally possible. As hole depth increases, cost increases. When hole depth becomes three times the hole diameter and more, cost increases exponentially.

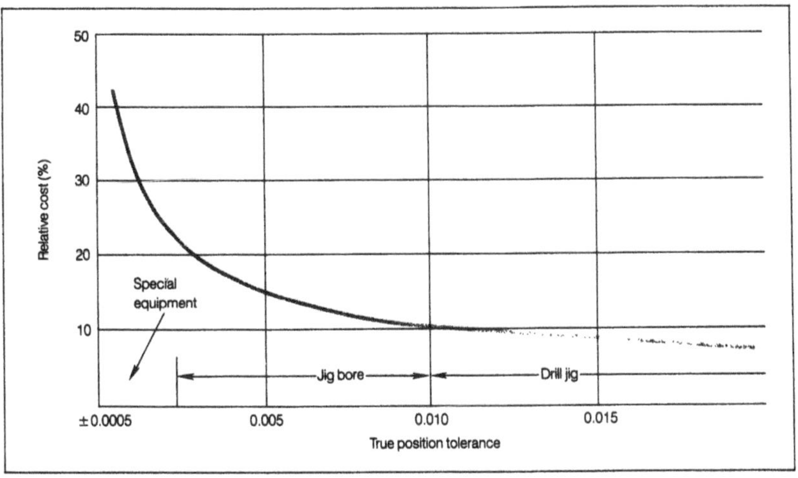

FIG. 2.8 Cost vs. tolerance.

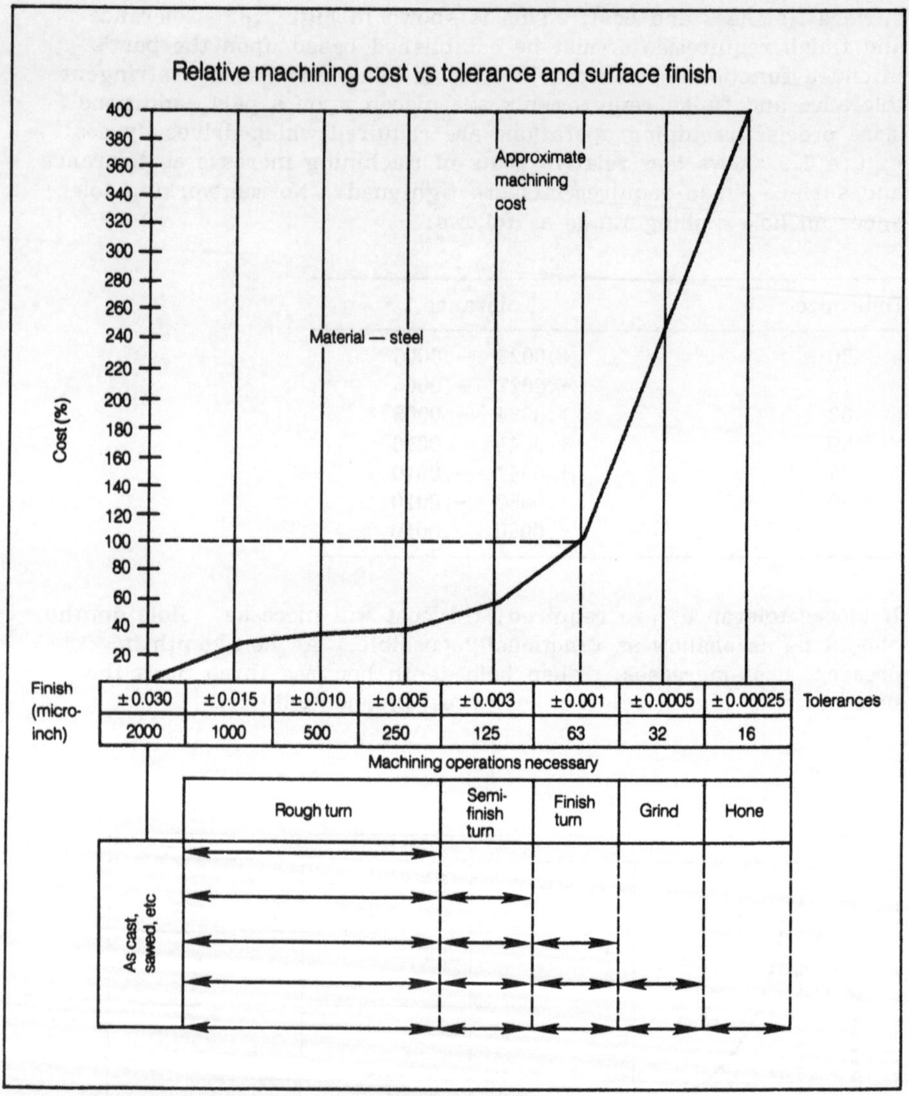

FIG. 2.9 Cost vs. tolerance and finish.

Much of what has been covered in this section on machined surface tolerances and finishes can be summarized in seven product design cautions, as follows:

1. Avoid surface finishes that cannot be achieved in the first operation of fabrication (for example, a forging that requires a 32 microinch finish will need a subsequent machining operation).
2. Generally, surface finishes of 63 microinches or less will require additional machining operations to meet specifications, and should be avoided.
3. Avoid tolerances on machined parts that require several operations to achieve. Use economical design tolerances indicated for that type of process.
4. Avoid hole depths which are three times or more greater than hole diameter.
5. Generally avoid tolerances tighter than .004 inch in drilling operations.
6. Generally avoid tolerances tighter than .0025 inch in milling operations.
7. Generally avoid tolerances tighter than .0015 inch in turning operations.

2.11 PRINTED CIRCUIT ASSEMBLY

Manufacturability in printed circuit assembly must address the citeria for electronic component selection and placement, mounting hole location and tolerances, and tooling hole requirements for double-sided, plated through-hole printed circuit cards, and for multi-layer, plated through-hole printed circuit cards.

The first consideration must be circuit card size, which should not exceed 16" × 18" if it is to fit on an automatic component insertion machine. When the total surface area of a circuit card is 36 square inches or less, "break-away" configuration should be considered. Figure 2.10 shows a break-away circuit card panel, containing four circuit cards retained by break-away tabs. This arrangement allows processing, handling, and testing as a single panel, rather than as separate circuit cards.

A major consideration in the design of printed circuit cards for ease of assembly is the proper location of tooling holes. Tooling holes may be drilled or punched depending upon material and board thickness. The following are design criteria for tooling holes:

1. Tooling holes should be separated by the maximum diagonal distance allowed by the size of the board or panel.

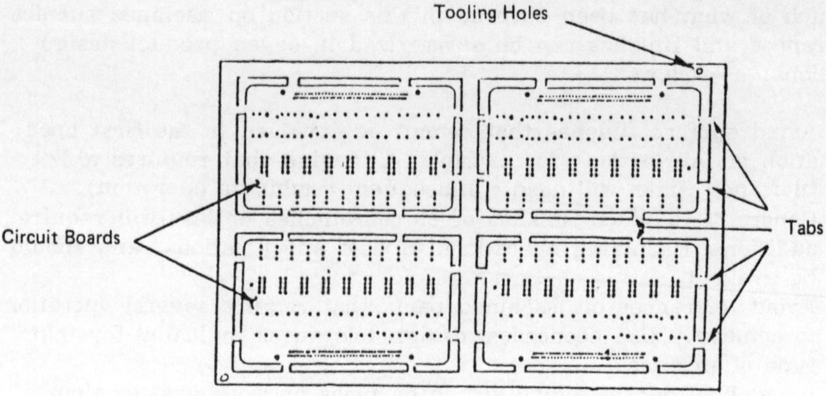

FIG. 2.10 Break-away PCB panel.

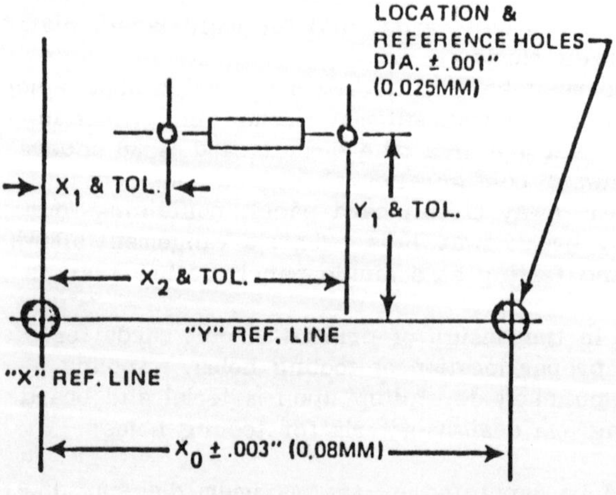

FIG. 2.11 Component location relative to tooling holes.

2. Tooling holes are to be placed in each panel of boards in a "break-away" configuration, as well as in each individual circuit board.
3. Tooling holes should be located within ±.003 inch.
4. Tooling hole size should be .125 inch diameter ±.001 inch.
5. Locations of all component mounting holes should be within ±.003 inch in relation to the tooling holes.
6. No components should be located within .050 inch radius around tooling holes.

Figure 2.11 shows proper location of components relative to tooling holes.

Orientation of axial leaded components to facilitate automatic insertion is another consideration which must be addressed. Components are to be located preferably in one axis orientation, however, two orthogonal axes are acceptable. Axial leaded components should be laid out on a .100 × .100 inch orthogonal grid. Figure 2.1 shows layouts that are preferred and acceptable for axial leaded components. Dual-in-line packages (DIPS) should be oriented as shown in Fig. 2.12, and lead spans should be as small as possible. Preferred DIP lead spans are as follows:

–.300 inch lead span, 2 to 20 leads
–.400 inch lead span, 22 to 40 leads
–.600 inch lead span, 22 to 40 leads

For DIPS, the minimum hole diameter is:

$$(.022)^2 + (\text{lead thickness})^2 + \text{Hole Location Tolerance} + .010 \text{ inch}$$

(But in no case less than .037 inch). Maximum hole diameter for DIPS is .041 inch.

Again, for axial leaded components, the range of component body sizes and lead diameters should be as small as practical. The positioning of a small body component immediately adjacent to a large body component should be avoided, as should the use of noncentered axial components as shown in Figures 2.13 and 2.14. Discrete components should not be mounted on both sides of a circuit card as shown in Figure 2.15.

For axial leaded components, the minimum hole diameter is calculated as follows:

$$\text{Lead Diameter} + .010 + \text{Hole Location Tolerance}$$

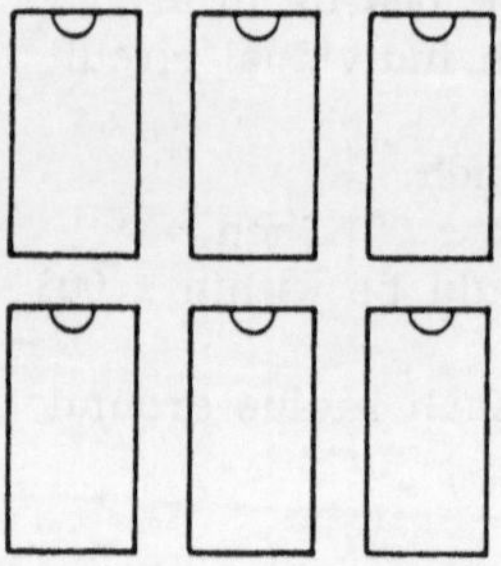

Preferred. All packages parallel to one axis with Pin-1 identically oriented.

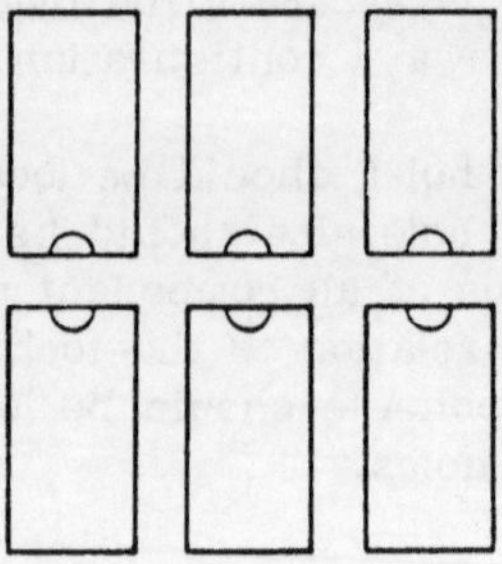

Acceptable as necessary. All packages parallel to one axis, but Pin-1 in both directions.

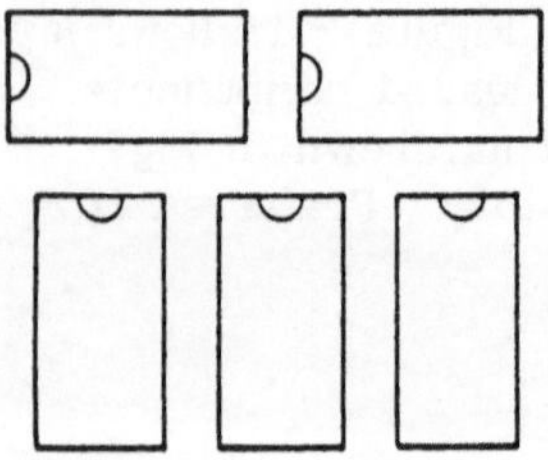

Preferred. Packages parallel to both axes, but Pin-1 identically oriented in each axis.

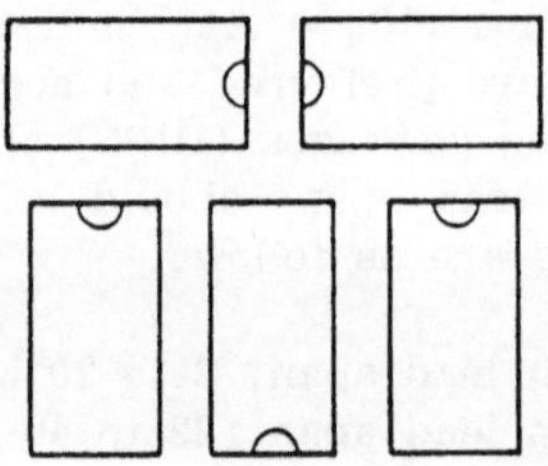

Acceptable as necessary. Packages parallel to both axes, and Pin-1 oriented in both directions in each axis.

FIG. 2.12 Dual-in-line package orientation.

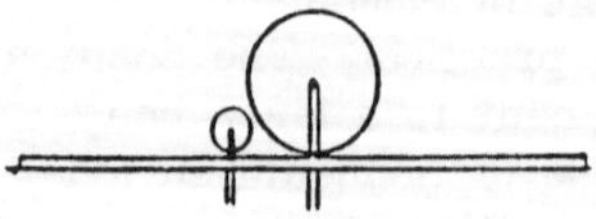

FIG. 2.13 Small vs. large component.

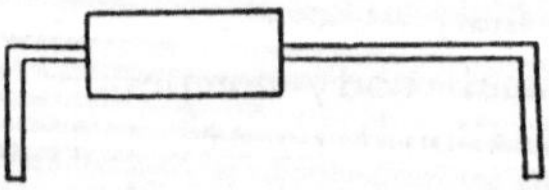

FIG. 2.14 Off-center component body.

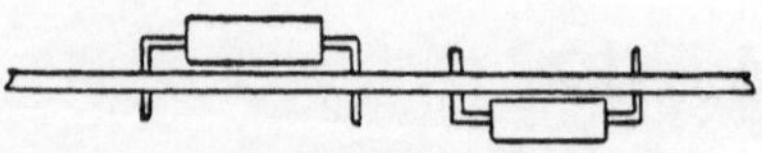

FIG. 2.15 Components on both sides.

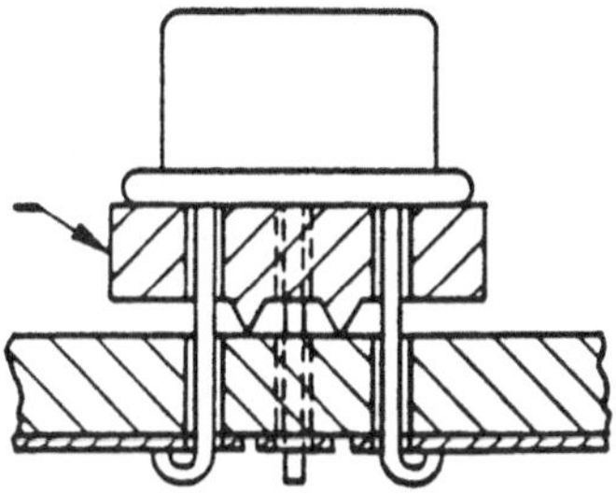

FIG. 2.16 Radial-leaded device.

For example, a 1/4 watt resistor with .025 inch lead diameter, and with a circuit board hole location accuracy within .003 inch of true position, we then have:

Minimum Hole Diameter = .025 inch + .010 inch + .006 inch

= .041 inch

Maximum hole diameter should never be more than .020 inch larger than the lead diameter.

As shown in Figure 2.16, radial leaded components should always be mounted so that the component base is parallel to the surface of the board. The minimum hole diameter would be calculated as follows:

Lead Diameter + .010 + Hole Location Tolerance

Maximum hole diameter requirements are the same as for axial leaded components.

Savings in the range of a five to one ratio can be achieved by designing a printed circuit card for automatic insertion versus the use of manual insertion.

2.12 ADHESIVE BONDING

Generally, adhesive bonding should be avoided in manufactured products if there is any other simple means of mechanical assembly. It is a time-consuming, usually messy, and often long cycle process due to cure times. If adhesive bonding is used, the use of single-part adhesives is preferred. In particular, hot melts are desirable due to their long shelf life and short cure times. Adhesives in tape or strip form are preferred over bulk or paste adhesives.

If multipart adhesives must be used, they should be comprised of no more than two components. Mix ratios should be as close as possible to one-to-one by volume or weight. Pot life of mixed adhesives should be sufficient to permit on-line mixing and dispensing as required by the process cycle time. Cure time should be matched to the operation cycle time. If cure time is expected to exceed the operation cycle time, the part should be designed to include a feature to prevent movement of the bonded area during subsequent operations.

Primers, if required, should have an unlimited life in an unprotected environment. If this is not possible, the primer life must be sufficient so as not to constrain the production process. Figure 2.17, shows types of joints where adhesive bonding is used. Two final cautions where adhesive bonding must be used in the manufactured product design:

1. Grit blasting, sanding or other abrasive action to prepare the surfaces to be bonded should be avoided.
2. The allowable amount of "squeeze out" or overapplication of the adhesive material should be clearly specified on the drawings.

Adhesives should be selected based upon materials involved, joint configuration, load requirements and production rate. The use of more than one type of adhesive should be avoided in a single assembly.

2.13 AUTOMATIC ASSEMBLY

Designing for automatic assembly must include all mechanical parts, as well as assemblies. The product must be designed so that it can be assembled as a stacked or layered unit, allowing installation of all parts from one direction only, usually from above. This allows gravity to assist in the feeding and placing of parts. Figure 2.18 shows an example of such an assembly.

Excessive lifting and rotating of parts during assembly should be minimized. The design should be such that only simple patterns of movement are required. Fasteners should be oriented vertically, or as a minimum, in one direction. Fasteners such as screws, bolts, clips and rivets should not be used. Assembly should be by snaps, interlocks, tabs, slots, and various press fits. Fasteners should be standardized. If possible, use the same size throughout the assembly. Various fastener configurations and their ease of use in automatic assembly are as follows:

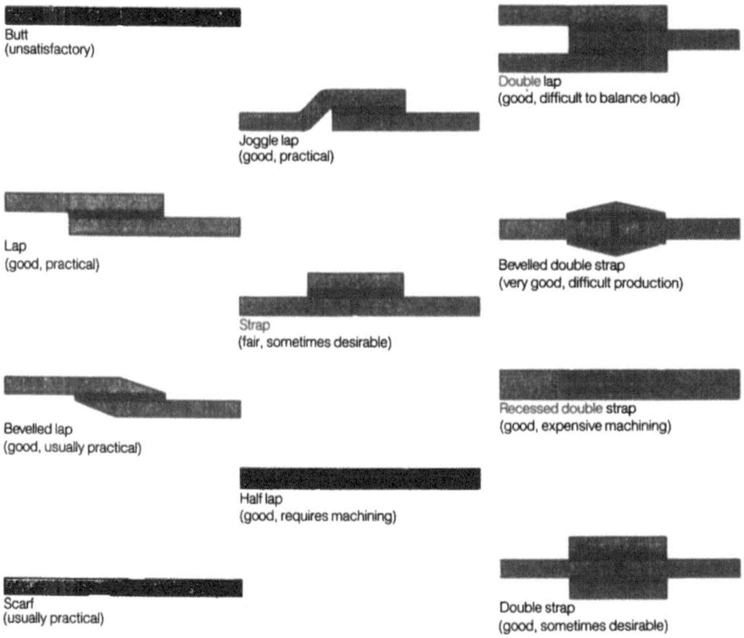

FIG. 2.17 Joint design for adhesive binding.

- Rolled thread point: difficult to locate, will not centralize without positive control on the outside diameter of the screw.
- Header points: slightly better than rolled thread points, but only if it is of the correct shape.
- Chamfer point: easy to locate.
- Dog point: easy to locate.
- Cone point: very easy to locate.
- Oval point: very easy to locate.

Figure 2.19 illustrates these various fastener configurations.

Component parts of the assembly should be designed with guide surfaces. These should be in the form of chamfers, ramps, or lead-ins to allow slightly misaligned parts to mate. Subassemblies should be self-contained units. Figure 2.20 illustrates good versus poor

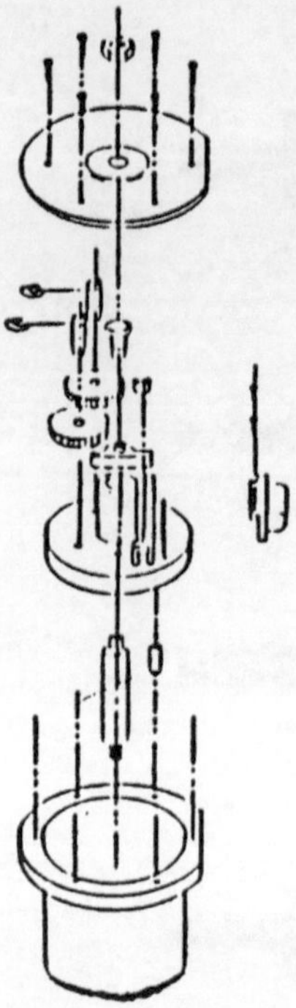

FIG. 2.18 Stacked design for automatic assembly.

Configuration	Fastener Type	Comments
	Rolled thread point	Difficult to locate; will not centralize without positive control on outside dia. of screws.
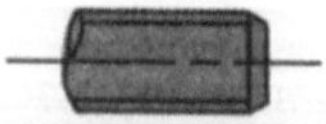	Header point	Slightly better, but only if it is of correct shape.
	Chamfer point	Reasonable to locate.
	Dog point	Reasonable to locate.
	Cone point	Very easy to locate.
	Oval point	Very easy to locate.

Fastener configurations.

FIG. 2.19 Fastener points.

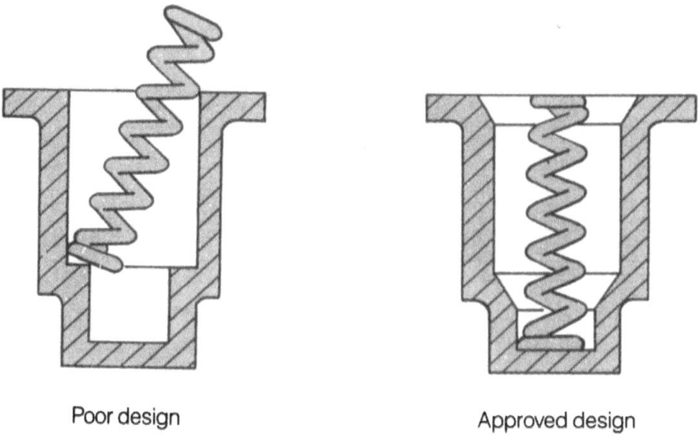

(a) (b)

FIG. 2.20 Guide surfaces.

design for a guide surface. To ensure that parts fit together without a secondary aligning and straightening operation, guide pins, D-shaped holes, and notched slots should be designed in. Figure 2.21 shows how this works. Parts located in this manner cannot be dislodged.

Parts should be designed to reduce handling needed to orient properly. This can often be accomplished by designing symmetry into the parts. Asymmetrical parts that have several orientation possibilities can be designed with an offset center of gravity, or a marked asymmetry that enables the part to correctly oiriented with automatic equipment.

Parts should be designed with features that facilitate orinetation for machine feeding, as illustrated in Figure 2.22. Parts should be designed for ease of moving and transporting, not only in grooves, chutes, channels, and on rails, but also for ease of transporting in systems such as pick-and-place units.

Parts should not be designed with open loops, sharp edges, acute angles or other features that might tangle and cause jamming in an automatic feeder. They should incorporate a flat base, low center of gravity, and mating surfaces that are generally parallel to the direction of feed travel. Figure 2.23 provides examples of this.

Components and subassemblies should be plugged directly together, eliminating connecting cables and wires. If this is not feasible, they should be connected directly to a circuit card or other rigid intermediary. Assemblies and subassemblies should be designed so that

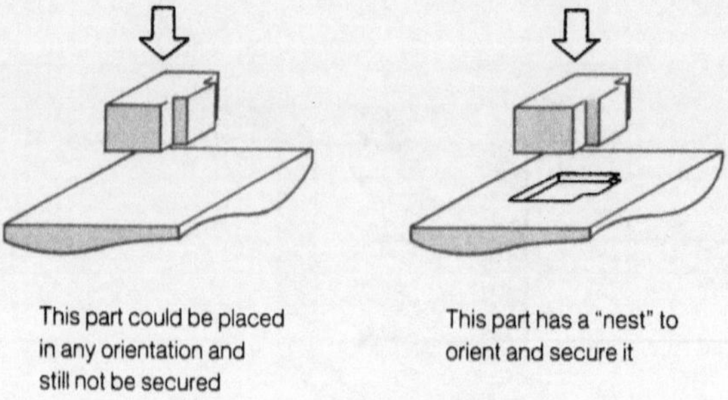

FIG. 2.21 Design aids orientation.

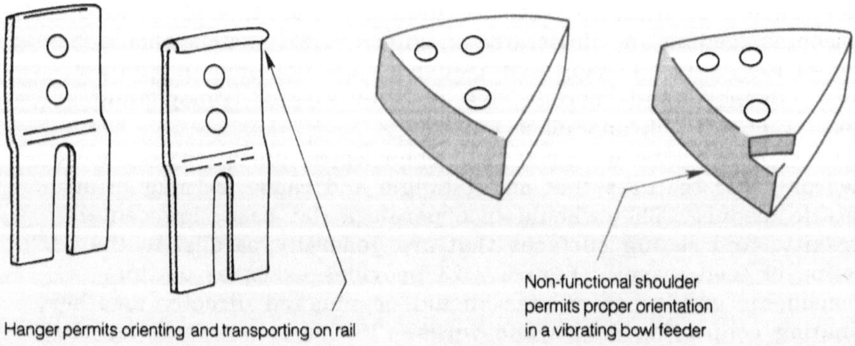

FIG. 2.22 Parts design for machine feed.

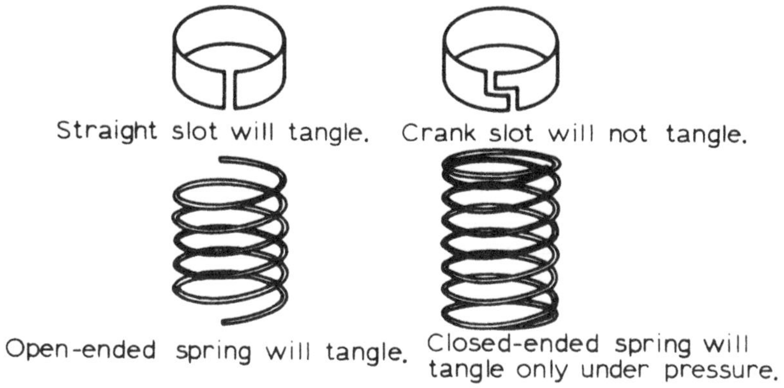

FIG. 2.23 Design for automatic feed.

any tool required for insertion or tightening can easily reach the required location and perform the required task. Products should be designed to provide unobstructed access to all parts and in such a way that parts can be assembled in one step or operation. Number of parts should be held to a minimum. One way to do this is to avoid subassemblies. Design the subassembly into one single part whenever possible.

The following is a rough guide to use when determining whether or not to design the product for automatic assembly:

Number of units	Recommended manufacturing process
100,000 or less	Manual with tooling and some mechanization
100,000 to 1,000,000	Mechanization with some robotics
1,000,000 and above	Full automation

Automatic assembly offers a multitude of advantages, not the least of which is the replacement of the human operator with automatic workheads where tasks are very simple. The human operator then is retained for the more complex tasks that would be uneconomical to automate. Other advantages of automatic assembly include:

- Reduction in the cost of assembly
- Increased productivity
- A more consistent product

BIBLIOGRAPHY

Boothroyd, G., Poli, C. R., and Murch, L. E., *Automatic Assembly*, Marcel Dekker, Inc., New York, 1982.

Gooch, J., George, M., and Montgomery, D., *America Can Compete*, Institute of Business Technology, Inc., Dallas, 1987.

Oxford, C., Causes of Twist Drill Breakage, *Tooling and Production*, January, 1978.

Tanner, J. P., Planning To Minimize Manufacturing Costs, *Production Engineering*, June, 1986.

Tanner, J. P., Estimating Manufacturing Costs, *Machine Design*, August 11, 1988.

QUESTIONS

1. How does the manufacturing engineer assist the design engineer in developing a producible product design?
2. Why is it so important for the manufacturing engineer to begin his work with the design team very early in the product develoment and design cycle?
3. Why are checklists such an important tool and aid to the manufacturing engineer in doing manufacturability analyses?
4. In selecting sheet aluminum for product design, what is the criteria that should be used?
5. What is the most expensive manufacturing operation in sheet metal fabrication, and how is it avoided?
6. Why should call-outs for hole depth on machined parts be as shallow as functionally possible?
7. What is the primary consideration from a manufacturability standpoint in specifying surface finishes on machined parts?
8. In the design of printed circuit cards, when should "break-away" configuration be considered?
9. Why should adhesive bonding be avoided in the design of most manufactured products?

3

Preproduction Planning

3.1 CONCEPTUALIZING THE MANUFACTURING PLAN

Preproduction planning defines the manufacturing plan. From work with design engineering on product manufacturability and from review and analysis of manufacturing drawings, the manufacturing engineer establishes the make or buy plan and a preliminary bill of material. Then he begins to conceptualize the steps and the sequences of manufacture, beginning with detailed parts, through the various levels of subassembly, to the final or top assembly. Figure 3.1 shows how this development takes place when a strong working interface exists with design engineering, and the product is newly designed and has never been manufactured.

This visualization process is aided by the various flow charting techniques, by the manufacturing engineer's knowledge and experience, and by the understanding of the product design and intended function. The visualization process is further enhanced by experience gained in fabricating, assembling and testing engineering models, prototypes, and development units to be used for further testing and evaluation.

The construction of a first article flow, with its time estimate for the manufacture of each part and assembly provides the baseline information needed for determining manufacturing lead times, preparing manufacturing schedules and build-lines, and establishing what outside supplied material is needed first and last to manufacture the product.

The operation process chart shows the sequence of operations required to manufacture the product and where the purchased parts and material flow into the process. The listing of tools, machines,

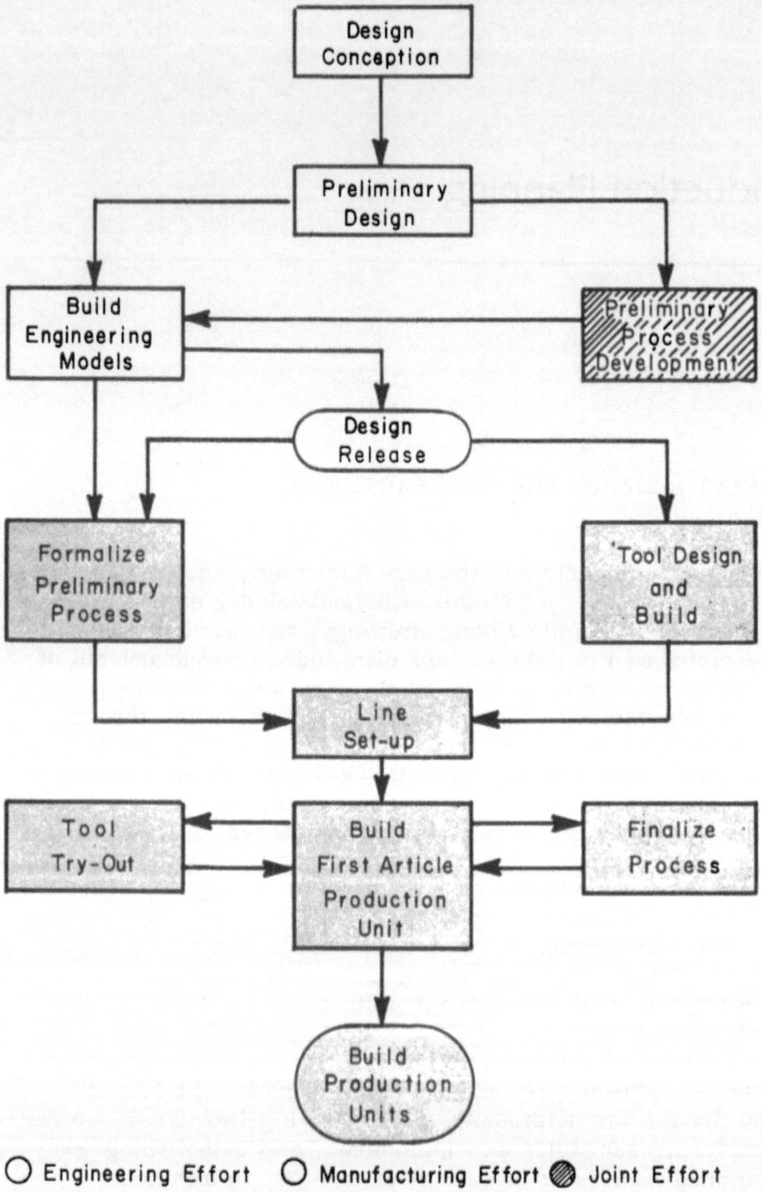

FIG. 3.1 Engineering and manufacturing engineering functions interfaced during design cycle. (With permission from *Industrial Engineering Magazine*, Vol. 2, No. 10, October, 1970, copyright © Institute of Industrial Engineers, Inc.)

and equipment required by operation then finish defining how the product will be manufactured. The next step is conceptualizing a design for special tools, test equipment and machines, followed by an estimation of manufacturing labor, floor space, and support facilities required.

3.1.1 Charting Techniques

The various process charts used by the manufacturing engineer are designed to help clearly visualize the manufacturing process in its entirety. The most useful chart for this purpose is known as the operation process chart. An operation process chart depicts graphically the points in the process where materials are introduced into the process, and the sequence of all manufacturing inspection, and test operations. The chart may contain such detail and information as standard times for each operation, production equipment and tooling required for each operation, and applicable process specifications. Symbols used in constructing operation process charts include:

○ *Operation:* Occurs when an object is intentionally changed in any of its physical or chemical features, is assembled from another object, or its arranged for another operation, inspection, or test.
□ *Inspection/test:* Occurs when an object is examined or verified for quantity or quality in any of its characteristics.

The principles of operation process chart construction are shown in Fig. 3.2. A preprinted chart format for operation process charting is not recommended, because of the wide range of use and application. All steps should be listed in proper sequence for each component or part, working vertically from top to bottom. The major component or assembly is always shown at the far right, and all other components are allotted space to the left of this main component or assembly. The presentation is similar to that of a mechanized assembly line, with components, parts, and subassemblies fed into the top assembly in proper sequence and at the proper point in the process. Operation and inspection descriptions should be brief, utilizing shop terminology such as drill, tap, ream, weld, assemble, solder, and test.

Figure 3.3 shows an operation process chart for a printed circuit card assembly. Time values, when assigned to each operation, test, or inspection, should be broken down into set-up and run times. Other equipment used, tools required, and special process provisions, then completes the operation process chart. The completed operation process chart highlights material, operations, inspections, tests, and time.

With the completed operation process chart, the concept of the entire process of manufacture can be visualized. All operations can

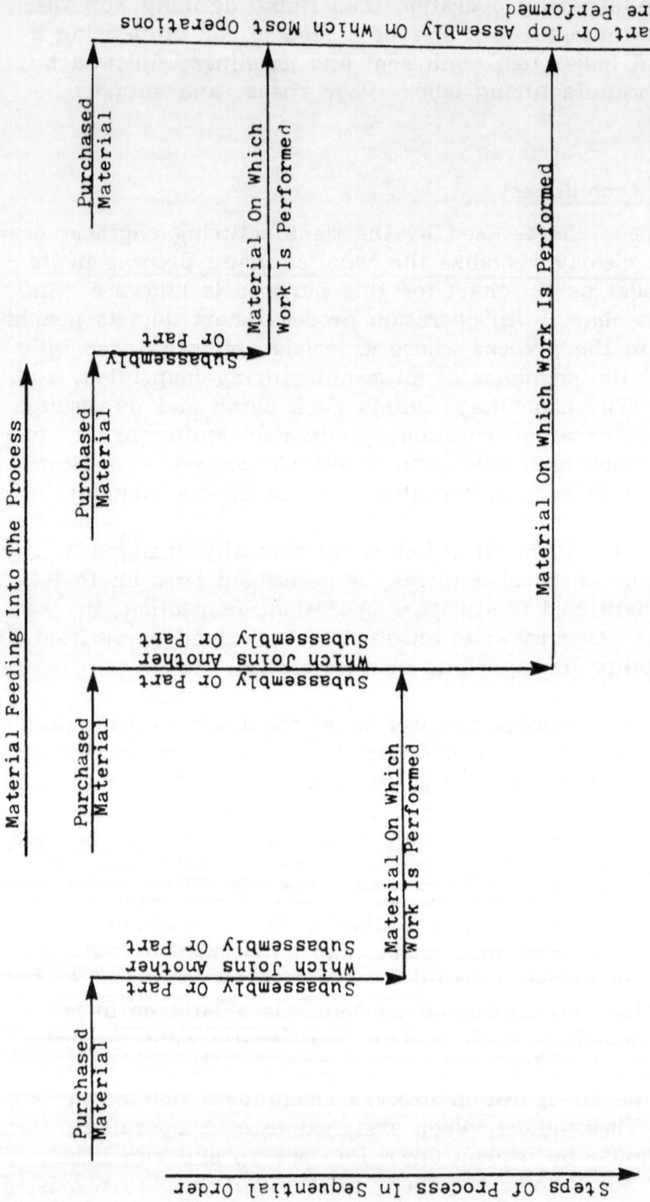

FIG. 3.2 Principles of construction of an operation process chart.

then be reviewed for optimum sequencing, alternative methods of processing, and most efficient methods. Equipment and tooling requirements at each step in the flow can be determined. Table 3.1 which lists all the basic manufacturing processes is an aid in constructing the operation process chart. It can readily be seen that every step on the chart will involve a manufacturing sequence, or an inspection or test.

The visualization process needed to construct the operation process chart is fairly straightforward for metal parts fabrication or for

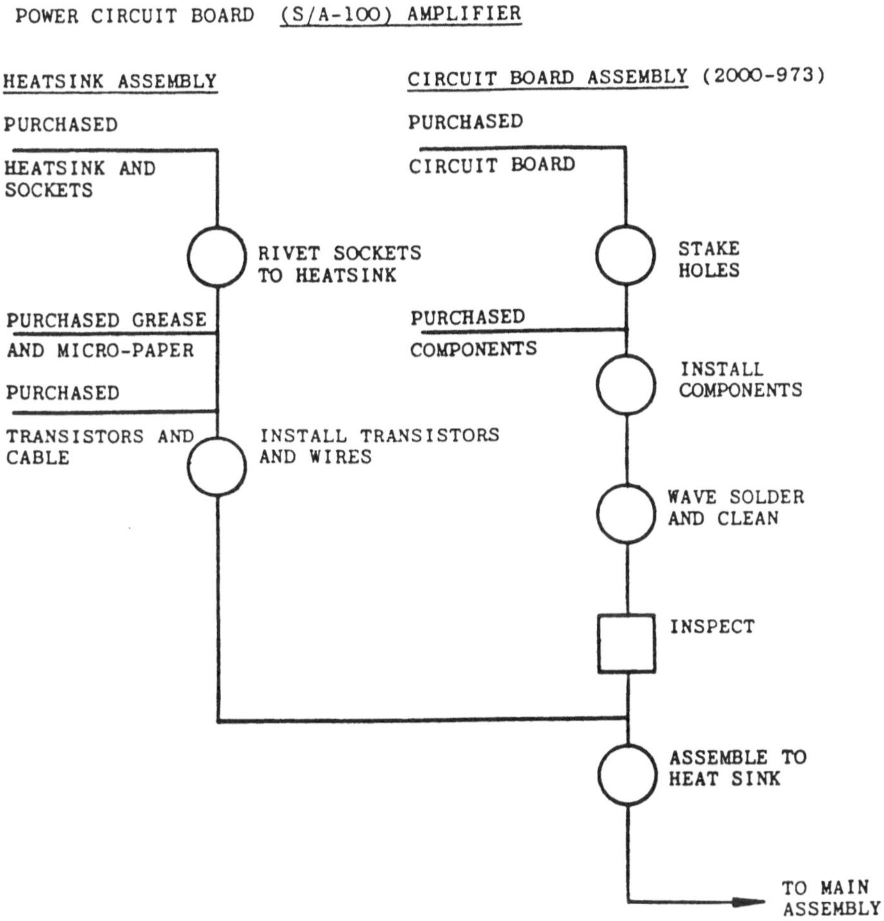

FIG. 3.3 Circuit card assembly.

machining. The primary requirement is a good working knowledge of the various processes involved and the ability to read and correctly interpret the requirements of the engineering drawings. Visualization of the manufacturing process and sequences in the final analysis defines the manufacturing plan. It is imperative that sufficient time be allowed for the manufacturing engineer to do a thorough job of preproduction planning. Many times the process of visualization for complex assemblies can be aided by exploded-view sketches or tooling isometrics of how the parts and assemblies actually

TABLE 3.1 Manufacturing Processes

	Casting and molding	
Sand casting		Compression molding
Investment casting		Transfer molding
Die casting		Extruding
Shell casting		Laminating
Permanent mold casting		Powdered metal molding
Injection molding		
	Machining	
Turning	Drilling	Broaching
Cutoff	Shaping	Honing
Milling	Grinding	Abrasive cutting
Routing	Etching	Sawing
	Forming	
Rolling	Forging	Forming
Coining	Swaging	Spinning
Drawing	Punching	Trimming
	Assembly	
Soldering	Brazing	Welding
Press fitting	Shrink fitting	Adhesive bonding
Mechanical fastening		
	Finishing	
Cleaning	Buffing	Polishing
Blasting	Plating	Deburring
Painting	Heat treating	Conformal coating

go together. Figure 3.4 shows a tooling isometric of the assembly breakdown of a small tactical missile.

3.1.2 Identifying Equipment and Tooling Requirements

Once the process of fabrication and assembly has been determined and the inspection and test requirements identified, then the manufacturing engineer must determine what production equipment and machines will be required in each step of the manufacturing process. In many cases this can be done quickly and easily, as the manufacturing process is centered around existing production machines and equipment, or standard equipment which can be readily obtained. In determining equipment required for metal forming or machining, the analysis of the drawing established the process and sequence of operations required to make the part, as well as identifying which machines are required. The analysis in Table 3.2 provides an example.

All product machined parts would be analyzed as shown in Table 3.2. This kind of analysis identifies machines and equipment required. From the basic analysis of the two example parts, it is apparent that we would need the following machines and equipment:

(1) Cutoff saw
(1) Engine lathe
(1) N/C lathe
(1) Three-axis, N/C mill
(1) Multiple spindle drill and tapping machine
(1) Vapor degreaser

TABLE 3.2 Analysis of Machined Parts

Item	Part number	Operation	Description
Nut, processor retainer	5257295	005	Saw
		010	Lathe, rough, and finish turn
		015	Multispindle drill
		020	Multispindle tap
		025	Degrease
Plate, hinge	5257345	005	N/C lathe
		010	Three-axis, N.C mill
		015	Degrease

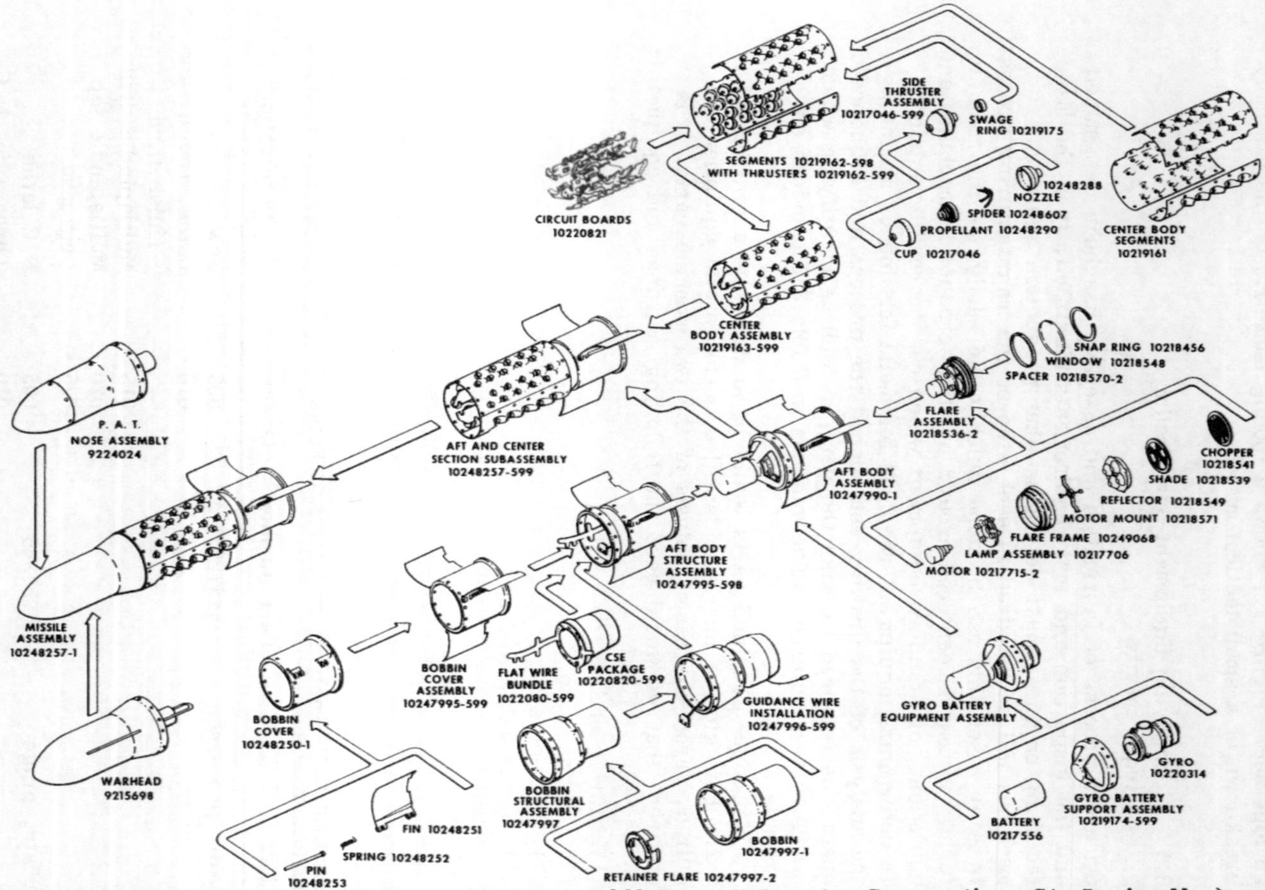

FIG. 3.4 Assembly exploded view. (Courtesy of McDonnell Douglas Corporation, St. Louis, Mo.)

The next step would be to determine the specific size and capability required for the machines and equipment identified. How large is the part that must be machined? What tolerances must we hold? Why an N/C machine rather than a conventional machine? What is the material to be machined? Why not a single-spindle drill rather than a multispindle machine? How many holes must be drilled in each part? How many are threaded holes and must be tapped? The answers to these and similar questions come from the analysis of the engineering drawing, the specifications, and the knowledge, experience, and professional judgement of the manufacturing engineer. Other very important considerations are the planned peak production rate and length of the production run.

Once we have fully defined the requirements the machines must meet, the next step is to find out if there are existing machines or equipment in-house that can do the machining required. Do these machines have the necessary rigidity, and can they hold the required tolerances? We must look at the age and condition of these machines, and whether they can meet the peak production rates planned. Another key consideration is work already on existing machines and work scheduled during the needed period of usage.

There could be several reasons for considering the purchase of a new machine. The first and most compelling would be that there is no machine in-house capable of running the part. Another might be that the existing equipment is behind state of the art in modern machine tool technology and would not be cost effective in the production of the new parts. Possibly subcontracting the parts to an outside shop that has the equipment and experience might be the most effective approach in the final analysis.

When considering the economics of buying new or additional production equipment, the job itself is a primary consideration. How long will it last? What is the potential for follow-up business? If the machine or equipment is general purpose, what are the possibilities of using it for other work in order to keep it fully loaded with work? If the proposed equipment will create a new manufacturing capability, is there a real possibility of bringing in more work with the new capability?

Tooling requirements are identified concurrently with the machine and equipment requirements. Working from the tooling philosophy that was established at the time the job was bid, we can determine what the tooling approach will be. We must know whether the product will be fabricated by using minimal shop-aid tooling or even a no-tooling approach, by fabricating on setup. The approach may be to invest in hard tooling and all that is involved in such a decision, including design and construction costs. It is essential that a tooling philosophy or approach be determined in advance, preferably in the proposal or estimating phase, and be used by the manufacturing engineer as his overall direction in determining the tools required to do the job.

Once the tooling philosophy has been determined, tools required at each step in the process are determined and listed. A drilling step, for example, may require a drill jig and a holding fixture, whereas an assembly operation may require weld fixtures, holding fixtures and/or solder fixtures. A careful and complete part print and assembly print analysis is required, with special attention given to the manufacturing step or operation, part or assembly shape and geometry, and production rate. An important part of the tooling for any product is the assembly tooling. Assembly tooling should include, in addition to holding fixtures, locating fixtures and the like, motion economy devices and other similar tools to facilitate the assembly operations from a methods standpoint.

On large jobs of major programs, it is often desirable to illustrate tooling concepts with isometric sketches or illustrations which show in a pictorial manner the planned production flow, equipment, and major tools exactly as they will appear when set up and operating. Figure 3.5 shows how the isometric for a moving belt assembly line used to assemble Dragon missile aft bodies would appear. Production from this line was 300 per day over two 8-hr shifts. The belt was set to move at 6 in. per minute; each station was a 3-min station, or about 18 in. in length. Similar moving belt assembly lines are used for the assembly of products ranging from television sets to small applicanes.

In planning tooling requirements, it should be kept in mind that tools fall into one of several categories:

1. Special-purpose tooling: tooling peculiar to the part or asembly it makes. This is tooling which must be designed and then built. It is good for the life of the product in production only.
2. General-purpose tools: tools which can be used in the production of any and all products. No special design is required. This kind of tooling is commercially available and is much less expensive than special-purpose tooling. Some examples would include tool balancers, small hand tools, power assembly tools, and standard assembly holding fixtures.
3. Expendable tocls: examples are cutters, end mills, drills, taps, broaches, and reamers. These tooling items should be stocked on a min-max basis. Provision should be made to identify the items by the manufacturing engineer during the preproduction planning phase.
4. Tooling components: included in this category would be standard tooling components such as quick-acting clamps and drill bushings.

The preproduction planning phase is most important when it comes to defining the tooling and fixtures required to build the product.

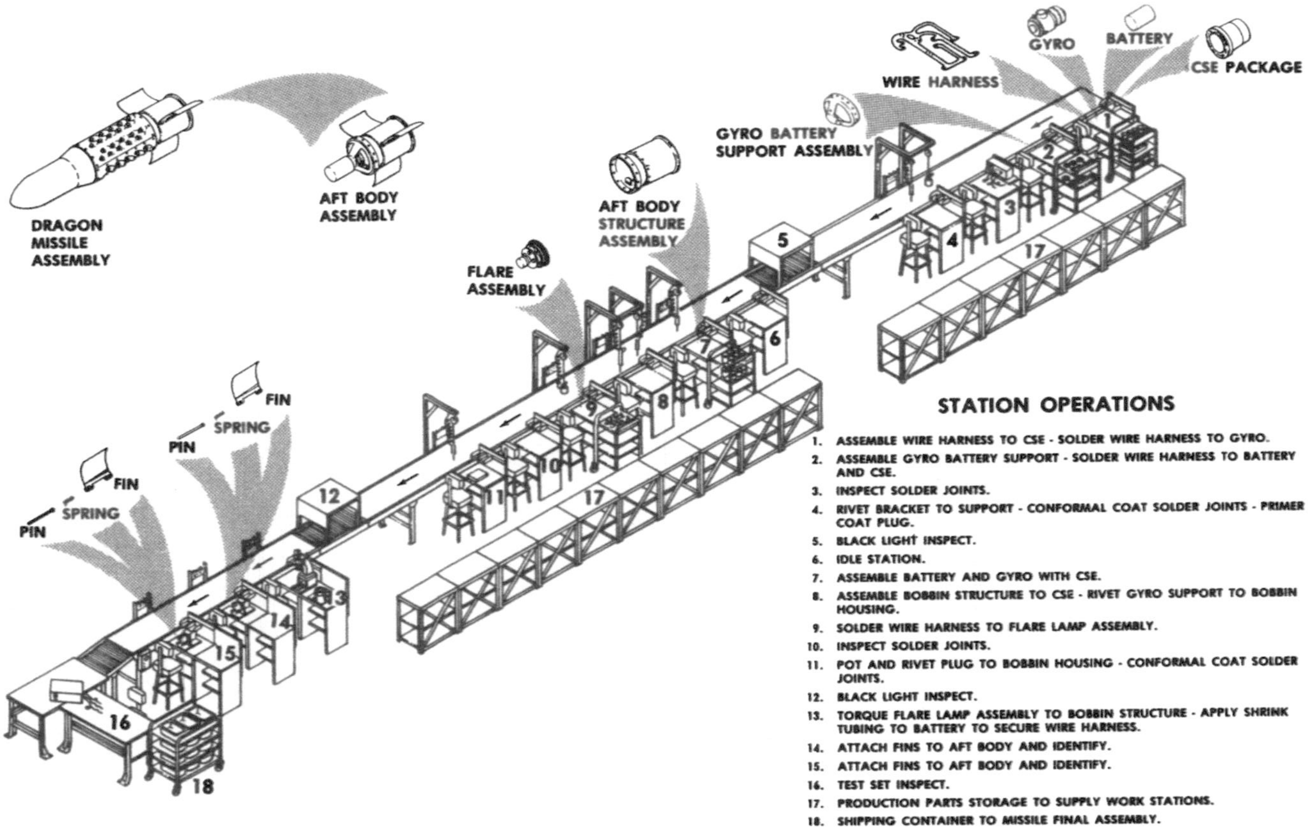

FIG. 3.5 Moving belt assembly line. (Courtesy of McDonnell Douglas Corporation, St. Louis, Mo.)

The manufacturing engineer must review each operation in the total manufacturing process and identify the tooling required. He must consider such factors as:

How the part or assembly will be held or positioned to perform the work required
What orientation and accessibility the fixture must provide the operator so that the work may be performed most efficiently
What relationship the tool will have to the other tools, parts, parts bins, and so on required to complete the operation
The spatial relationship between the operator and the workpiece to provide an efficient and comfortable work position for the operator

When special-purpose tooling is required, the manufacturing engineer should utilize design concepts of previously well-engineered tools and fixtures. A manual which depicts these previously successful tool design concepts should be available to the manufacturing engineer during the preproduction planning phase. The importance of having such a manual cannot be emphasized strongly enough.

3.1.3 Manpower and Floor Space

Manpower requirements are determined from the standard times that were established for each operation, factored by a realization factor and multiplied by the planned production rate. These figures should be checked against the original estimate to make certain that there is substantial agreement between both sets of calculations. If there is a discrepancy, it should be investigated to determine what is different or has changed since the original estimate was made.

The ground rules, assumptions, and constraints in effect at the time of the original estimate should be reviewed to determine whether there have been any changes now that the job is in-house. In particular the product design or configuration at the time of the estimate should be checked against the current configuration. Has the complexity or parts count increased? If so, by how much, and what does this do to the labor required in manufacturing? Any changes to the original base-line at the time of the estimate can be crucial to completing the job within the cost that was negotiated with the customer. Any differences should be brought up by the manufacturing engineer to management during the preproduction planning phase.

In addition to product design changes, which can be critical in a product that is making the transition from development to production, there can be other changes even in a stable or previously built product that can mean substantial differences from the original cost estimate, especially in the area of production labor. A change in the performance requirements or in the product acceptance criteria can cause major cost changes from the original estimate, in additional inspection or test requirements. It can also increase rework costs. In addition, particular attention should be given to the small and otherwise subtle changes which, taken individually, have a quite small effect, but, when taken together, can have a major cost impact.

The operation process chart, which described the manufacturing process used to build the product, also defines the layout of the factory to support product manufacturing. The amount of floor space required is a function of the size of machines and equipment required in the manufacturing process, the area required to work and service the machines and equipment, and the number of units of product in flow and in process at the peak planned production rate. Labor-intensive assembly areas, where shop benches and storage racks would constitute the equipment in the area, are sized by multiplying the number of operators in the area at peak rate on a single work shift by 50 square feet per operator. Floor space requirements determined in this way should be increased by 15% for aisles, material drop areas, and other support functions.

If the product is to be run on existing machines and equipment, the requirement for additional floor space would be minimal and would include only dedicated area for the new product, if any, and whatever expansion of the common processing areas would be required due to the new product causing an overall capacity increase. Once floor space requirements have been determined, the next step is to identify special facilities requirements. These may include such items as clean rooms, requirements for hoods, vents, and other similar openings in roofs or walls, special power requirements such as 28-volt direct current lines for testing aircraft instrumentation, and requirements for sound and noise protection.

As mentioned earlier, the operation process chart determines product flow and how dedicated areas would be laid out. In the preproduction planning phase, the overall block layout plan and most of the detailed area layouts should be developed. Figure 3.6 shows an overall block layout plan for a small electronics manufacturing plant. Overall flow of work is shown, which helps in visualizing the total factory layout plan.

The final phase in preproduction planning is the development of detailed layouts for product production areas. The overall block layout plan has defined the space requirements, the relationship of one area or department to another, and the overall flow of work and material through the facility. The operation process chart has defined in detail the product flow. The manufacturing engineer must now decide the arrangement of equipment and facilities that is needed in each of these areas of the block layout plan.

It should be emphasized at this point that this is a preproduction plan and, as such, is subject to more detailed analysis as the total production plan is finally detailed. In the preproduction phase,

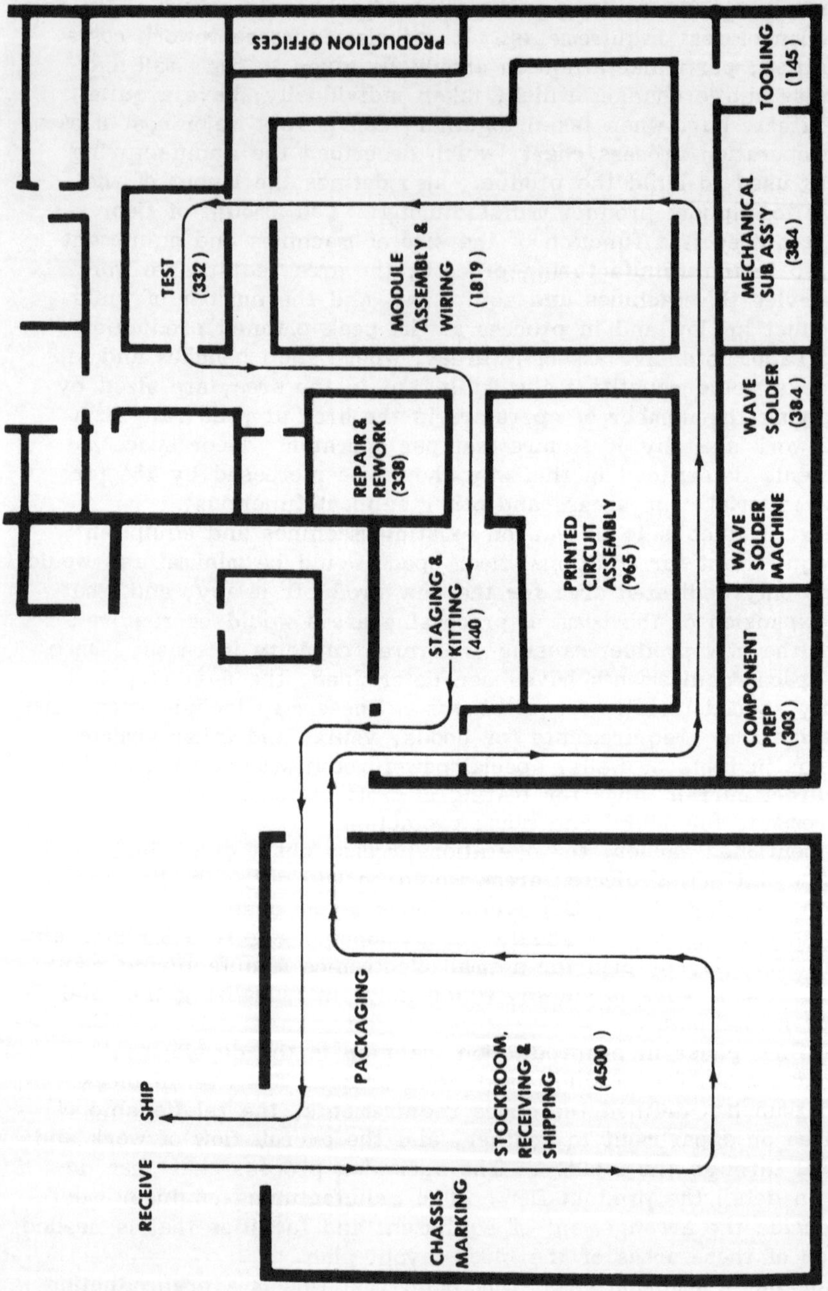

FIG. 3.6 Overall block layout plan.

layouts of production lines and manufacturing equipment may be conceptualized and roughed out, based on planned production rates, the number of work stations needed, and where in the process flow the equipment should be placed to ensure optimum support to the production operations. This layout would be an idealized layout, which does not consider the constraints imposed by the building, by other products that may be in production at the same time, and so on. Figure 3.7 shows such an idealized layout plan in isometric for the manufacture of wire harnesses. This idealized layout would then be worked into a detailed layout for one of the block areas on the overall block layout plan.

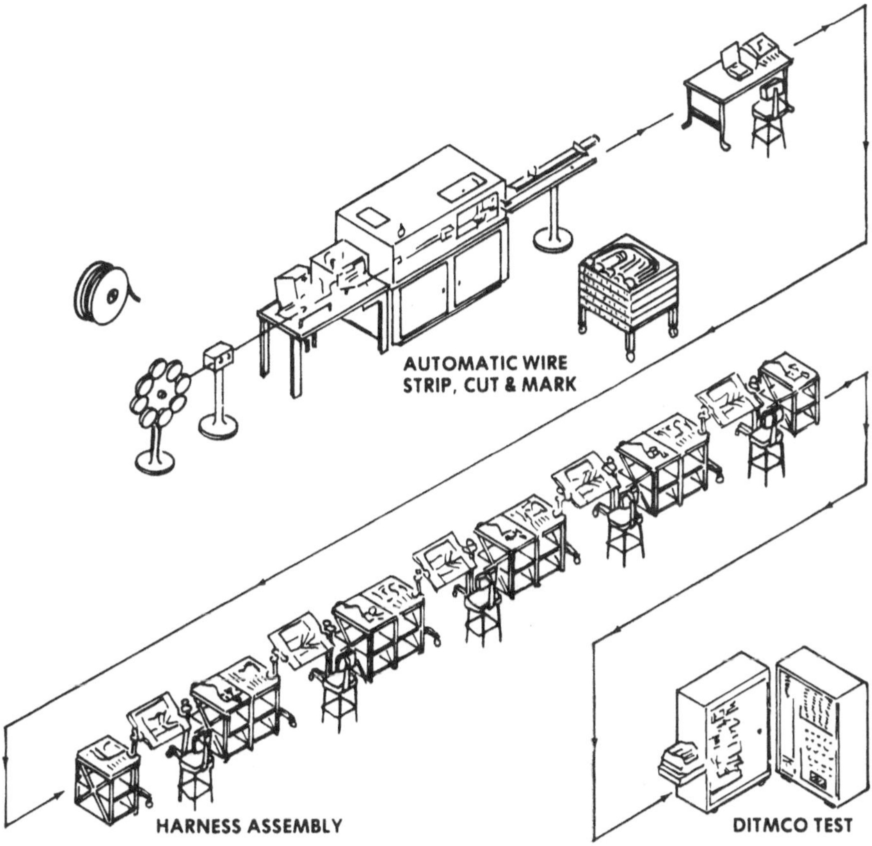

FIG. 3.7 Conceptual layout and work flow for wire harness manufacturing. (Courtesy of McDonnell Douglas Corporation, St. Louis, Mo.)

In a small company or plant, a shop that specializes in machining, metal forming, plating, and finishing, a foundry, an investment casting house, or where a single product line is run on a continuing basis, the preproduction planning for a new job, or the continuation of an existing job, may require a totally different approach than the one described above, and it may be far less formal. The point to be emphasized, however, is the need for a preproduction plan developed before the product goes into production. This planning may have been done when the estimate was prepared and the work was quoted to the customer. When the order is placed, the plan is then executed. However it is done, a preproduction plan is a necessity.

3.2 FIRST ARTICLE FLOWS AND MANUFACTURING BUILD LINES

First article flows of the product and manufacturing build lines which show the time-phased, cumulative buildup of production to the end of the program are basic planning tools for the preproduction plan. The first article flow depicts how the product goes together, in what sequence, the elapsed time in days to do each build sequence, what fabrication and subassembly operations can be done in parallel, leading to the total elapsed time to build one complete unit of the product. This information from the first article flow chart provides what is needed to construct the manufacturing build line, by providing the needed setback times to begin production to meet delivery requirements.

It is essential that the manufacturing engineer understand and be able to apply these important planning tools. Time is always a key element in any preproduction plan and failure to consider actual shop flow times, rather than standard times, is why many preproduction plans often go astray. Shop flow times are the actual floor-to-floor times it takes to physically move the parts and materials into place, do the work required to fabricate, or assemble, inspect, test, and stage for the next operation, and then to physically move the part or subassembly to the next series of operations. It includes allowances for delays, waiting time, and other factors that enter into the "real world" of production in your plant or facility.

3.2.1 First Article Flow Chart and What It Tells

Figure 3.8 shows a typical first article flow chart used in the Aerospace Industry. It graphically depicts the elapsed time in working days to accomplish the sequential final assembly of the Tomahawk Cruise Missile. Also included on the chart, but not always a requirement, is the lead time needed to order the long-lead aluminum plate required for the machining of the forward and aft cylinder skins.

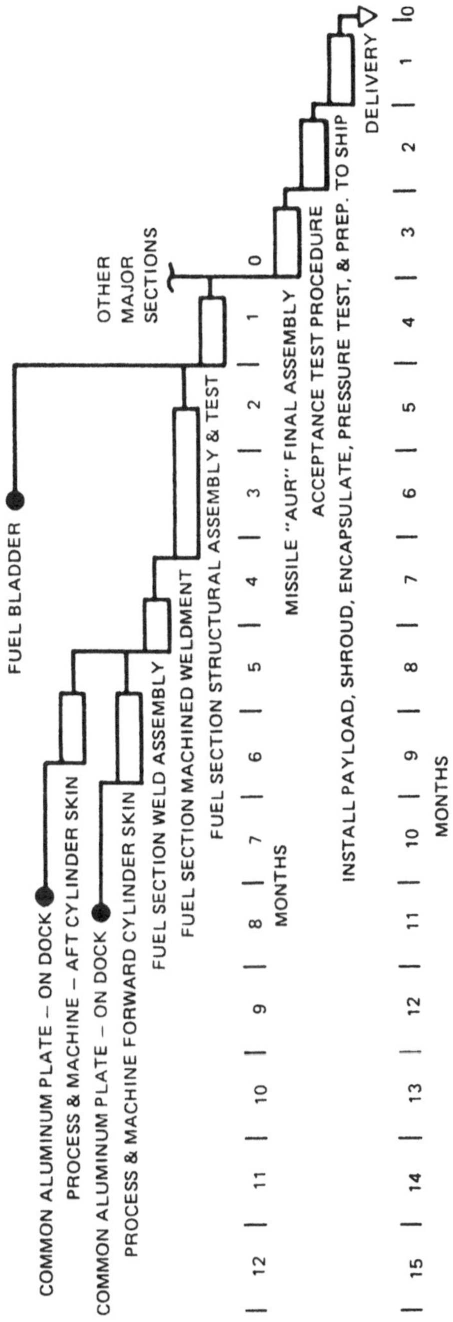

FIG. 3.8 First article flow for Tomahawk Cruise Missile final assembly and test.

The time scales on the chart are in months of elapsed time, the upper time scale representing time through final assembly, and the lower scale time through delivery to the Navy. The time scale can be in any unit of time that is appropriate for the product, from days to months.

It is readily apparent that the first article flow chart becomes the basis for development of production schedules, the manufacturing build lines, and the total time-phasing of production start-up and buildup to peak rate production. It tells when parts and all materials must be available, when lower level fabrication and subassembly build must begin, when tooling and test equipment must be available, when manpower for the production lines must be in-place and available, and more, in order to meet planned product delivery schedules. The importance of this time-phased setback information to the preproduction planning cannot be emphasized strongly enough. Only the manufacturing engineer, with his technical and product knowledge, plus his knowledge of shop operations and flow times, can construct this important document.

3.2.2 Manufacturing Build Lines

Figure 3.9 shows a manufacturing build line for the assembly and test of electronic assemblies and subassemblies for the Mark 46 Torpedo. The setback time spans for constructing this chart are taken directly from the first article flow charts for these units. In this particular case, the plan is to build in block releases in order to more readily accommodate the required delivery to the Navy. Since the chart shows product build and delivery in cumulative terms, the build lines themselves are curves as shown. Note that as the buildup gets closer to peak rate production, the spans between start of lower level subassembly build and completion of final assembly increases to almost four months, indicating added time as a cushion for unforeseen impacts and contingencies.

3.2.2 Peak Rate Units in Flow Chart

Another valuable tool in developing the preproduction plan is the peak rate units in flow chart. An example is shown in Fig. 3.10. This chart defines product flow at any specific time during peak production. The chart in Figure 3.10, is based upon a peak production rate of 53 units per day, and a work station time of 1.5 hours. This shows the number of units that should be in-work at each of the work stations, inspection stations, and test stations to maintain the peak production rate of 53 units per day.

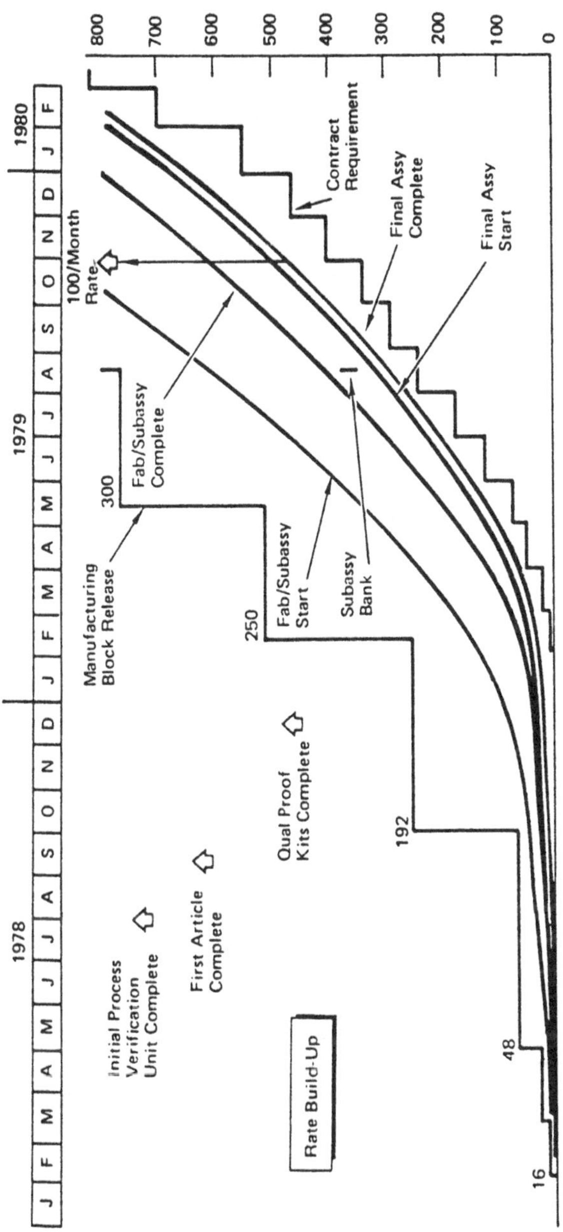

FIG. 3.9 Manufacturing build line for Mark 46 NEARTIP, Torpedo program.

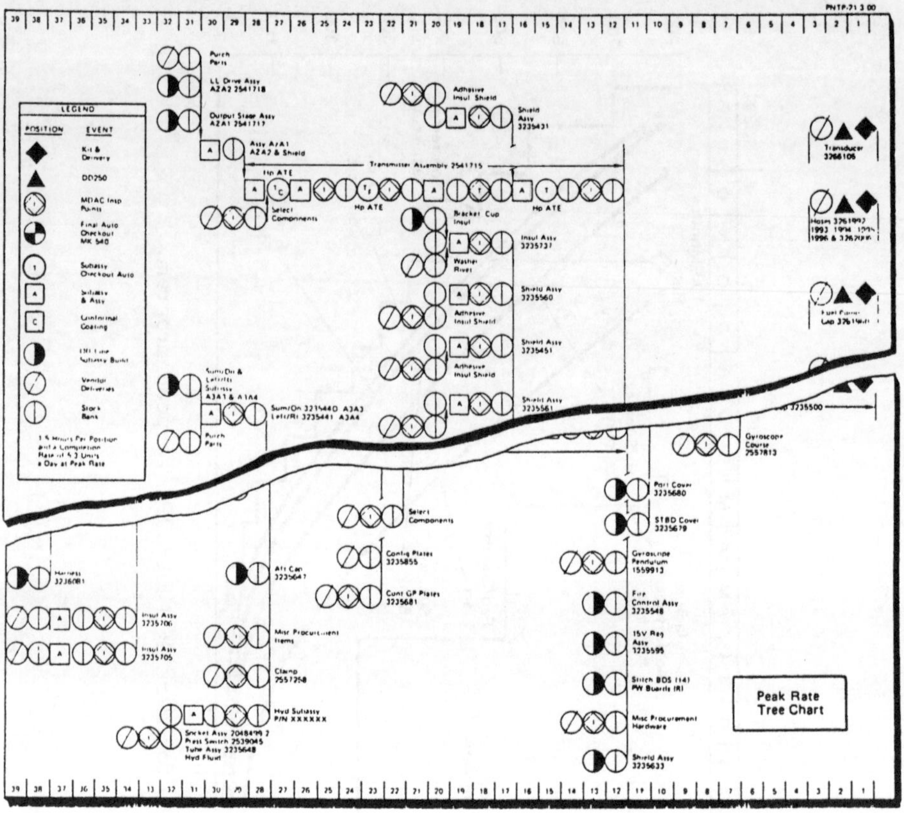

FIG. 3.10 Peak rate units in flow for a transmitter assembly.

3.3 MECHANIZATION AND AUTOMATION PLANNING

Mechanization and automation are no longer the wave of the future; they are here with us today. As stated in Modern Machine Shop in the January 1984 issue, "Numerical control has been the major concept among forces driving the metalworking industries across the threshold into a new era. Closely coupled with the booming technologies of microelectronics and computers, it has truly revolutionized manufacturing." Flexible manufacturing systems are beginning to revolutionize the assembly plant. Flexible assembly systems allow work to be selectively routed to certain assembly stations, bypassing stations where the work does not have to go. Flexible assembly systems can reduce work in process, improve product quality, maximize the use of floor space, and minimize capital expenditures for equipment. They also offer the flexibility to change or relocate the assembly system at minimum cost.

The manufacturing engineer today must understand and be able to plan for these new technologies, as well as understand the steps that lead from manual assembly to mechanization, and ultimately the automation of manufacturing processes. He must have a working knowledge of computers and computer control. Knowledge and skill in the fundamentals of numerical control of machines are essential for the manufacturing engineer in the metalworking industry today.

3.3.1 Numerical Control and Computerized Numerical Control

The Electronics Industries Association defines numerical control (N/C) as a system in which motions are controlled by the direct insertion of numerical data at some point. The system must automatically interpret at least some portion of these data. Numerical control can be considered as an extremely versatile method of operating machines automatically through discrete numerical values which are sent to the machine from some form of stored input media. In machining, N/C directs the relative motion of the workpiece and the cutting tool.

Numerical control machines can produce parts quickly and with great accuracy. It should, however, be point out that numerical control has its limitations, particularly when it comes to applications involving the production of a large number of parts. Numerical control is especially suited to small lot sizes, and the ability to automate the manufacture of small batch or lot sizes.

A numerical control system normally consists of five basic components:

1. Data input devices
2. Machine control unit
3. Controlled equipment or machine
4. Servo drives for each axis of motion
5. Feedback devices for each axis motion

These components operate as a system to perform all of a machine operator's function. In semiautomatic systems, some manual machine operator functions such as tool changing would be done by a human operator, but the system would inform the operator when to do so.

Numerically controlled machine tools are classified as either point-to-point or contouring systems. In a point-to-point system, the tool moves from location to location, performing all the required operations. A salient feature of a point-to-point system is that the tool does not contact the workpiece while moving from one location to another. In this kind of system, either the spindle moves around a fixed workpiece or the table moves beneath a fixed spindle. The tool does not contact the workpiece until the planned coordinate location has been reached. The path to reach this position must avoid any fixture or other tools. The need for positioning tools and clamps or other tools is greatly diminished by the system. System programming is relatively simple and can be done by technicians who have the ability to read engineering drawings and possess a basic knowledge of machining principles and practices.

In a contouring or continuous-path numerical control system, there is contact between the tool and workpiece as movement occurs. Such systems are used for milling and turning operations that can profile and shape the workpiece. Contour systems can control up to five axes simultaneously. These systems can also be programmed to do point-to-point work. In continuous-path systems the posittional relationship between the workpiece and tool must be maintained at all times, as the path of the cutter in moving from position to position ultimately determines workpiece shape.

Numerical control is not the panacea for all manufacturing problems. In many instances conventional, manually operated machines or mass-production, transfer-line equipment may offer a better solution. The answer is more often than not based upon lot size. Numerical control is recommended for lot sizes in the range 1 to 100. Numerical control should be used only where it is more economical or will do the work faster, more accurately, and better than other methods.

In addition to batch or lot size, the following criteria can be used in deciding which parts should be produced by numerical control:

1. Parts which, if run on conventional equipment, would incur high tooling costs relative to total manufacturing cost
2. Parts requiring lengthy setup time relative to run time on conventional machines
3. Parts which are machined in small batches or lots
4. Parts with anticipated design changes
5. Parts with complex configurations and tight tolerances
6. A wide range of parts requiring frequent machine setup changes and a large tooling inventory, if machined on conventional equipment
7. Parts which have mathematically defined, complex contours
8. Very expensive parts where operator error could be very costly, especially as the part nears completion
9. Parts which require repeatability from part to part and from lot to lot.

Numerical control is so attractive to many companies because of the lack of availability of trained machinists and the very few young people today who are willing to work under machine shop conditions and to put in the months and years needed to become a skilled machinist. With numerical control, the part processing knowhow, the mathematical skill, and the other skills required to machine a part have moved from the shop floor to the office, from machinist to programmer.

With numerical control, the machine runs automatically except for the loading and unloading of the workpiece; the complete machining cycle per piece is automatic. There is little possibility for operator intervention or operator error. If the machining cycle is sufficiently long, one operator can tend two or more machines. With numerical control piece after piece will be identical, thus reducing inspection time significantly. Changing the setup is much less difficult for a numerically controlled machine than for a conventional machine. It therefore takes much less time to change setup for a different part configuration.

Numerical control has given rise to the multifunction machining center. The machining center can, for example, mill, drill, and bore. It may have 90 or more tools stored on the machine, ready for immediate automatic changeover to the machine spindle as required. Through multiple operations in one setup of the workpiece, the part stays on a single machine and as a result spends less time in the machine shop.

Perhaps the biggest gain to be realized from numerically controlled machines over conventional machines is the significant increase in machine utilization. Through the control of wasted, redundant, and nonproductive motions and waiting times, utilization as high as 70% is not uncommon. Also, numerical control has made it possible to drive cutting tools more efficiently and much closer to their theoretical maximum capacity.

3.3.2 Automatic Assembly

When considering whether automatic assembly of the product would be feasible, the manufacturing engineer should look at a few key factors or considerations:

1. The cost of both manual and automatic assembly
2. Production rates
3. Labor availability
4. Length of the production run
5. Whether the product configuration lends itself to automatic assembly

The above considerations apply to new products. For existing manual assembly lines, consideration must be given to the problem caused by the elimination of human assembly operators. Capital investment in automatic assembly equipment must be amortized over the life of the product in production, unless the equipment is general purpose and may be adapted for other products. If the product production run is short, automatic assembly is not feasible unless there are other extenuating circumstances.

Automatic assembly clearly offers some very real advantages:

1. Reduced assembly cost with a corresponding increase in productivity
2. A more uniform product that is consistently repeatable
3. Elimination of the human operator and potential safety problems

The typical automatic assembly machine consists of the following:

1. Transfer system for moving the assembly from workstations to workstation
2. Automatic workheads to perform single assembly operations
3. Vacant workstations for human operators to perform the more complex assembly operations
4. Inspection stations which check that the various operations have been completed successfully

The automatic workheads can be fed manually with loose or magazine-staged parts, or can be supplied from an automatic feeder, such as a vibratory bowl with track feed. The workhead would consist of a parts placing mechanism, or a fastening-type device. Some examples would be screw and nut running heads, riveting heads and guide placing mechanisms.

One final word of caution on automatic assembly systems. When considering the cost of such a system, be sure to include the cost of system shakedown and debugging. This can be as much as 30% of the cost of system design and construction. In the usual automatic assembly system, each workstation must:

1. Handle the work completed from the previous station.
2. Present this work in a usable form to the following station.
3. Handle the range of tolerances characteristic of incoming piece parts.
4. Operate at quoted rates, which might be as high as 2000 to 3000 cycles per hour.
5. Be able to operate on a continuing production basis (which might be as long as 5, 10, 15, or more years with cycles in the 100 million and higher range).

First and foremost, the automatic assembly system represents a production capability, not hardware. All of the manufacturing engineer's evaluations must consider whether such a system has the capability to give the production needed. If your plant cannot service and maintain the system, if your manufacturing people do not want it, or if your piece parts are of such quality that the system will not handle them, the automatic assembly system will not work.

3.3.3 Robotics

Robots may be used for assembly, material handling, painting, welding and a variety of other tasks requiring a high degree of precision, repeatability, or work that would be hazardous to human operators. Basic assembly or material-handling robots, which are most often used in industry today, consist of a single mechanical arm under computer control. A number of different size, capacity, and capability choices are available in robots today. In the area of assembly robots, there are small assembly robots with x, y, z linear motion, a four-position turret equipped with an electro-optical position centering system, gripper, screwdriver, and nutrunner. The individual tools are servo-driven under computer control with feedback signals

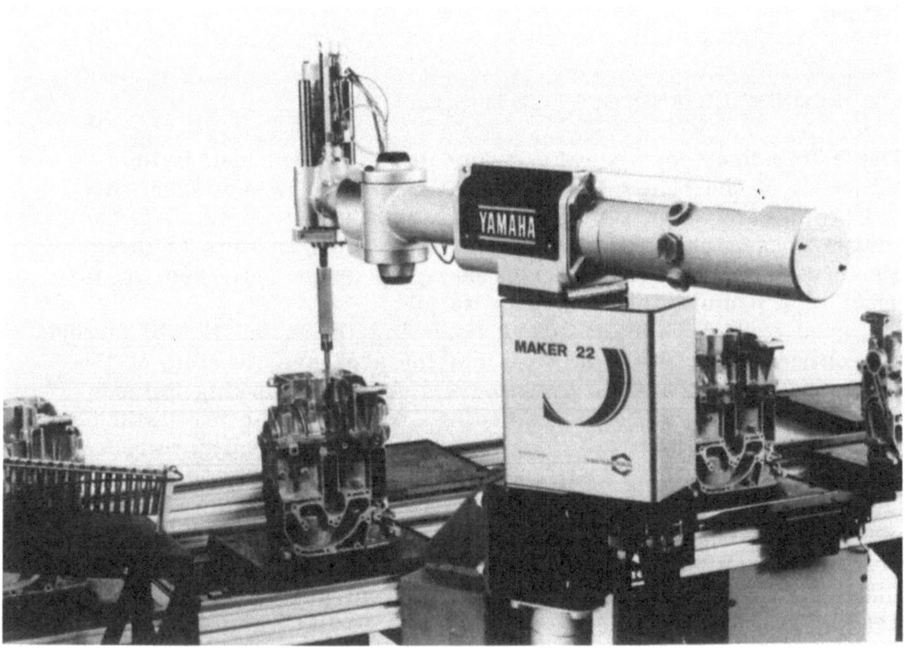

FIG. 3.11 Assembly robot driving bolts into engine castings. (Courtesy of United States Robot Corporation, King of Prussia, Pa.)

for torque and force sensing. Such a robot would be used in small precision assembly applications such as component insertion into printed circuit cards. Figure 3.11 shows such a system, which is a special purpose precision assembly robot. Computer control is through a programmable microprocessor position controller.

Similar robots feature six axes of motion (three linear and three rotary). A robot assembly cycle may start with the robot picking up a new printed circuit board from a supply rack and transporting the board to a special holding fixture in the work area. Next, the robot, according to a systematic program, acquires components from component racks and inserts each component in position on the printed circuit board. Finally the robot removes the completed board from the holding fixture and places it in a finished board rack. The robot can also perform adjusting tasks such as setting a potentiometer to a desired output level; this is done by feeding a signal to the controller, which can instruct the robot to rotate the potentiometer until the desired power level is achieved.

Over 100,000 robots are in use in industry in the United States today. The use of robotics should be a consideration of the manufacturing engineer as he does his preproduction planning. Again, as in automatic assembly, cost and productivity should be the deciding factors.

3.4 ECONOMICS AND COST JUSTIFICATION

The bottom line, so to speak, in any preproduction plan is the economics of the plan. Is it the most cost-effective approach that can be taken? Will it result in delivering a quality product to the customer, on time, and at optimum cost? If the answers to these questions are not clear or cannot be easily quantified, then the preproduction planning has not done its job.

A good example is a company considering the installation of numerical control. Numerical control is not for everyone, in spite of the many advantages it offers over conventional metalworking methods. Companies which are likely candidates for numerical control still face important and difficult decisions as to which numerically controlled machines to buy. The large capital investment, involving hundreds of thousands of dollars, is ample reason alone to conduct a careful cost analysis before ordering the equipment. Most companies, however, have neither the time nor the data required for a thorough analysis. The cost impacts of numerical control which require investigation in a thorough analysis are numerous and often ambiguous. Simple answers do not exist.

Most of the literature on numerical control justification is of a qualitative nature. Advantages of numerical control are usually listed along with lists of manufacturing conditions necessary to achieve these advantages. While this is helpful, the attributes of numerical control must be quantified and measured. Claims made for numerically controlled manufacturing range from 0 to 99% savings in costs. It should be noted that some companies have actually experienced increased costs in going to numerical control. Because of this wide variance in costs, each numerical control procurement decision should be based on data and on the circumstances relevant to it alone.

The same comments apply, more or less, to decisions involving automatic assembly, flexible manufacturing systems, and robotics. Each procurement decision must be a stand-alone analysis based on quantifiable data relevant to it alone.

BIBLIOGRAPHY

Boothroyd, G., Poli, C. R., and Murch, L. E., *Automatic Assembly*, Marcel Dekker, Inc., New York, 1982.

Knox, Charles S., *CAD/CAM Systems, Planning and Implementation*, Marcel Dekker, Inc., New York, 1983.

1988 N/C CAM Guidebook, *Modern Machine Shop*, Gardner Publications, Cincinnati, January, 1988.

Nutt, M. C., *Functional Plant Planning, Layout and Materials Handling*, First Edition, Exposition Press, New York, 1979.

Maynard, H. B., *Industrial Engineering Handbook*, Third Edition, McGraw-Hill, Inc., New York, 1971.

QUESTIONS

1. Why is preproduction planning so important?
2. What is an Operation Process Chart, and why is it considered the basic outline of the manufacturing plan?
3. In planning tooling requirements, what four categories are used to classify tools?
4. What are shop flow times and how do they differ from standard times?
5. What are the five basic components of any numerical control system?
6. What are some of the criteria used in deciding which parts to produce by numerical control?
7. List the principal advantages to be realized from numerical control in the machine shop?
8. What are some of the reasons for selecting automatic assembly when planning production?

9. What would be some reasons for not selecting automatic assembly?
10. If robotic assembly is selected by the manufacturing engineer, what are the primary considerations?
11. What should a good preproduction plan provide?

4

Work Measurement and Standards

4.1 ESTABLISHING THE STANDARDS BASE

Manufacturing engineers, as much as or more than industrial engineers, deal with time standards day in and day out in their work with methods, processes, line balancing, cost trade studies, plant layout, and manpower staffing. Standards are the common denominator, the baseline used for measurement and comparison. These standards are an important tool of the manufacturing engineer, and must be as accurate and reliable as economics and conditions will permit. Labor standards can be derived or developed by using a variety of methods and techniques ranging from engineered work measurement to the averaging of historical cost data. The more labor-intensive the plant or factory operations, the greater the need for accurate time standards. Labor performance measurement and incentive payment plans almost always require fully engineered work standards. Engineered standards are the most accurate and the most reliable standards, and also the most expensive to establish. Engineered standards require full-time maintenance. Every time the method for an operation changes, the standard must be changed. If there is no continuing maintenance program, initially established engineered time standards soon become obsolete.

4.1.1 Engineered Work Measurement

Engineered work measurement includes all of the accepted industrial engineering techniques for establishing time standards. Basically, this includes time study, work sampling, and all of the predetermined standards systems.

Predetermined Time Standards

Some of the better known predetermined time systems include methods time measurement (MTM), work factor (WF), Maynard operation sequence technique (MOST), motion time analysis (MTA), basic motion time study (BMT), dimensional motion time (DMT), and many others. In this chapter, we will describe how MTM, WF, and MOST work, and how the manufacturing engineer can become proficient in these systems through self-study, formal training, and certification as a licensed, qualified practitioner.

Methods time measurement: Methods time measurement (MTM) is a procedure which analyzes any manual operation or method into the basic motions required to perform it and assigns to each motion a predetermined time standard which is based on the nature of the motion and the conditions under which it was made. This is the definition of MTM given by Maynard et al. in their book *Methods Time Measurement* published in 1948.

The basic objective of MTM is to optimize methods as well as to set standards. It offers the great advantage of making it possible to plan and develop accurate methods before production starts, eliminating the need for changes in methods in the middle of the production run. The MTM system allows the establishment of engineered time standards without the use of a stopwatch, and subsequent performance rating of the operator. Operator performance rating, as used in a time study, can vary significantly between observers, and is one of the weaknesses of stopwatch time study. The MTM system overcomes this problem. MTM can be applied by observation or by visualization. The first method involves the analysis of actual work on the production floor, and the second involves the analysis of jobs to be put into production.

MTM is a system of established basic work motions and the time allowed for their performance. The MTM system also established the laws and concepts of how and why motion patterns are made by operators with normal mental and physical qualifications. The MTM procedure consists not only of data tables which establish basic work motions under a variety of conditions, but also of the times for their performance which serve as the basis of measurement of any manual operation. It defines the path about which the sequences of motions will follow, much as the laws of physics explain mathematically the expected material results that will be encountered under varying physical conditions.

The MTM system is built around eight manual movements, nine pedal and trunk movements, and two ocular movements. There are 19 basic motions to consider in establishing any motion pattern. The time for each of these motions is a function of the physical conditions that are involved in the motion's performance, and the nature of the conditions under which it is made. To put it another way, the time for a given motion is affected by a combination of physical and mental conditions.

The law governing the sequences and combinations of motions in MTM is called the principle of the limiting motion. This principle states that a physically and mentally qualified operator can perform some motions simultaneously, while others cannot. This law is just as important as the time data tables shown in Fig. 4.1, because it turns out that

TABLE I – REACH – R

Distance Moved Inches	Time TMU A	Time TMU B	Time TMU C or D	Time TMU E	Hand In Motion A	Hand In Motion B	CASE AND DESCRIPTION
3/4 or less	2.0	2.0	2.0	2.0	1.6	1.6	A Reach to object in fixed location, or to object in other hand or on which other hand rests
1	2.5	2.5	3.6	2.4	2.3	2.3	
2	4.0	4.0	5.9	3.8	3.5	2.7	
3	5.3	5.3	7.3	5.3	4.5	3.6	B Reach to single object in location which may vary slightly from cycle to cycle.
4	6.1	6.4	8.4	6.8	4.9	4.3	
5	6.5	7.8	9.4	7.4	5.3	5.0	
6	7.0	8.6	10.1	8.0	5.7	5.7	C Reach to object jumbled with other objects in a group so that search and select occur
7	7.4	9.3	10.8	8.7	6.1	6.5	
8	7.9	10.1	11.5	9.3	6.5	7.2	
9	8.3	10.8	12.2	9.9	6.9	7.9	
10	8.7	11.5	12.9	10.5	7.3	8.6	
12	9.6	12.9	14.2	11.8	8.1	10.1	D Reach to a very small object or where accurate grasp is required
14	10.5	14.4	15.6	13.0	8.9	11.5	
16	11.4	15.8	17.0	14.2	9.7	12.9	
18	12.3	17.2	18.4	15.5	10.5	14.4	
20	13.1	18.6	19.8	16.7	11.3	15.8	
22	14.0	20.1	21.2	18.0	12.1	17.3	E Reach to indefinite location to get hand in position for body balance or next motion or out of way
24	14.9	21.5	22.5	19.2	12.9	18.8	
26	15.8	22.9	23.9	20.4	13.7	20.2	
28	16.7	24.4	25.3	21.7	14.5	21.7	
30	17.5	25.8	26.7	22.9	15.3	23.2	
Additional	0.4	0.7	0.7	0.6			TMU per inch over 30 inches

TABLE II – MOVE – M

Distance Moved Inches	Time TMU A	Time TMU B	Time TMU C	Hand In Motion B	Wt. Allowance: Wt. (lb.) Up to	Wt. Allowance: Dynamic Factor	Wt. Allowance: Static Constant TMU	CASE AND DESCRIPTION
3/4 or less	2.0	2.0	2.0	1.7				
1	2.5	2.9	3.4	2.3	2.5	1.00	0	
2	3.6	4.6	5.2	2.9				A Move object to other hand or against stop.
3	4.9	5.7	6.7	3.6	7.5	1.06	2.2	
4	6.1	6.9	8.0	4.3				
5	7.3	8.0	9.2	5.0	12.5	1.11	3.9	
6	8.1	8.9	10.3	5.7				
7	8.9	9.7	11.1	6.5	17.5	1.17	5.6	
8	9.7	10.6	11.8	7.2				B Move object to approximate or indefinite location.
9	10.5	11.5	12.7	7.9	22.5	1.22	7.4	
10	11.3	12.2	13.5	8.6				
12	12.9	13.4	15.2	10.0	27.5	1.28	9.1	
14	14.4	14.6	16.9	11.4				
16	16.0	15.8	18.7	12.8	32.5	1.33	10.8	
18	17.6	17.0	20.4	14.2				
20	19.2	18.2	22.1	15.6	37.5	1.39	12.5	
22	20.8	19.4	23.8	17.0				C Move object to exact location.
24	22.4	20.6	25.5	18.4	42.5	1.44	14.3	
26	24.0	21.8	27.3	19.8				
28	25.5	23.1	29.0	21.2	47.5	1.50	16.0	
30	27.1	24.3	30.7	22.7				
Additional	0.8	0.6	0.85		TMU per inch over 30 inches			

TABLE III A – TURN – T

Weight	30°	45°	60°	75°	90°	105°	120°	135°	150°	165°	180°
	Time TMU for Degrees Turned										
Small – 0 to 2 Pounds	2.8	3.5	4.1	4.8	5.4	6.1	6.8	7.4	8.1	8.7	9.4
Medium – 2.1 to 10 Pounds	4.4	5.5	6.5	7.5	8.5	9.6	10.6	11.6	12.7	13.7	14.8
Large – 10.1 to 35 Pounds	8.4	10.5	12.3	14.4	16.2	18.3	20.4	22.2	24.3	26.1	28.2

TABLE III B – APPLY PRESSURE – AP

FULL CYCLE			COMPONENTS		
SYMBOL	TMU	DESCRIPTION	SYMBOL	TMU	DESCRIPTION
APA	10.6	AF + DM + RLF	AF	3.4	Apply Force
			DM	4.2	Dwell, Minimum
APB	16.2	APA + G2	RLF	3.0	Release Force

TABLE IV – GRASP – G

TYPE OF GRASP	Case	Time TMU	DESCRIPTION	
PICK-UP	1A	2.0	Any size object by itself, easily grasped	
	1B	3.5	Object very small or lying close against a flat surface	
	1C1	7.3	Diameter larger than 1/2"	Interference with Grasp on bottom and one side of nearly cylindrical object.
	1C2	8.7	Diameter 1/4" to 1/2"	
	1C3	10.8	Diameter less than 1/4"	
REGRASP	2	5.6	Change grasp without relinquishing control	
TRANSFER	3	5.6	Control transferred from one hand to the other.	
SELECT	4A	7.3	Larger than 1" x 1" x 1"	Object jumbled with other objects so that search and select occur
	4B	9.1	1/4" x 1/4" x 1/8" to 1" x 1" x 1"	
	4C	12.9	Smaller than 1/4" x 1/4" x 1/8"	
CONTACT	5	0	Contact, Sliding, or Hook Grasp	

EFFECTIVE NET WEIGHT			
Effective Net Weight (ENW)	No. of Hands	Spatial	Sliding
	1	W	$W \times F_c$
	2	W/2	$W/2 \times F_c$

W = Weight in pounds
F_c = Coefficient of Friction

TABLE V – POSITION* – P

CLASS OF FIT		Symmetry	Easy To Handle	Difficult To Handle
1—Loose	No pressure required	S	5.6	11.2
		SS	9.1	14.7
		NS	10.4	16.0
2—Close	Light pressure required	S	16.2	21.8
		SS	19.7	25.3
		NS	21.0	26.6
3—Exact	Heavy pressure required	S	43.0	48.6
		SS	46.5	52.1
		NS	47.8	53.4
SUPPLEMENTARY RULE FOR SURFACE ALIGNMENT				
P1SE per alignment >1/16 ≤1/4"		P2SE per alignment ≤1/16"		

*Distance moved to engage—1" or less

TABLE VI – RELEASE – RL

Case	Time TMU	DESCRIPTION
1	2.0	Normal release performed by opening fingers as independent motion
2	0	Contact Release

TABLE VII – DISENGAGE – D

CLASS OF FIT	HEIGHT OF RECOIL	EASY TO HANDLE	DIFFICULT TO HANDLE
1—LOOSE—Very slight effort, blends with subsequent move	Up to 1"	4.0	5.7
2—CLOSE—Normal effort, slight recoil	Over 1" to 5"	7.5	11.8
3—TIGHT—Considerable effort, hand recoils markedly	Over 5" to 12"	22.9	34.7

SUPPLEMENTARY

CLASS OF FIT	CARE IN HANDLING	BINDING
1—LOOSE	Allow Class 2	————
2—CLOSE	Allow Class 3	One G2 per Bind
3—TIGHT	Change Method	One APB per Bind

TABLE VIII – EYE TRAVEL AND EYE FOCUS – ET AND EF

Eye Travel Time = $15.2 \times \frac{T}{D}$ TMU, with a maximum value of 20 TMU

where T = the distance between points from and to which the eye travels.
D = the perpendicular distance from the eye to the line of travel T.

Eye Focus Time = 7.3 TMU.

SUPPLEMENTARY INFORMATION

– Area of Normal Vision = Circle 4" in Diameter 16" from Eyes

– Reading Formula = 5.05 N Where N = The Number of Words.

TABLE IX – BODY, LEG, AND FOOT MOTIONS

TYPE		SYMBOL	TMU	DISTANCE	DESCRIPTION
LEG–FOOT MOTION		FM	8.5	To 4"	Hinged at ankle.
		FMP	19.1	To 4"	With heavy pressure.
		LM__	7.1	To 6"	Hinged at knee or hip in any direction
			1.2	Ea. add'l inch	
HORIZONTAL MOTION	SIDE	SS__C1	*	<12"	Use Reach or Move time when less than 12". Complete when leading leg contacts floor.
			17.0	12"	
			0.6	Ea. add'l inch	
		SS__C2	34.1	12"	Lagging leg must contact floor before next motion can be made
			1.1	Ea. add'l inch	
	TURN	TBC1	18.6	——	Complete when leading leg contacts floor
	BODY	TBC2	37.2	——	Lagging leg must contact floor before next motion can be made
	WALK	W__FT	5.3	Per Foot	Unobstructed
		W__P	15.0	Per Pace	Unobstructed
		W__PO	17.0	Per Pace	When obstructed or with weight
VERTICAL MOTION		SIT	34.7	——	From standing position
		STD	43.4	——	From sitting position
		B,S,KOK	29.0	——	Bend, Stoop, Kneel on One Knee
		AB,AS,AKOK	31.9	——	Arise from Bend, Stoop, Kneel on One Knee
		KBK	69.4	——	Kneel on Both Knees
		AKBK	76.7	——	Arise from Kneel on Both Knees

TABLE X – SIMULTANEOUS MOTIONS

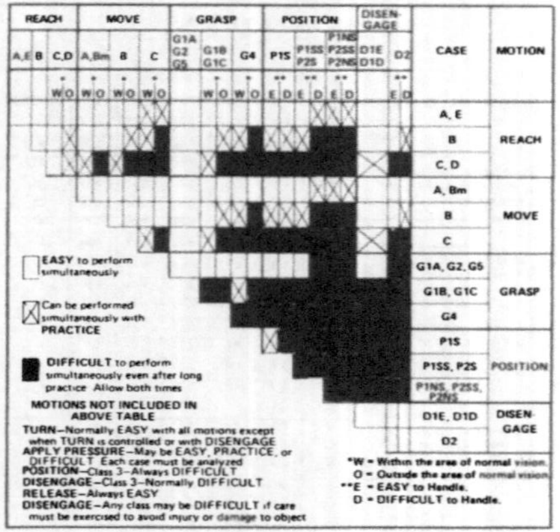

SUPPLEMENTARY MTM DATA

TABLE 1 – POSITION – P

Class of Fit and Clearance	Case of † Symmetry	Align Only	Depth of Insertion (per ¼") 0 >0≤1/8"	2 >1/8≤¾	4 >¾≤1¼	6 >1¼≤1¾
21 .150" – .350"	S	3.0	3.4	6.6	7.7	8.8
	SS	3.0	10.3	13.5	14.6	15.7
	NS	4.8	15.5	18.7	19.8	20.9
22 .025" – .149"	S	7.2	7.2	11.9	13.0	14.2
	SS	8.0	14.9	19.6	20.7	21.9
	NS	9.5	20.2	24.9	26.0	27.2
23* .005" – .024"	S	9.5	9.5	16.3	18.7	21.0
	SS	10.4	17.3	24.1	26.5	28.8
	NS	12.2	22.9	29.7	32.1	34.4

*BINDING—Add observed number of Apply Pressures.
DIFFICULT HANDLING—Add observed number of G2's.

†Determine symmetry by geometric properties, except use S case when object is oriented prior to preceding Move.

TABLE 1A – SECONDARY ENGAGE – E2

CLASS OF FIT	DEPTH OF INSERTION (PER 1/4") 2	4	6
21	3.2	4.3	5.4
22	4.7	5.8	7.0
23	6.8	9.2	11.5

TABLE 2 – CRANK (LIGHT RESISTANCE) – C

DIAMETER OF CRANKING (INCHES)	TMU (T) PER REVOLUTION	DIAMETER OF CRANKING (INCHES)	TMU (T) PER REVOLUTION
1	8.5	9	14.0
2	9.7	10	14.4
3	10.6	11	14.7
4	11.4	12	15.0
5	12.1	14	15.5
6	12.7	16	16.0
7	13.2	18	16.4
8	13.6	20	16.7

FORMULAS

A CONTINUOUS CRANKING (Start at beginning and stop at end of cycle only)

$$TMU = [(N \times T) + 5.2] + F + C$$

B INTERMITTENT CRANKING (Start at beginning and stop at end of each revolution)

$$TMU = [(T + 5.2)F + C] N$$

C = Static component TMU weight allowance constant from move table
F = Dynamic component weight allowance factor from move table
N = Number of revolutions
T = TMU per revolution (Type III Motion)
5.2 = TMU for start and stop

1 TMU = .00001 hour	1 hour = 100,000.0 TMU
= .0006 minute	1 minute = 1,666.7 TMU
= .036 seconds	1 second = 27.8 TMU

FIG. 4.1 MTM application data tables. (Copyright by the MTM Association for Standards and Research. No reprint permission without written consent from the MTM Association.)

regardless of the accuracy of the standard times that are established for the individual motions, the standard established from their application will not be accurate if the principle of the limiting motion is not considered.

As indicated earlier, the MTM system may be applied by visualization:

1. Create the operation.
2. Visualize and organize information.
3. Plan the methods of operation.
4. Analyze operation details and establish the time.

The visualization technique requires complete attention to details to avoid error. If we use the observation method, we must:

1. Observe the operation.
2. Broadly analyze and record existing information.
3. Record the operation method.
4. Analyze the method and establish the time.

The MTM data tables shown in Fig. 4.1, are expressed in terms known as time measurement units (TMUs). These are leveled or normal times and contain no allowances for personal needs, fatigue, or delays. The time measurement unit conversion to decimal hours, minutes, and seconds is as follows:

1 TMU = 0.00001 hr
1 TMU = 0.0006 min
1 TMU = 0.036 sec

MTM-1 is the basic MTM system from which all other systems of the MTM family have been developed. It is a procedure which analyzes any manual operation or method into the basic motions required to perform it and assigns to each motion a predetermined time standard based on the nature of the motion and the conditions under which it is made.

The basic motions that make up the MTM-1 system can best be described as follows:

Reach—The basic hand or finger motion employed when the primary purpose is to transport the hand or fingers to a destination. The variables that determine reach would be the nature of the ending or destination, distance, and type of reach, depending on the presence or absence of acceleration and deceleration.

Grasp—The basic hand or finger element employed to secure a degree of control of an object. The variables of grasp consist of the type of grasp and the nature of the object and its location.

Move—The basic hand or finger motion employed when the primary purpose is to transport an object to a destination. The variables are similar to those for reach, with the added variable of weight being a factor.

Position—The basic hand or finger motions employed to align, orient, and engage one object with another to obtain a specific relationship. This primarily involves assembling one part into another. The variables for position are the class of fit, the object's symmetry, and the depth of insertion.

Release—The basic hand or finger motion employed to relinquish control of an object.

Turn—The basic motion employed to rotate the hand about the long axis of the forearm.

Crank—The motion employed when the hand follows a circular path to rotate an object, with the forearm pivoting at the elbow and the upper arm essentially fixed. The variables of crank include the number of revolutions, whether the motion is continuous or intermittent, the size of crank, and the resistance.

Apply pressure—An application of muscular force during which object resistance is overcome in a controlled manner, with little or no motion.

Disengage—The basic hand or finger element employed to separate one object from another object where there is a sudden ending of resistance.

Body motions—These motions contain such elements as walk, sit, stand, bend, stoop, turn body, kneel, foot motions, and leg motions.

Eye action—Eye focus is the basic visual and mental element of looking at an object long enough to determine a readily prominent characteristic. Eye travel is the basic motion of shifting vision from one location to another.

MTM-2 is a compact, self-contained, predetermined time system synthesized from basic MTM data. The speed of MTM-2 application has been found to be twice that of MTM-1 application. The accuracy is within plus or minus 5% of the 95% confidence level when compared to MTM-1 analysis with cycles of approximately 1 min exclusive of limiting process time. The motions of MTM-2 have been well defined and, as with MTM-1, precise rules of application have been determined. The categories of MTM-2 are as follows:

Get—A motion with the predominate purpose of reaching with the hand or fingers to an object, grasping the object, and subsequently releasing it.

Put—This is the other principal motion on the MTM-2 data card, and is a combination of the basic MTM move motion with that of crank.

Other MTM-2 categories include *Apply pressure*, *regrasp*, *eye action*, *crank*, *step*, *foot motion*, *bend*, and *arise*. The definitions of these are the same as in MTM-1.

MTM-3 is a system of synthesized MTM data and is the third generic level of the family of MTM data. MTM-3 was designed with the same criteria as MTM-2. The accuracy of MTM-3 is within plus or minus 5% at the 95% level of confidence when compared to MTM-1 analysis in cycles of approximately 4 min, exclusive of limiting process times. MTM-3 is seven times as fast to apply as MTM-1.

The MTM-3 system consists of only four categories of manual motions, which include:

Handle—A motion sequence with the purpose of getting control over an object with the hand or fingers and placing the object in a new location.

Transport—A motion with the purpose of placing an object in a new location with the hand or fingers.

The other categories, including step and foot motions and bend and arise, have the same basic definitions as in MTM-1.

MTM-V is a special, high-level MTM system designed for use in application of the manual portion of machine shop-related work. It is indended primarily to fill the need to determine and develop manual time data for machine shop-related work at all levels of manufacturing, from single units to continuous production, and the intermittent manufacture of any lot size.

The MTM-V system offers significant advantages as an accurate base for projecting production. The system provides capabilities for the control and maintenance of time standards and methods for individual machines. The design of the MTM-V system permits it to be used together with other MTM systems when necessary. Other advantages are that it is easy to maintain, designed for standard data construction, traceable to MTM, combinable with other systems, and easy to apply.

MTM-M is a second-level, functional system of original data designed for the analysis and measurement of manual assembly work performed under stereoscopic magnification from 5 to 30 power. The system was developed by a consortium of industrial members of the MTM Association working in conjuction with the University of Michigan.

The MTM-M system has four major tables and one subtable which contain all the data necessary for application. These tables are related to the direction of movement:

II—Inside to inside
IO—Inside to outside
OO—Outside to outside
OI—Outside to inside
IF—Infield to final target

Selection of the appropriate data tables is determined by a four-variable consideration:

1. Type of tool
2. Condition of tool
3. Terminating characteristic of the motion
4. Distance/tolerance ratio

Notations for MTM-M are designed to describe the motion and respective variables involved. The notation is alphanumeric and was designed for eventual computer application.

MTM-MEK fills a need for predetermined standards in small-batch and one-of-a-kind production. The system defines seven different time elements, including *get* and *place*, *handle tool*, *place*, *operate*, *motion cycles*, *body motion*, and *visual control*. In addition to the generic analyzing system, special-purpose data or continuation stages have been developed covering such activities as bolting and assembling standard parts.

Work factor: Work factor is defined as an elemental time system for compiling time standards to establish the expected productivity of a human performing useful manual and mental work. Stopwatches and other timing devices are not used with work factor except for machine or process times. There are four levels or integrated systems of work factor to accommodate the industrial situation. These include:

Detailed work factor—For highly repetitive, short-cycle operations. Work factor time unit equals 0.0001 min.

Ready work factor—For longer cycle and less repetitive operations. Work factor time unit equals 0.0010 min.

Abbreviated work factor—For custom operations with very little repetition, such as toolroom operations. Work factor time unit equals 0.0050 min.

Work factor standard data—For rapid standards and rate setting in all production operations.

Work factor time is defined as the time required for the average experienced operator, working with good skill and good effort and under standard working conditions, to perform one work cycle or operation, on one unit or piece, according to a prescribed method and specified quality. The work factor time does not include allowance for personal needs, fatigue, environmental unavoidable delays, or incentive payment. Work factor time is not comparable with times referred to as normal, day-work performance, 60-minute-hour performance, or other terms used to indicate the work pace expected of the average worker who performs without incentive or at a level of productivity commensurate with base rate output.

Allowances for fatigue, environmental unavoidable delays, and personal needs are added to the work factor time in accordance with the working conditions involved. Once they are established, they should not be varied except in special cases involving unusual working conditions. When this occurs, work sampling studies are recommended for determining appropriate changes to the established allowances.

Adequate task physique is a work factor term which describes the physical ability of a worker to perform a given task at rates commensurate with work factor time. When the worker has an adequate task physique, he performs the work without undue fatigue, hazard, or physical strain and maintains physical well-being throughout the working day.

Adequate task intelligence indicates that the average experienced operator has sufficient intelligence to perform the task involved at a rate commensurate with work factor time. This assumes that the subject, intent, procedures, tools, work instructions, and all other items related to the work are within the operator's understanding and experience.

Adequate task vision is the work factor term for sufficient visual ability to perform properly the visual task involved at a rate that is commensurate with work factor time. Visual tasks with close detail, fine color distinctions, poor illumination, or weak contrasts may require better than normal vision, whereas those involving large, clearly defined shapes, bright colors, and contrasts can be adequately performed with poor or less than normal vision.

In its simplist form, work factor is essentially a method of determining the time for a given motion pattern by:

1. Making a detailed analysis of each motion based on the identification of the four major variables of work and the use of work factors as the unit of measurement
2. Applying to each motion the proper work factor time value from the motion time table (see Fig. 4.2)

A basic motion in work factor is any motion whose performance involves the least amount of diffiuclty or precision for any given distance and body member combination.

Work factor is a unit used as the index of additional time required over and above the basic time when motions are performed involving the work factor variables of manual control or weight/resistance.

Work factor standard element is the term applied to the basic key divisions of work, including transport, grasp, and assemble. All manual work consists of one or more work factor standard elements, which are also composed of one or more motions.

The motion analysis, then, is the combination of work factor symbols used to identify a motion for direct reference to the motion time table (Fig. 4.2) in selecting the proper work factor time value.

The work factor predetermined time system is structured around the basic premise that four major variables affect the time it takes to perform manual motions. These are:

Work-Factor Motion-Time Table for Detailed Analysis

Time in Work-Factor Units

Distance moved, inches	Basic	Work Factors 1	2	3	4
(A) Arm, measured at knuckles					
1	18	26	34	40	46
2	20	29	37	44	50
3	22	32	41	50	57
4	26	38	48	58	66
5	29	43	55	65	75
6	32	47	60	72	83
7	35	51	65	78	90
8	38	54	70	84	96
9	40	58	74	89	102
10	42	61	78	93	107
11	44	63	81	98	112
12	46	65	85	102	117
13	47	67	88	105	121
14	49	69	90	109	125
15	51	71	92	113	129
16	52	73	94	115	133
17	54	75	96	118	137
18	55	76	98	120	140
19	56	78	100	122	142
20	58	80	102	124	144
22	61	83	106	128	148
24	63	86	109	131	152
26	66	90	113	135	156
28	68	93	116	139	159
30	70	96	119	142	163
35	76	103	128	151	171
40	81	109	135	159	179
Weight, pounds:					
Male	2	7	13	20	Up
Female	1	3½	6½	10	Up
(T) Trunk, measured at shoulder					
1	26	38	49	58	67
2	29	42	53	64	73
3	32	47	60	72	82
4	38	55	70	84	96
5	43	62	79	95	109
6	47	68	87	105	120
7	51	74	95	114	130
8	54	79	101	121	139
9	58	84	107	128	147
10	61	88	113	135	155
11	63	91	118	141	162
12	66	94	123	147	169
13	68	97	127	153	175
14	71	100	130	158	182
15	73	103	133	163	188
16	75	105	136	167	193
17	78	108	139	170	199
18	80	111	142	173	203
19	82	113	145	176	206
20	84	116	148	179	209
Weight, pounds:					
Male	11	58	Up		
Female	5½	29	Up		

Distance moved, inches	Basic	Work Factors 1	2	3	4
(L) Leg, measured at ankle					
1	21	30	39	46	53
2	23	33	42	51	58
3	26	37	48	57	65
4	30	43	55	66	76
5	34	49	63	75	86
6	37	54	69	83	95
7	40	59	75	90	103
8	43	63	80	96	110
9	46	66	85	102	117
10	48	70	89	107	123
11	50	72	94	112	129
12	52	75	97	117	134
13	54	77	101	121	139
14	56	80	103	125	144
15	58	82	106	130	149
16	60	84	108	133	153
17	62	86	111	135	158
18	63	88	113	137	161
19	65	90	115	140	164
20	67	92	117	142	166
22	70	96	121	147	171
24	73	99	126	151	175
26	75	103	130	155	179
28	78	107	134	159	183
30	81	110	137	163	187
35	87	118	147	173	197
40	93	126	155	182	206
Weight, pounds:					
Male	8	42	Up		
Female	4	21	Up		
(F, H) Finger-hand, measured at fingertip					
1	16	23	29	35	40
2	17	25	32	38	44
3	19	28	36	43	49
4	23	33	42	50	58
Weight, pounds:					
Male	⅔	2½	4	Up	
Female	⅓	1¼	2	Up	
(FT) Foot, measured at toe					
1	20	29	37	44	51
2	22	32	40	48	55
3	24	35	45	55	63
4	29	41	53	64	73
Weight, pounds:					
Male	5	22	Up		
Female	2½	11	Up		
(FS) Forearm swivel, measured at knuckles					
45 degrees	17	22	28	32	37
90 degrees	23	30	37	43	49
135 degrees	28	36	44	52	58
180 degrees	31	40	49	57	65
Torque, pound-inches:					
Male	3	13	Up		
Female	1½	6½	Up		

Work-Factor* symbols:

- W = Weight or Resistance
- S = Directional Control (Steer)
- P = Care (Precaution)
- U = Change of Direction
- D = Definite Stop

Walk Time

Type	30-inch paces 1	2	Over 2
General	Analyze from Table	260	120 + 80 per pace
Restricted	Analyze from Table	300	120 + 100 per pace
Add 100 for 120- to 180-degree turn at start or finish			
Up steps (8 inches rise, 10 inches flat) 126			
Down steps 100			

Head turn:
45 degrees 40
90 degrees 60

1 Time Unit = 0.006 second
= 0.0001 minute
= 0.00000167 hour

FIG. 4.2 Work factor motion time tables for detailed analysis. (Copyright by the Wofac Company, Moorestown, N.J.)

1. Body member used
2. Distance moved
3. Manual control required
4. Weight or resistance involved

Body members used are identified by observation and the application of an exact definition. Work factor motion times have been compiled for them, and include:

Finger-hand	Arm
Forearm swivel	Trunk
Foot	Leg
Head	

Distance is the easiest of the variables to recognize and determine because it is easily measured. All distances, except those with a change in direction, are measured as a straight line between the starting and stopping points of the motion arc described by the body member. This is just the opposite of MTM. The actual motion path is measured only when a direction change is involved.

Manual control is the most complex of the variables. Its evaluation is difficult because there is no physical measuring device, such as a scale or balance, which can be used as a check.

Weight or resistance in manual motions is measured in pounds for all body members except forearm swivel. Since this is a rotary motion, it is measured in inch-pounds of torque.

As with MTM, the manufacturing engineer should be familiar with work factor if it is used in his plant or facility. He should attempt to use and apply detailed work factor only after approved training and certification. Several simplified versions of work factor have evolved since work factor first made its appearance in 1947. One version, known as ready work factor, was developed for use by persons not engaged in work study as an occupation. It is based on detailed work factor and is suitable for establishing day work and incentive time standards. Figure 4.3 shows the ready work factor time table. Again, the manufacturing engineer should be cautioned that training in ready work factor is a prerequisite for correct use or application in standards setting.

Abbreviated work factor provides a work measurement procedure where rapid standards setting is a requirement. The application procedures are quite different from either detailed or ready work factor. All time values are part of the analysis form, eliminating the need for reference to tables. The abbreviated time unit is 0.005 min. Minimum writing is required for descriptive purposes, and many other shortcuts have been incorporated in this procedure.

Standard data developed from detailed or ready work factor are the preferred approach, as mentioned earlier. Generally, most manufacturing engineers have not had the training or experience in predetermined

All times (boldface numbers) are expressed as Ready Work-Factor Time Units. One Ready Work-Factor Time Unit = 0.0010 minute.

SYMBOLS: – indicates "up to and including." > indicates "greater than."

	Work Factors				
	0	1	2	3	4
	Very Easy	Easy	Average	Difficult	Very Difficult
TRANSPORT	Weight Limits, lbs.				
Finger, Hand	–1	–2	–3	–5	>5
Arm	–2	–4	–6	–10	>10
Foot	–3	–8	>8	—	—
Leg	–5	–16	>16	—	—
Trunk	–7	–32	>32	—	—
Motion Distance,[1] in.					
Very Short –4 A[2]	**2**	**3**	**4**	**5**	**6**
Short –10 B	**4**	**5**	**6**	**7**	**8**
Medium –20 C	**5**	**7**	**9**	**11**	**13**
Long –30 D	**7**	**9**	**11**	**13**	**15**
Very Long –40 E	**9**	**11**	**13**	**15**	**17**
GRASP Simple	Pinch[3]	Wrap and Transfer[3]	Average No. Motions		
Manipulative	—	—	–2	–3	–4
Complex[7]	Major Dimension, in.		>¼		–¼
	Diameter, in.		>¼	–¼	
	Thickness, in.		>3/64	–3/64	–¼
Visual / Blind	**1**	**2**	**3** / **4**	**5** / **6**	**8**

PICK-UP[4] Distance, in.	0	1	2	3	4
–4 A	**8**	**9**	**10**	**12**	**15**
–10 B	**12**	**13**	**14**	**16**	**19**
–20 C	**17**	**18**	**19**	**21**	**24**
–30 D	**21**	**22**	**23**	**25**	**28**
–40 E	**25**	**26**	**27**	**29**	**32**
Distance, in.	Add-on for Weight				
–10	—	**1**	**2**	**3**	**4**
>10	—	**2**	**4**	**6**	**8**
Add-on for:					
Blind	—		**1**		—
Simo	—		**2**[5]		

PRE-POSITION	One Hand		Two Hands		
	Major Dimension, in.				
% Occurrence	>3/8–4	–10	–3/8	–10	>10
25	**1**		**2**		
50	**2**	**3**		**4**	
75	**3**	**4**	**5**	**5**	**6**
100	**4**	**5**	**6**	**7**	**8**
Simo	Add 50% to time			—	—

RELEASE	Gravity	Unwrap	—
	1	**2**	

ASSEMBLE

	Mechanical[6]					
	Open			Close		
	Ratios					
Target, in.	–0.4	–0.9	>0.9	–0.4	–0.9	>0.9
>3/8	**2**	$\mathbf{3}^{1}$	$\mathbf{7}^{1}$	**2**	$\mathbf{3}^{1}$	$\mathbf{7}^{1}$
–3/8	$\mathbf{3}^{1}$	$\mathbf{4}^{2}$	$\mathbf{8}^{2}$	$\mathbf{5}^{3}$	$\mathbf{6}^{4}$	$\mathbf{10}^{4}$
–1/8	$\mathbf{6}^{4}$	$\mathbf{6}^{4}$	$\mathbf{10}^{4}$	$\mathbf{9}^{7}$	$\mathbf{9}^{7}$	$\mathbf{13}^{7}$
Tolerance, in.	Surface[6]					
>3/8	**3**			**3**		
–3/8	$\mathbf{5}^{2}$			$\mathbf{6}^{3}$		
–1/8	$\mathbf{8}^{5}$			$\mathbf{12}^{9}$		

% Add-ons for Assemble

Distance, in.	–1	–2	–3	–5	–7	–15	>15
Gripping Distance	—	—	10	20	30	50	70
Distance Between Temporary Blind	—	20	30	50	70	2 Asy / 150	2 Asy +5 / —
Permanent Blind	30	50	70	150	250	500	—

Miscellaneous

Index: Mechanical **3** Surface **4**

Simo: 50% of (Aligns + Add-ons)

Seat: Turn; Use Transport Rules

Dislodge; Use Assemble Rules

Weight, lbs.	—	–2	–4	–6	–10	>10
% Add-on to Total Asy Time		0	30	50	70	100

MENTAL PROCESS Focus **2**, Inspect **3**, React **2**

BODY TURN Head Turns: –45° **4**, –90° **6**
Body Turns: –90° (0 Foot) **10**
–90° (2 Feet) **20**, –180° **26**

WALK (in terms of 30-inch Paces)
General **12** + **8** per Pace
Restricted **12** + **10** per Pace
Up and Down Steps **10** per Step
Stand-up **13** Sit Down **9**

RELEASE PRESSURE **1**

CIRCULAR MOTIONS Diameter, in.	No. Work-Factors: Single Circle	Multiple Circles
–1	**1**	**2**
–2		**1**
>2		**0**

1 For Trunk Motion multiply Distance by 2.
2 Use class A for Apply Pressure and for Forearm Swivel.
3 For weight over 3 pounds multiply time by 2.
4 Times based on Visual Grasp. Work-Factors in Pick-up are considered to be identical to those in Grasp. Times include Reach, Grasp and Move, not Release.
5 Add Blind value also.
6 Small numbers are Align times, large numbers are total Assemble times.
7 For Complex Grasp: add 2 Units for Simo and 1 Unit each for Entangled, Nested and Slippery.

FIG. 4.3 Ready work factor table. (Copyright by the Wofac Company, Moorestown, N.J.)

time systems to enable the direct establishment of standards from basic, detailed MTM or work factor. Also, such an approach would not be cost-effective, or the best use of the time and talent of the manufacturing engineering staff. Industrial engineers, certified practitioners, and/or consultants should be employed to develop the standard data program and to maintain it.

Maynard operation sequence technique (MOST): No discussion of predetermined time systems would be complete without examining the Maynard operation sequence technique (MOST). It is a unique system that is simpler to apply than other systems, yet does not sacrifice accuracy of results. MOST is constructed around three basic motion sequences: general spatial move, controlled move, and tool use. It is up to 50 times faster to apply than MTM, for example, and is now in the form of a computer program with the attendant advantages of mass data update, rapid file access, and, of course, greater speed of application. The theory underlying MOST is that all work basically equates to the movement of objects. Since MOST is a system for measuring work; it concentrates on the movement of objects. The primary work units then become activities involved in moving objects, and not basic motions. The activities then are described as subactivities for basically moving an object.

In the MOST system, objects can be moved in only one of two ways; either they are picked up and moved freely through space, or they are moved and maintain contact with another surface. An example would be a box that can either be picked up and carried from one end of a work bench to the other, or can be pushed across the top of the work bench. For each type of move, a different sequence of events takes place, giving rise to a separate MOST activity sequence model for each.

Tool usage is analyzed through a separate activity sequence model. This model allows the analyst the opportunity to follow the movement of a hand tool through a standard sequence of events, which is a combination of the two basic sequence models.

Three activity sequences cover all manual work. These are as follows:

1. General move sequence for the spatial movement of an object freely through the air
2. Controlled move sequence for the movement of an object when it remains in contact with a surface or is attached to another object during movement
3. Tool use sequence for the use of common hand tools

The sequence model concept is the basic premise on which the MOST predetermined time system is built. The sequence model forces the analyst to focus on the movement of objects, instead of operator body motions. This allows a larger unit to work with, and a clearer, more understandable method description. It also forces the analyst to consider all of the subactivities possible in moving an object. This results in more consistency and less application error.

MOST is based upon fundamental statistical standard deviation concepts. This gives the system its speed, ease of application, accuracy, and, most important, its outstanding consistency. As with all work measurement techniques, the time values that result from performing a measurement should be based upon a defined, well-engineered method. With MOST, the difference between the calculated time and the actual time lies between the engineered method and the method actually being used.

As with all predetermined time systems, training and certification are a requirement with MOST, before the manufacturing engineer or anyone else should attempt to use the system for standards setting. However, of all the predetermined time systems, MOST is the easiest and fastest to apply. MOST systems have proved to be accurate and consistent. The technique is used worldwide by many companies with very good results.

Stopwatch Time Study

Barnes defines time study as used to determine the time required by a qualified operator, working at a normal pace, to do a specified task. Stopwatch time study is the most commonly used method for measuring work in industry today. The decimal minute stopwatch and time study board are the primary pieces of equipment required (see Fig. 4.4). The procedures for making a time study may vary slightly, but basically the steps are as follows:

1. Obtain and record information about the operation and the operator being studied.
2. Divide the operation into elements of work, and record a detailed and complete method description.
3. Improve the method, the tools, and the workplace layout.
4. Observe and record the time taken by the operator to perform the operation.
5. Rate the operator's performance.
6. Determine the number of cycles to be timed, and make certain that sufficient cycles are timed for statistical validity.
7. Determine allowances for personal needs, fatique, and delay times.
8. Determine the standard time for the operation.

It is very important that step 3 above take place before the time study is made. The job or operation is not ready for time study until the method has been analyzed and necessary improvements have been made. Every manufacturing engineer should have had some training in time study, or in one of the predetermined time systems. Time and motion

FIG. 4.4 Decimal minute stopwatches and time study board. (Courtesy of the Meylan Corporation, New York, N.Y.)

training is essential for the successful performance of the manufacturing engineer's job. He must, as a minimum, know how to subdivide an operation into the elements of work and write a good elemental description. He should be familiar with the decimal minute stopwatch, and how to break his elements when making a time study. In addition, he should have training in performance rating so that he can distinguish the various degrees of operator proficiency and effort when making a time study. Proficiency in motion and time study can be gained by formal classroom study and by self-study, but, in addition, it requires practice to become fully proficient. Stopwatch time study is used to establish engineered time standards, but it is also used to check existing standards, to make production studies, and to determine process and equipment cycle times.

Basically there are two ways of making a time study. The first technique is called snapback time study, and involves timing each element of the operation separately and snapping the watch back to zero each time the element is timed. The snapback method is recommended for obtaining a sufficient number of readings on short-cycle operations, or where an element was missed in a continuous time study. The second method is known as continuous time study simply because the stopwatch, once started, is never stopped until the entire study is complete. This kind of study is the most difficult to perform because the time study observer must read the watch and record the time for each element while the hand is moving on the watch, and also rate operator performance for the element and record it.

Continuous time study is recommended for all time study standards work, as it is the only method which picks up everything the operator does in performing the task. This enables identification of "foreign elements" where time is lost. Continuous time study is also used in performing an all-day production study of an operator to determine lost time, nonproductive time, waiting time, as well as the time spent working productively. Figure 4.5 shows an example of a continuous time study, and Fig. 4.6 shows an example of a snapback time study.

As in predetermined time systems, stopwatch time should be used to establish standard data, which can be applied by relatively untrained standards applicators simply by counting the number of parts or components and filling in a standard data work sheet. The procedure for determining the number of observations needed for a given time study involves first determining the standard error of the average for each individual element as follows:

$$\sigma\bar{x} = \frac{\sigma'}{\sqrt{N}}$$

where

$\sigma\bar{x}$ = standard deviation of the distribution of averages

σ' = standard deviation of the universe for a given element

N = actual number of observations of the element

Standard deviation is the root-mean-square deviation of the observed readings from their average:

Observation	Hamilton Standard Job	SHEET	One
Observer	Roy J. Tanner	DATE	8/17/85

Production Study Observations

No.	Element	Clock time	Actual time
111	Break time	10.09	—
112	Setup for next assembly	10.24	15.00
113	Leave area to obtain molds and return	10.27	3.00
114	Prep mold braces	10.28	1.00
115	Add insulator, braces, and screws to mold	10.29	1.00
116	Add mating mold halves with screws	10.30	1.00
117	Add inserts, assemble mold, check and verify fit	10.33	3.00
118	Prep mold braces	10.34	1.00
119	Add insulator, braces, and screws to mold	10.35	1.00
120	Add mating mold halves with screws	10.36	1.00
121	Add inserts, assemble mold, check and verify fit	10.39	3.00

FIG. 4.5 Continuous study observation sheet.

$$\sigma = \sqrt{\frac{(X_1 - \bar{X})^2 + (X_2 - \bar{X})^2 + \cdots + (X_n - \bar{X})^2}{N}}$$

$$= \sqrt{\frac{\Sigma X^2}{N} - \bar{X}^2}$$

where

X = each stopwatch reading or individual observation

$\bar{X}$ = average or mean of all readings of an element

Σ = sum of individual readings

A 95% confidence level and ±5% precision are commonly used in time study. This means that the chances are at least 95 out of 100 that the

PART NUMBER	OPERATION DESCRIPTION	
PART NAME	OPERATION NUMBER	MACHINE
MATERIAL SPECIFICATION	MATERIAL OR PART SIZE	OBSERVER
MACHINE TIME	FIXTURE OR DIE NUMBER	DATE
START TIME	STOP TIME	SHEET OF

	ELEMENT DESCRIPTION		OBSERVATIONS										TOTAL MIN.	AVG. MIN.	L F	NEAT MIN.
			1	2	3	4	5	6	7	8	9	10				
		T														
A		R														
		T														
B		R														
		T														
C		R														
		T														
D		R														
		T														
E		R														
		T														
F		R														
		T														
G		R														
		T														
H		R														
		T														
I		R														
		T														
J		R														
		T														
K		R														
		T														
L		R														
		T														
M		R														

REMARKS	TOTAL MINUTES	
	PERSONAL FATIGUE AND DELAY %	
	STANDARD TIME IN MINUTES	
	STANDARD HOURS	
	CYCLE HOURS	

FIG. 4.6 Snapback time study observation sheet.

AMETER OF TOOL OR WORK__________ TYPE WORK__________

EED__________ FEED__________ TOOL LIFE PER GRIND__________

DTH OF CUT__________ DEPTH OF CUT__________ LENGTH OF CUT__________

IIP LOAD PER TOOTH ☐ OR FEED PER REVOLUTION ☐__________ SURFACE FT. PER MIN.__________

TOOL COST AND SKETCHES

FIG. 4.6 (Continued)

average value for the element will not be in error more than ±5% of the true element time. The required number of observations to predict the true time within a ±5% precision and 95% confidence level can be calculated as follows:

$$N' = \left(\frac{40 \sqrt{N\Sigma^2 - (\Sigma X)^2}}{\Sigma X} \right)^2$$

Work Sampling

Work sampling is a measurement technique for the quantitative analysis, in terms of time, of the activities of men, machines, or any observable state or condition of operation. It is especially useful in the analysis of nonrepetitive or irregularly occurring activity, where no complete methods and frequency description are available. A work sampling study consists of a large number of observations taken at random intervals. In taking the observations, the condition of the object of study is recorded, and this condition is categorized into predefined groups of activities peculiar to the particular work situation. From the proportions of observations falling into each of the categories, interferences are drawn concerning the total work activity under study. As an example, if a group of production operators are observed to be waiting for material in a third of the observations of their work activity, we might draw the inference that better material planning and scheduling, rather than better work methods, represent the most obvious area for improvement.

The underlying theory of work sampling, then, is that the percentage of observations recording either a man or machine as idle, or as working, or in any other condition reflects to a known degree of accuracy the average percentage of time actually spent in that state or condition. If the observations are randomly distributed over a sufficiently long period of time, this theory is held to be true, regardless of the nature of the observed activity.

Suppose we wish to make a machine utilization study in our plant machine shop. Several vital steps would be necessary to carry out the study. These steps would include:

1. *Defining the project*: What do we want to know? In or case we want to know whether a machine is idle, is in work, is being set up, or is being serviced or repaired, and we want to know how much of the available time is spent in each state or condition.

2. *Making a preliminary study*: This may be as few as 100 observations taken at random intervals to see what percentages we obtain for each of the states or conditions for possibly one or two of the key or principal shop machines.

3. *Redefining the project*: We would have some actual data at this point, and are now familiar with the work sampling method, the machines,

and the procedures employed in the machine shop. In this case, we have found that we pretty much want to stay with the study as it was originally conceived.

4. *Planning the actual study*: We have now defined the study. We know where we will gather the data, and what categories we will use (see Fig. 4.7, machine utilization observation sheet). We also know how the observations will be made and who will make them. A clearly defined

OBSERVER: ____________
WEEK ENDING: ____________

MACHINE NUMBER: ____________ LOCATION: ____________

ELEMENTS	MONDAY	TUESDAY	WEDNESDAY	THURSDAY	FRIDAY	TOTAL
In Use						
Set Up						
Service & Repair						
Not in Use						

MACHINE NUMBER: ____________ LOCATION: ____________

ELEMENTS	MONDAY	TUESDAY	WEDNESDAY	THURSDAY	FRIDAY	TOTAL
In Use						
Set Up						
Service & Repair						
Not in Use						

MACHINE NUMBER: ____________ LOCATION: ____________

ELEMENTS	MONDAY	TUESDAY	WEDNESDAY	THURSDAY	FRIDAY	TOTAL
In Use						
Set Up						
Service & Repair						
Not in Use						

FIG. 4.7 Machine utilization observation sheet.

path has been laid out on a layout of the machine shop for the work sampling observers to follow, and the points along the route where they will stop and record their observations are shown. To next determine the number of observations, the easiest recommended approach is one in which we do not have definite tolerance requirements and are not really certain how close these total tolerances should be, and we do not want to run the risk of taking more observations than we really need. We take a block of observations from the preliminary study, calculate the percentage occurrences of the several categories, and then move ahead to determine the tolerances we actually have in each category. We then have two options:

Accept the average percentage occurrences in each of the categories, and their associated tolerances, and discontinue the study at this point.

Add successive blocks of additional observations and determine the combined percentage occurrences and the tolerances after adding each block, until we have the reliability that we feel we need.

5. *Taking the observations*: Observations must be taken at random intervals if the study is to have statistical validity. Times for observation trips can be constructed from a table of random numbers (see Fig. 4.8). A timetable of random trip times covering each day can be set up, or the random times can be stratified by hours in the workday.

It should be pointed out that observation trips into the area require the greatest amount of time, not the actual observations made. It is extremely important that the trips be planned in such a way that the maximum number of observations is taken per trip. Get everything you can out of each trip.

It is also advantageous to split up the trips among several observers. No one individual should have to spend too much time in taking work sampling observations. A schedule of about 10 trips per day at about 10 min per trip will require only 2 hr. Thus it is possible and desirable to use other manufacturing engineers and personnel not immediately involved in the design and direction of the study. By using several observers, it is possible to schedule observations closer together, and it also tends to balance the effect of observer bias, since it is well known that we all see the same situation a little differently.

6. *Analyzing the data*: You should now have a large amount of data. The most important thing to remember at this point is not to destroy original data. Maintain the identity of each individual observation, so that you will always know the day, the approximate time, and the machine involved. You may need to support your calculations at some later date, and the original data could prove invaluable.

No one can tell you exactly how to interpret your data. It all depends on what you want to know. You should be able to use an entire block

93734	01115	93668	98210	74351	74511	59558	13946	84653	26619
78046	36556	52983	63314	63428	62772	81475	82588	04863	67553
13856	90217	62338	07427	88600	51436	85819	35322	98239	04479
60629	34594	94757	23990	53285	89621	26727	89836	00132	73218
40534	96125	89065	17245	23000	40886	13436	70122	39038	47038
08123	31849	50177	27095	87986	51022	33837	87284	86968	03021
88032	02832	87185	31683	41726	49981	03380	78201	49734	90899
46721	30942	83460	49679	45815	13050	33796	94097	88317	90395
75684	64787	54889	33308	77806	24991	12767	68376	44835	58774
41644	78813	31303	00101	60401	48080	43670	09732	42827	43263
08533	40604	72071	19546	91230	41711	08097	88032	35547	88263
62267	83359	76489	88492	67103	53240	38754	22448	79734	86600
45188	40211	31521	67605	64452	61671	95814	14328	15551	66260
83708	97234	39653	13413	58656	99094	37725	17625	97754	38730
47682	11510	39769	96286	45477	67789	53056	90585	98860	10947
47274	24009	09586	79738	26668	98156	43251	70613	18510	27580
47764	93802	59122	36233	74756	14387	33719	24862	65268	23591
65398	95217	44823	58024	02229	39160	99509	23795	23549	23725
90200	03072	16926	00779	74304	70657	99731	95659	39473	00534
76116	41731	93522	40654	13201	76237	51760	93859	16655	42936
10500	82529	70573	60023	36437	91822	19897	83175	92020	40368
73129	51104	11001	72445	35826	48508	58795	83000	04848	91628
12789	39825	58290	28556	08074	47142	45621	26831	69227	64547
49431	48619	20684	98122	05325	59878	08651	15627	11928	71035
83861	41642	26414	03780	59875	05511	84753	81327	46369	96872
12191	41142	70423	10673	19083	53688	57121	16186	16009	64075
68794	14525	09448	63406	57620	03452	63541	80623	23772	32250
05621	47985	05287	07703	12624	33067	68297	40827	41467	31132
49624	17243	86462	90171	97999	65911	59913	05648	26039	01568
41909	87188	24948	94285	85120	45938	05733	02121	10397	31038
05343	98714	36298	16914	42719	55448	96582	26976	02142	97862
98512	64106	76781	27644	07210	92564	91955	40493	92088	16851
98960	08043	80681	83743	67085	09666	31387	13099	62719	33629
34583	79520	65947	58060	93289	53229	27319	34884	38461	62561
89098	26561	26153	44042	99546	25669	73824	02044	77748	83543
63754	15419	51882	97195	17248	95121	42889	06868	34246	32235
84061	02526	86747	54370	07161	11729	15078	27129	93875	98234
69684	40905	06480	49353	68698	37329	10905	65737	20836	25374
93257	79590	52694	53238	08085	13045	99055	65876	78512	98754
55160	45972	83651	12782	11737	85403	09291	51762	52977	17206
29250	69956	99529	44047	06044	98492	27429	35026	40592	62936
49305	68764	83680	34372	11988	79542	20655	54004	03527	71832
18033	69707	99173	67768	80685	51423	66920	70093	75441	34691
01160	33167	78354	56227	73914	53317	90068	19860	16457	12128
76715	76806	12275	14069	35637	57161	61396	72119	29033	29258
73247	65102	51278	73727	95424	64525	80920	30584	26087	31038
66151	49015	44649	75095	00621	78527	34183	08291	93119	43512
92260	24470	53500	79066	65867	03494	51468	05934	58966	84999

FIG. 4.8 Table of random numbers.

of data, or to break it down by day of the week, individual machine, or any other way. Calculate percentage occurrences and tolerances, and compare the results by variable (Tuesday idle time versus Wednesday idle time, for example), or against what you think you want (for instance, 15% idle time seems higher than it should be).

7. *Doing a follow-up*: If a work sampling study resulted in certain changes in an area, a study of the new condition might be necessary to prove that the changes did, in fact, produce the results desired.

Work sampling is a very valuable tool for the manufacturing engineer. It is one of the least expensive ways to gather accurate information quickly. It requires a rudimentary knowledge of statistics and the theory of probability. No formal training or certification is required to use work sampling. It can be learned by self-study or by formal classroom training. Every manufacturing engineer should understand and have some proficiency in this technique.

Work sampling can be used to set standards, but to do so requires that an accurate record be maintained of the work turned out during the course of the study, and with each observation there must be a recording of operator performance rating. The work sampling observations will identify the percentage occurrence of each of the elements of the job, and the idle and unproductive time can be identified and pulled out. It is not recommended that the manufacturing engineer attempt to set standards with work sampling unless he is quite proficient in work sampling as used to gather information.

The use of work sampling to identify machine utilization and machine downtime can be of tremendous assistance to the manufacturing engineer in justifying new or additional machines, or in surplusing old and worn-out equipment. It provides factual information which cannot be disputed and questioned by management. Work sampling can be used to identify areas and operations in the factory with low productivity and performance. It can highlight poor methods, inefficient tooling, and material handling problems, and can be used to determine P.F.&D (personal, fatique, and delay) allowances. Work sampling is especially useful in industries such as shipbuilding where operations are spread out over large areas and are extremely difficult to coordinate and control with any degree of efficiency. Again, it highlights where problems exist, so that they can be worked out.

4.1.2 Standard Data Development

In the final analysis, for standard times to be cost-effective, as well as for ease of application, they should be in the form of standard data. Standard time data can be defined as a compilation of all the elements that are used in performing a given class of work with normal elemental time values for each element. The data are used as a basis for determining time standards for work similar to that for which the data were determined without making actual time studies.

'ART NAME ______ OPERATION ______
'ROCESS PLAN DATE ______ MATERIAL ______
ALLOWANCE ______ STD. S/U ______ RUN TIME ______
STIMATOR ______ DATE ______

ARNESS ASSEMBLY EH

DESCRIPTION	MINUTES	FR	TOTAL MINUTES
ET-UP			
IT-MCO HARNESS TEST TE			
Prepare for Test/Lot	6.0		
UN TIME			
'REPARE WIRE ENDS MANUALLY			
Clock in on DPI/Step	.60		
'ill out C.A.T. Card & Stamp/Job	1.50		
:.A.T. Card, Stamp Only/Stamp	.04		
'ind Wire Number on Plan/Wire	.11		
:heck off Wire Number on Plan/Wire	.10		
:ut with Diagonal Pliers/Cut	.07		
:ut with Nail Clippers/Cut	.05		
'repare Pigtail Normal Wire/Pigtail	.85		
'repare Pigtail "H" Film Wire/Pigtail	.63		
:omb out Shield (Incl.Cover Removal)/Shld.	1.27		
Intwist Conductors (2 & 3 Cond.)/Wire	.07		
leasure Conductor & Cut/Conductor	.21		
leasure Strip Lgth.& Mark/Conductor	.15		
land Strip-Auto Strippers/Conductor	.07		
land Strip-Knife/Conductor	.43		
'wist Strands/Conductor	.06		
land Tin with Soldering Iron/Cond.	.15		
'Hand Crimp Lug, Taper Pin, etc./Crimp	.20		
'Hand Crimp (2) Wires, 1 Lug/Crimp	.25		
'Hand Crimp (3) Wires, 1 Lug/Crimp	.29		
Install Ferrule/Job, Set-Up	.27		
Install Ferrule/Ferrule, Run Time	1.15		
Install Solder Sleeve/Sleeve	.60		
Install Vartex Over Pigtail/Pigtail	.59		
Install Vartex over Floating Shld./Shld.	.10		
Install Vartex over Ferrule/Ferrule	.61		
Install Wire Marker (Non-Shrink.)/Marker	.20		
'lark Conductor, Cut, Strip & Twist	.38		
Wire Identification (Masking Tape)	.17		
'For Wires smaller than 10 gage			
'or 10 Gage and Larger, Use Values Below:			
:rimp Lug-Hydraulic Lugger/Lug	.69		
:rimp Lug-Bnch. Lugger 6-14 Gage/Lug	.77		
:rimp Lug-Bnch. Lugger 00-4 Gage/Lug	.80		
HANDLING			
Get Parts, Folder, etc./Job	1.8		
Get, Locate & Secure Plug or P/A to Harness Board/Plug or Plug Assembly	.70		
Get, Locate & Secure Sub-Assy. to Harness Board/Sub Assembly	1.02		
Get, Locate & Fit Item to Harness Assy./Item	.12		
Get, Locate & Secure Lacing Bar with 2 Nuts	.58		
Get & Locate Loose Wire to Harn. Bd./Wire	.14		
Remove Wired Plugs from Harn.Bd./Plug	.44		
Remove Wired Lug from Stud/Lug	.05		
Remove Wired Sub-Assy. from Harn.Bd./S-Assy.	.74		
Remove Lacing Bar from Harness Board	.55		
Remove Harn.Assy. from Bd./Junct.or 10" Lgth	.09		
Place & Remove Complete Harness from Board for Fit Check/Junct. or 10" Length.	.18		
Bag and Box Harness/Job	2.03		
Search Out Wire in Bag of - 30 Wires	9.91		
60 Wires	33.56		
90 Wires	70.95		
120 Wires	122.08		
ROUTING			
Route Wires Straight/Each 10" Length	.02		
Route Thru Bends or Breakout/Bend or Brk.	.06		
Route Thru Vartex/Inch	.04		
Route Thru Transition/Wire	.18		
Route Thru Grommet or Hole, Open/Wire	.05		
Route Thru Grommet or Hole, Congested/Wire	.09		
Route to Clip Spring or Equiv./Wire	.03		
Dress Wires/Wire	.03		

RUN TIME DESCRIPTION		MINUTES	FR	TOTAL MINUTES
Spot Tie/Each	Open	.26		
	Congested	.76		
Lacing, Starting Knot/Each		.26		
Lacing, Continuous Lock Stitch/Inch		.33		
Tywrap/Each	Open	.22		
	Congested	.31		
Install Tywrap thru Anchor/Each		.34		

SEARCH NUMBERED WIRES

NO. OF WIRES	MINUTES	NO. OF WIRES	MINUTES	NO. OF WIRES	MINUTES	NO. OF WIRES	MINUTES
1	.03	17	2.58	33	8.67	49	18.37
2	.08	18	2.85	34	9.17	50	19.09
3	.15	19	3.14	35	9.68	51	19.82
4	.24	20	3.45	36	10.21	52	20.57
5	.33	21	3.77	37	10.75	53	21.33
6	.44	22	4.10	38	11.30	54	22.10
7	.57	23	4.45	39	11.87	55	22.89
8	.71	24	4.81	40	12.45	56	23.70
9	.86	25	5.18	41	13.05	57	24.51
10	1.02	26	5.57	42	13.66	58	25.35
11	1.20	27	5.97	43	14.28	59	26.19
12	1.40	28	6.38	44	14.92	60	27.02
13	1.61	29	6.81	45	15.62	61	27.89
14	1.83	30	7.26	46	16.29	62	28.78
15	2.06	31	7.71	47	16.97	63	29.68
16	2.31	32	8.19	48	17.66	64	30.59

For a trunk containing more than 120 wires:

$$\text{TIME} = \left[\frac{(N+2)\quad(N-1)}{4}\right] + N \quad .0277 \text{ Min.}$$

Where N = Number of wires in breakout

WIRE PREP - ENDS 12 GA & UP

MEASURE, CUT & STRIP, PLUS / SINGLE CONDUCTOR WIRE:	TIN	FREQUENCY	CRIMP (1) WIRE PER LUG OR PIN	FREQUENCY	CRIMP (2) WIRES PER LUG OR PIN	FREQUENCY	CRIMP (3) WIRES PER LUG OR PIN	FREQUENCY
Common Wire End - Open	.64		.70		.74		.79	
- Cong.	.82		.81					
Pigtail Shld. Norm.	2.08		2.13		2.18		2.22	
Float Shld. Norm.	1.58		1.64		1.68		1.73	
Pigtail Shld."H" Film	1.86		1.92		1.97		2.01	
Float Shld."H" Film	1.37		1.43		1.47		1.52	
TWO CONDUCTOR WIRE:								
Pigtail Shld. Norm.	2.78		2.90		2.99		3.08	
Float Shld. Norm.	2.29		2.41		2.50		2.58	
Pigtail Shld."H"Film	2.57		2.69		2.78		2.87	
Float Shld."H"Film	2.08		2.19		2.28		2.37	

MARK, CUT STRIP & TIN ONLY - SINGLE CONDUCTOR WIRE:

Common Wire End - Open	.37
- Congested	.55
Pigtail Shield Normal	1.61

Values are per wire, vartex over shields included.

FIG. 4.9 Wire harness fabrication standard data.

RUN TIME

DESCRIPTION	MINUTES	FR	TOTAL MINUTES
HOOK UP AND SOLDER WIRES			

DESTINATION	SOLDER TIME ONLY		1 WIRE/ TERMINAL		2 WIRES/ TERMINAL		3 WIRES/ TERMINAL	
	MIN.	FR	MIN.	FR	MIN.	FR	MIN.	FR
P.C. Pad	.05		.31					
Congested			.73					
Plug Pin 10-26G W/Weep Hole	.27		.64					
Plug Pin 10-26G	.27		.56		.85			
Congested			.92		1.54			
Plug Pin 4-8 G	.38		.84					
Congested			1.25					
Plug Pin 4/0-2G	1.00		1.45					
Congested			2.74					
Turret, Island	.09		.48		.71		.94	
Congested			.79		1.33		1.89	
Turret, Buss	.89		1.28		1.41		1.54	
Congested								
Bifurcated.Isl.	.26		.64		1.08		1.49	
Congested			1.08		2.00		2.87	
Bifurcated, Bus	.89		1.28		1.41		1.54	
Congested								
Eyelet Terminal	.11		.50		.65		.81	
Congested			.76		1.08		1.41	
Hook & Other	.07		.45		.59		.73	
Congested			.64		.90		1.16	
Use Assist Tool			.06		.06		.06	
Tin Plug Pin	.18		.31					
Hand Solder Circuitry/Sq."	.09		.20					
Touch-up Circuitry/Sq."	.09		.19					

DESCRIPTION	OPEN MINUTES	FR	CONGESTED MINUTES	FR
ROUTE, PREP, TIN, HOOK-UP, SOLDER, CLEAN & DRESS WIRES				
Turret Terminal (Island) 1W	.90		1.39	
Turret Terminal (Island) 2W	1.56		2.53	
Turret Terminal (Island) 3W	2.21		3.68	
Bifurcated Term.(Island) 1W	1.07		1.68	
Bifurcated Term.(Island) 2W	1.93		3.20	
Bifurcated Term.(Island) 3W	2.77		4.67	
Eyelet Terminal (Island) 1W	.92		1.36	
Eyelet Terminal (Island) 2W	1.50		2.28	
Eyelet Terminal (Island) 3W	2.08		3.21	
Hook Term. or Otner 1W	.88		1.24	
Hook Term. or Other 2W	1.57		2.10	
Hook Term. or Other 3W	2.00		2.96	

DESCRIPTION	MINUTES	FR	TOTAL MINUTES
VARTEX & MARKER INSTALLATION			
Cut, Install & Tape Wire Marker (Shrinkable)	.23		
Cut & Install Wire Marker (Non-Shrinkable)	.20		
Install Bundle Marker (Shrinkable)	.08		
Install Bundle Marker (Non-Shrinkable)	.57		
Install Marker Plat/Marker	.62		
Slide Short Vartex Over 1 Wire/Piece	.04		
Slide Short Vartex Over Solder Connection	.05		
Install Short Vartex, Under 36", Over Wires (Non-Shrinkable)/Inch of Wire	.08		
install Long Vartex, Non-Shrinkable/Piece	8.07		
Install Long Vartex, Non-Shrink./In.of Wire	.13		
Install Long Vartex,Shrinkable/Piece	.14		
Shrink Short Vartex, Per Piece	.15		
Shrink Long Vartex, Gun Handling/Piece	.14		
Shrink Long Vartex, Per Inch	.23		
Shrink Boot or Transition, Per Piece	11.36		
Wrap Vartex for Filler on Strain Relief, Per Piece	.17		
Cover Exposed Shield with Vartex & Tie/ Piece	1.67		

RUN TIME

DESCRIPTION	MINUTES	FR	TOTAL MINUTES
COAXIAL CABLES			
Prepare One End	1.69		
Prepare One End & Install Connector	3.69		
Prepare Two Ends & Install Connectors/ Cable	7.37		
MISCELLANEOUS			
Insert Pinned Wire in Connector,Per Wire	.22		
Install Plug Seals, Per Seal	.09		
Install Prepared Lug Over Stud, Per Lug	.03		
Install Dust Cap, Each	.21		
Install Strain Relief & Tighten, Each	1.56		
Remove & Replace Potting Boot, Each	.36		
Install Plug in Vise & Remove, Per Lug	.16		
Install Common Points, Each	3.23		
Install Rubber Bands on Harness Bd.,Each	.08		
Install Environmental Seal on Plug, Each	.54		
TORQUE MECHANICAL FASTENERS			
Change Socket on Torque Wrench/Socket	.16		
Set Required Torque/Gage	.05		
Torque Fastener/Fastener	.15		
Stripe Fastener/Stripe	.23		
POSITION HARNESS IN CHASSIS			
Harness Installation:			
Normal/Harness	.13		
Confined/Harness	.15		
Locate Straight Run:			
to 4" Per Run	.04		
Each additional 4" to 6" Per Run	.02		
Locate Bend of Breakout, Per Breakout	.06		
Reform & Locate Bend, Confined/Breakout	.12		
Dress Branched Conductors/Conductor	.06		
DIT-MCO HARNESS TEST- TE			
Constant, Per Harness	3.51		
Connect and Disconnect, Per Connector	.51		
Test, Per Conductor	.10		
Self Check Equipment, Per Harness	.031		
x Total R/T			

SUB TOTAL RUN TIME

PAGE #1 RUN TIME

ALLOWANCE

TOTAL STD. RUN TIME

FIG. 4.9 (Continued)

Figure 4.9 shows a standard data work sheet for wire harness fabrication. It is only necessary that the standards applicator work down the data sheet to find the elements of work that apply to the wire harness the standard is being developed for, and write in the count or frequency at which each element occurs in the fabrication of that particular wire harness. When he has gone over the entire list of elements, selected those that apply, and written in the frequency of occurrence of each, he then multiplies the standard time for each element by the frequency to obtain the total standard time for that element for the particular harness. The next step is to add all of the extended element times to arrive at the total wire harness leveled time. The allowances for personal, fatigue, and delay are added to the total leveled time to give the total standard time for the fabrication of the wire harness.

The analyst or engineer who uses standard data in work sheet form to set standards does not require extensive training or experience in work measurement. He can set a standard quickly and accurately for a variety of different product configurations, as long as they fall into the general class of work covered by the standard data. Since the standard data are used over and over again, for a variety of product configurations, the initial cost of making the studies and developing the standard time data can be prorated over many product lines and programs. Industrial engineers with experience and expertise in work measurement can do the engineered standards work, either on a one-time consulting basis or as specialists on the company staff. The net result is that the manufacturing engineer can quickly establish the accurate standard times he needs, without making expensive original studies.

4.2 STANDARDS APPLICATION

Time standards are the common denominator for the calculation of costs and the measurement of factory labor performance. Factory labor is production labor which can be reasonably and consistently related directly to a product, or unit of work being manufactured, processed, or tested. It involves work affecting the composition, condition, or production of a product. It does not include the functions of engineering, production control, and production planning. It does include such functions as machining, welding, fabricating, cleaning, painting, assembling, and functional testing of production articles.

4.2.1 The Job Shop

Job order manufacturing is the norm in most of the industrialized world today. Job order manufacturing is characterized by short production runs, frequent setups and teardowns, minimum or shop aid type tooling, and labor costs that are at the high end of the learning curve. Time standards established by original studies for this kind of work are impractical and prohibitively expensive. It makes little difference whether the standards are set by predetermined time systems or by time study. The only practical approach is to use standard data, assuming they are available.

The worst-case situation, often found in the smallest companies or shops which have only a handful of employees, is that the only thing available is historical cost data. The manufacturing engineer in this situation should use these data to establish standards for new work coming in, and as a basis for any cost trade studies he may be required to make. Such data should also be used bidding on new work. Historical cost data often contain delay and waiting time, rework time, and low operator or worker production time caused by poor methods or tools that do not work properly. Such data should be screened very carefully before they are used as a yardstick for measuring future performance.

An example of how a standard might be established in a job shop environment is shown below, with particular attention to the establishment of correct allowances for heavy work. In this example a crew is unloading boxes from a truck and placing them on a pallet or skid. The other factors or conditions in effect are as follows:

1. The operation is performed at a warehouse ramp.
2. The boxes weigh 25 lb each and the employee is under load 25% of the time. The boxes are being taken from stacks slightly higher than his waist and are placed on pallets resting on the truck bed.
3. The work is purely routine.
4. The employee walks approximately 5 ft with each box.
5. The cycle time (per box) is 0.500 min; actual under-load elements equal 0.125 min (if per pallet the percentage may be somewhat less).
6. No restrictive safety devices are required.
7. A forklift operator is considered a part of the unloading crew.

The computation of allowances is as follows:

Personal	
Base	4.2%
Class B slightly disagreeable, exposed to weather	3.0%
Fatigue	
Physical—25lb handled 25% of the time (total under-load element time, 0.125, divided by cycle time, 0.500, equals 25%)	4.0%
Mental—class A—work committed to habit	0

Fatigue (continued)	
Position—class C (walking)	1.0%
Monotony—class C (0.500 min)	2.0%
Delays	
Class A, little coordination with adjacent jobs	1.0%
Total allowance	15.2%

Allowance factor:

$$AF = \frac{100\%}{100\% - 15.2\%} = \frac{100}{84.8} = 1.179$$

Computation of standard: If this operation is studied and the normal time is determined to be 0.500 min, the standard time would be computed as follows:

$$0.500 \times 1.179 = 0.590 \text{ standard minutes}$$

The number of decimal places used would depend on the time increments used in the man-hour accounting system and the volume of production.

4.2.2 Production

Continuous or high-volume production such as automobile production, the production of television sets, or a long production run of automatic screw machine parts is characterized by expensive planning, hard tooling, few setups or teardowns, and labor costs that track predictably and consistently down the learning curve. Labor standards and costs are calculated to several decimal places, and standards must be truly accurate engineered standards, as product costs must be kept low to realize the benefits of production.

Here again, standards would be established from standard data. Original studies would be limited to new operations, problem areas, or choke points and validation studies of standards set from standard data. The key role played by accurate standards in production should be highlighted. Balanced production lines require the best work measurement techniques to ensure balance in work content between stations. If, for example, a moving-belt assembly line moves at a rate of 6 in. per minute, and each work station on the line is a 3-min station, then the operator would have an 18-in.-wide work area on the belt to accomplish his or her work, before the product or assembly moves in front of the next operator. There can be no more or no less than 3 min of work to perform for each operator on the line. If an operator must leave the line for any reason, there must be a relief operator available to step in until the original operator returns.

Once established, production standards are relatively easy to maintain, as methods, tooling, and other factors of production are closely controlled.

4.2.3 Automated Processes

Automation and automatic assembly processes are machine-controlled and direct factory labor plays a minimal part in the operation. Times for these operations are process or machine times for the most part, and there may or may not be time for an operator to tend or service the process or machine during its operation cycle. If an operator is required, as during a semiautomatic operation, and is part of the total process cycle (e.g., an operator loading parts into an automatic assembly machine), the operator is part of the total process cycle, and standard times must be established for the direct labor part of the operation.

Analysis of which parts of the cycle are machine-controlled and which parts are operator-controlled can best be done by using a multiple activity chart, similar to the one shown in Fig. 4.10. The cycle time, usually in minutes, is plotted in the center of the chart; the activity of the machine is logged on one side opposite the time scale; and the activity of the operator is plotted or logged on the other side against the center time scale. Such an analysis provides the needed visibility in showing how the machine and operator would interface.

4.2.4 Realization Factors

A realization factor is a calculated factor (exclusive of personal, fatigue, and delay allowances) by which labor standards are modified when developing actual man-hour requirements. If, for example, it is desired to determine what it will actually cost to build the first five units of a new product in production for the first time, then the realization factor applied to the standard time would provide this. It would include learning time, time lost for scrap and for rework, and time lost due to tooling and methods which do not function as envisioned and must be corrected. Realization is a fact of life in manufacturing; the standard is the goal which is attained (if at all) at the appropriate number of units as defined by the learning curve. Manpower staffing in manufacturing is determined by labor standards factored by an appropriate realization factor. Fig. 4.11, shows the elements that make up realization factors at various stages of production.

4.2.5 Labor Performance Measurement

Once an ongoing standards program has been established, and at least half of the plant or factory manufacturing operations are covered by

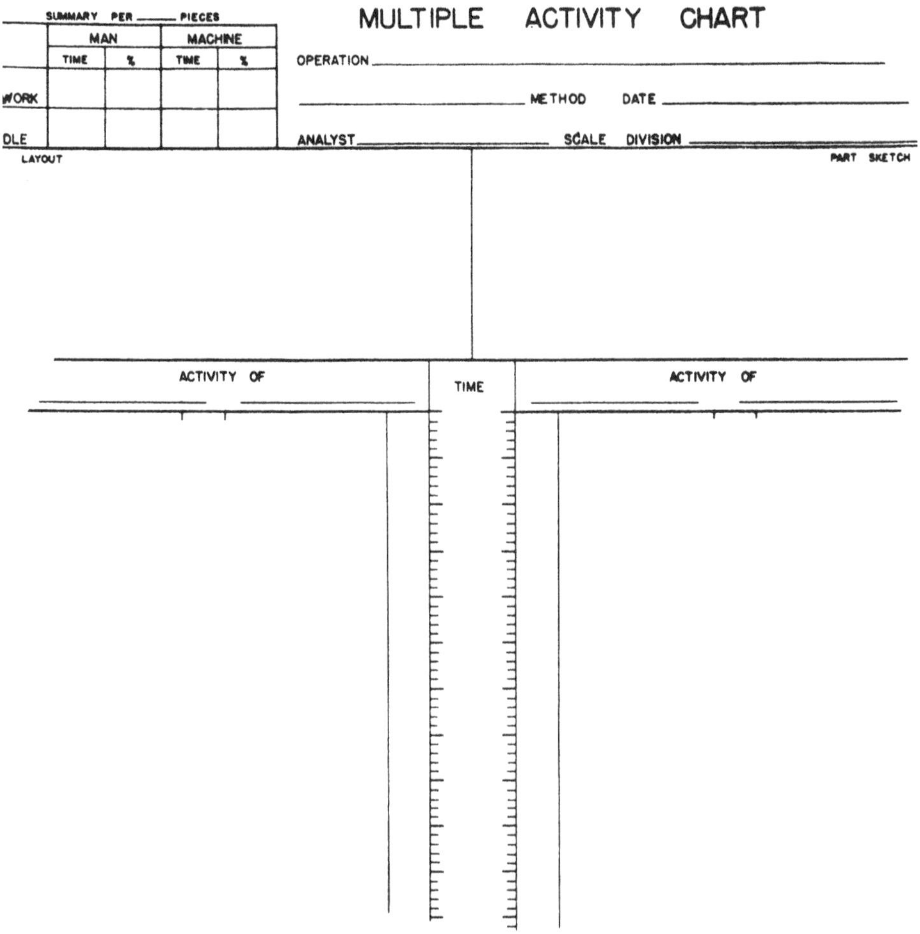

FIG. 4.10 Multiple activity chart.

standards, then labor performance against standard should be reported to management. The reporting system should provide performance reporting by department, by product line, and by individual operator. For a performance reporting system to have any real value, it must be based upon standard earned hours, determined by number of good units of product produced in a given time versus hours actually expended in producing those units.

Labor performance reports should be issued weekly in summary form, and daily to report operator performance and departmental performance.

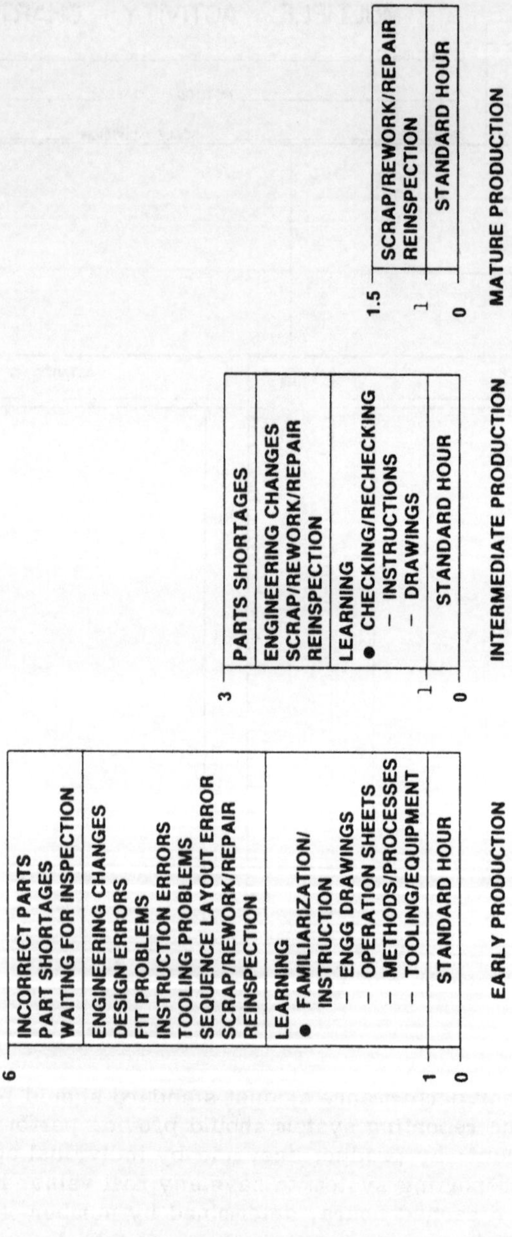
6
INCORRECT PARTS
PART SHORTAGES
WAITING FOR INSPECTION
ENGINEERING CHANGES
DESIGN ERRORS
FIT PROBLEMS
INSTRUCTION ERRORS
TOOLING PROBLEMS
SEQUENCE LAYOUT ERROR
SCRAP/REWORK/REPAIR
REINSPECTION
LEARNING
● FAMILIARIZATION/ INSTRUCTION
– ENGG DRAWINGS
– OPERATION SHEETS
– METHODS/PROCESSES
– TOOLING/EQUIPMENT
1
STANDARD HOUR
0
EARLY PRODUCTION
3
PARTS SHORTAGES
ENGINEERING CHANGES
SCRAP/REWORK/REPAIR
REINSPECTION
LEARNING
● CHECKING/RECHECKING
– INSTRUCTIONS
– DRAWINGS
1
STANDARD HOUR
0
INTERMEDIATE PRODUCTION
1.5
SCRAP/REWORK/REPAIR
REINSPECTION
1
STANDARD HOUR
0
MATURE PRODUCTION

FIG. 4.11

To be used as a management tool by shop supervision, the reports must be timely to enable corrective action to be taken. The time to correct a problem is when it first happens, or is happening. Labor performance information reported a month after it happens is too late for any action to be taken until small problems have become large problems. As a minimum, labor performance reporting must be timely, adequate, and accurate.

4.2.6 Other Uses of Standards

Labor standards should be used by the manufacturing engineer and others as an input in the development of budgets, plans, and schedules, in addition to their use in measuring labor performance. Also included would be the use of labor standards for:

1. *Estimating*: Standard hours should be the basis for all cost estimating and proposal work. Standards are the base against which learning curves and other realization factors are applied to arrive at estimated labor costs in man-hours. Labor estimates with standards as a base can usually withstand any audit, and are much easier to defend and support in customer negotiations. In making comparisons of different approaches while the estimate is being prepared, using standard times is the only way to ensure that the cost comparison has meaning and that all factors have been removed.
2. *Budgeting*: All direct labor budgets should be based upon standards factored by a realization factor. This is true whether the company has a standard cost system, a job order cost system, or a derivative of both. If there are no standards, then direct labor budgets would be based upon estimates and cost history.
3. *Cost trade studies*: One of the important duties of the manufacturing engineer is making cost analyses and cost trade studies to select the most cost-effective of several alternative approaches. This might include an analysis to determine which of two or three possible choices of equipment would yield the greatest return in manufacturing, or it might center around which design approach for a new product would be more producible in manufacturing. Cost trade-offs based upon standard times bring all of the labor factors involved down to a common basis for easy comparison and subsequent presentation and selling to higher management. When the cost savings are in terms of standard hours saved, there is little room for doubt. Savings are not inflated by any factors and, in reality, are in terms of real, quantifiable savings, or cost avoidance.

QUESTIONS

1. Why are standards important to the manufacturing engineer?
2. What are the requirements of engineered standards?
3. What are the three basic methods used in industry today to set standards, and what are the advantages and disadvantages of each?
4. What is standard data, and how is it used?
5. What are the four levels or integrated systems of Work Factor?
6. What is the underlying theory of MOST, and what are its primary activity sequences?
7. What statistical confidence level and degree of precision are commonly used in stop-watch time study?
8. What is the underlying theory of work sampling?
9. Why is work sampling a very valuable tool for the manufacturing engineer?
10. Describe one important use of work sampling that is of major assistance to the manufacturing engineer?
11. What is a realization factor, and what does it include?

5

Capital Equipment and Facilities

5.1 IDENTIFYING THE REQUIREMENTS

The manufacturing engineer is the key individual in the identification of facilities and capital equipment for manufacturing or production. His knowledge and expertise are critical in the determination of area and facilities for manufacturing, and only he can identify the machines and equipment required. The manufacturing engineer is a key participant in long-range planning for future requirements, in near-term preproduction planning where immediate requirements are identified, and in planning for the replacement of worn and obsolete production facilities and equipment. In many instances, it is highly economical to refurbish and rebuild existing machines and equipment rather than to buy new. Here again, the manufacturing engineer has a key role to play in making this decision.

5.1.1 Long-Range Planning

A long-range facilities plan must be based upon a long-range sales forecast and a determination by management of where the company wants to go in the future, what new capabilities will be developed, and what existing or old capabilities or product lines will disappear. As part of this planning, it is necessary to plan what facilities and equipment will be required, and approximately what it will cost. This type of plan looks ahead 5, 10, and even 15 years. Major new construction is planned at least 5 years in advance. The time phasing is very important, for it will show when funds must be available, when funds must be committed, how long it will take to complete construction, when new equipment and machines must be placed on order, when they will be delivered, and when installation and checkout for production will be completed.

The long-range plan will identify what facilities, machines, and items of equipment are required. It will indicate the cost for each, and when action must be initiated to start construction or to order machines and equipment. All of the detailed milestones for each item will be scheduled through start-up of production. In addition, the long-range plan may contain preliminary floor layouts for facilities, and brief descriptions of the operating features of machines and equipment. Also, a brief justification statement may be included with each item stating why it is needed. The long-range plan, once approved by management, becomes the "road map" which guides near-term and other operating plans. The manufacturing engineer must define the size and ultimately the cost of new production facilities. He must identify which machines and other production equipment are needed and what specifications they must meet, which ultimately determines what they will cost.

5.1.2 Preproduction Planning

The preproduction planning phase for new products, or previously built products going into production, reveals the firm, near-term requirements for facilities, equipment, and machines. How these requirements develop from this process is described in Chapter 3. Essentially, the definition of the manufacturing process, which evolves from product definition as described in the drawings and specifications, determines what equipment and machines are needed and what their capabilities must be, and the planned peak production rate determines equipment and machine capacity as well as ultimate facility floor space requirements.

As the need for a specific machine or piece of equipment is identified, the manufacturing engineer should begin contacting manufacturers of machines and equipment to determine availability, cost, and features and capabilities of the equipment and machines which could meet his needs. He should begin to build a literature file for machines and equipment from different manufacturers which meet his requirements. As soon as possible the manufacturing engineer should prepare a detailed machine or equipment specification for each different machine or piece of equipment that he will need. This specification should then be sent to all of the prospective machine and equipment manufacturers for their quotations of cost and delivery. The machine or equipment specification is an absolute must prior to requesting quotes. If a manufacturer is asked to quote on providing a machine without a specification from the requester, the manufacturer will quote cost and delivery for an item that highlights the features he has available, which may or may not be what you need to satisfy your company's requirements. The equipment or machine manufacturer is in business to make money, and without an appropriate equipment specification, he will include features and extras which you might not need, and overlook those that you do need, because he does not really know your operation.

Suppose a need has been identified for a hydraulic press, as an example. The manufacturing engineer should begin to build a file on hydraulic presses which as a minimum should contain the kind of information shown in Fig. 5.1 and Table 5.1.

In addition to determining hydraulic press tonnage requirements, it is also necessary to obtain information on machine action such as the required number of strokes per minute for the intended application. Table 5.2 provides this information. Other specifications in selecting a hydraulic press would include press type, such as open gap or four column, accessory requirements, and details of press construction including the frame, the cylinder, maximum system pressure, horsepower, and speed.

Information acquired should also explain any hydraulic press limitations such as shock after breakthrough in blanking, that a hydraulic press requires external or auxiliary power to feed stock, and that stroke depth cannot be controlled closer than .020 inch.

A machine tool or a major piece of capital equipment represents a major investment of company resources with long-term consequences. It is essential that the manufacturing engineer do a thorough job in researching the requirements and writing the specification. The machine or equipment specification should thoroughly define the performance requirements the machine or equipment must meet and the warranty requirements, spare parts and service needs, as well as acceptance criteria.

5.1.3 Planned Replacement

As machines and production equipment accumulate many hundreds and even thousands of hours of run time, even with planned and preventive maintenance, moving parts begin to wear and break down. Often advances in the state of the art in manufacturing technology make continued use of a machine or piece of equipment uneconomic, even though it is in reasonably good condition. Often a key machine or item of production equipment can be built or refurbished to "like new" condition at a fraction of the cost of a new machine. Without exception, the manufacturing engineer must make the necessary studies and analysis and recommend the appropriate course of action to management.

Obsolescence

Suppose that we are considering the replacement of a conventional engine lathe with a numerically controlled (N/C) lathe. In any capital

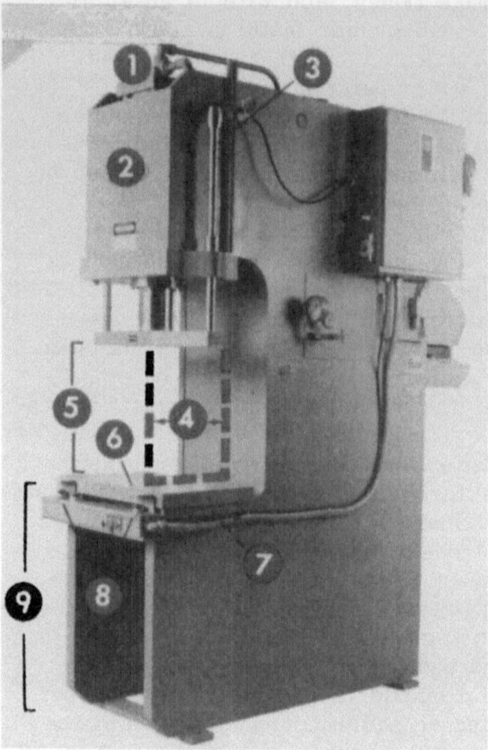

FIG. 5.1 Hydraulic press nomenclature. (Courtesy of Greenerd Press & Machine Company, Inc., Nashua, New Hampshire.)

1. *Cylinder* The piston diameter and the oil pressurization determine the force (tonnage) of a given press. A cylinder assembly consists of housing, piston, ram, packing and seals.
2. *Frame* The main structure of the press containing the cylinder(s) and the working surfaces.
3. *Stroke Control Rod* Stroke length can be set for any distance within the stroke capacity of a cylinder. The rod carries adjustable cams which actuate fixed limit switches to set stroke depth or return length. Other functions can also be set on the control rod.
4. *Throat Clearance* The distance from the vertical centerline of the ram to the frame member behind the bed. This distance determines the largest diameter piece that can be positioned with the part centerline under the center of the ram.
5. *Daylight* The vertical clearance from the top of the bed to the underside of the ram in its maximum up position. This term

equipment evaluation analysis, two series of costs must be considered. The first is known as the capital recovery costs (the recovery of the initial investment for the new lathe). The second series is the annual operating and maintenance costs. Normally, these costs increase each year as the maintenance costs increase with the age of the machine. The rate at which capital costs are recovered depends upon the method of depreciation used.

Figure 5.2 shows a graph with the two series of costs plotted against time. The goal is to determine whether the net savings in operating and in maintenance costs with numerical control are enough to more than offset the high initial outlay. If they are, the minimum in the graph for a numerically controlled machine will be lower than the minimum for the conventional machine it is being compared with.

The initial investment includes all the costs required to procure and install the N/C lathe and to bring it on-line in production. These initial investment costs include purchase of the lathe, shipping and installation costs, the cost of training operators and maintenance and programming personnel. Also included would be the cost of tooling and support equipment.

The net savings between the two machines can be divided into direct savings and indirect savings. Direct savings consist of savings in machining time through use of numerical control (the hours it takes to complete work on a conventional machine minus the time it would take to do the same work by using numerical control).

FIG. 5.1 (Continued)

sometimes is confused with the mechanical press term "shut height." Shut height is the clearance over the bed with the ram full down. "Daylight" described the maximum vertical capacity of the press.

6. *Bolster* A plate or structure mounted on the bed. Hydraulic press manufacturers provide a removable bolster on most models.
7. *Bed* The flat, stationary machined surface that supports the bolster or dies.
8. *Dual Palm Button Controls* A common method of actuating hydraulic presses. Both buttons must be depressed together to bring the ram down. . . thus using both of the operator's hands. Control circuits include non-repeat and anti-tie-down features.
9. *Work Height* The distance from the floor to the top of the bolster.

Table 5.1 Hydraulic Press Tonnage Requirements Calculations

1. *General*• When pressure per square inch is known

$$\frac{\text{psi} \times \text{area of work}}{2000} = \text{tons of ram force}$$

Example: Where it is known that 100 psi is needed to do a job on a 5" × 8" wide piece.

$$\frac{100 \times 5'' \times 8''}{2000} = 2 \text{ tons}$$

2. *Press fit*• To determine the pressure required to press fit two round pieces together such as a shaft pressed into a bushing, use the following formula: $P = D \times \Pi \times L \times i \times F$

Where P = pressure required in tons
D = diameter of the part to be pressed in inches
Π = 3.14
L = length of part to be pressed in inches (Note: the length of the interference fit only.)
i = interference in inches - (usually .002" to .006")
F = A constant pressure factor = 250 ÷ D

Example: A steel shaft 2" in diameter pressed into a hole 3" long. The interference fit between the two diameters is .006".

2" × 3.14 × 3" × .006 × (250 ÷ 2) = 14.13 Tons

3. *Punching pressure*• A quick guide to determine pressure requirements for punching in steel is as follows:

Diameter × thickness × 80 = TONS

80 in this formula is a constant for steel. Use 65 for brass.

Example: A 3" hole in .250" stock: 3" × .250" × 80 = 60 TONS

For noncircular holes, instead of the diameter use 1/3 of the total length of cut.

Example: A rectangular hole 4" × 6" in .250" stock.

(4" + 6" + 4" + 6" ÷ 3) × .250" × 80 = 133.3 TONS

4. *Deep Drawing*• Deep drawing calculations can be complex. The press, dies, material, radius, part shape, all have a bearing. A simple guide is the following formula for drawing and round shells.

$$P = \frac{\pi d t S \left(\frac{D}{d} - K\right)}{2000} \qquad \sqrt{D = d^2 + 4db} \quad \left(\text{for blank dia.}\right)$$

(Continued)

Table 5.1 (Continued)

Where P is tons of ram force; d is the cup diameter, D is the blank diameter; t is the metal thickness; S is tensile strength in psi; and K is a constant. (Use 0.7)

Example: A 3" cup from 9" blank in .065" stock with a tensile of 80,000 psi.

$$\frac{3.14 \times 3'' \times .065'' \times 80{,}000 \left(\frac{9''}{3''} - 0.7\right)}{2000} = 56.3 \text{ TONS}$$

Note: A 75-ton press would be recommended.

(Courtesy of Greenerd Press & Machine Company, Inc., Nashua, New Hampshire).

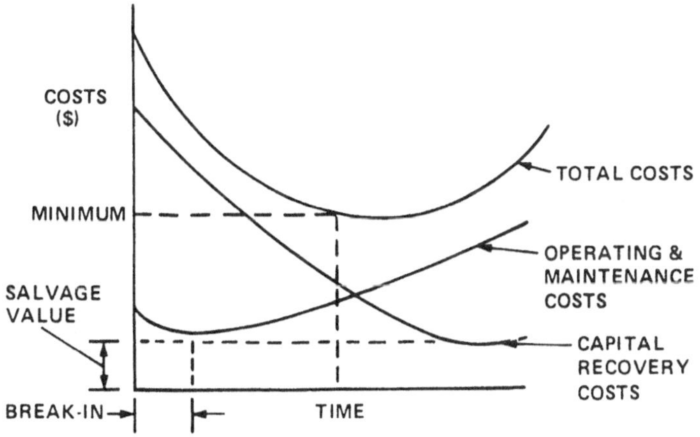

FIG. 5.2 Graphical plot of numerical control machine costs versus time. (Reprinted by courtesy of the Society of Manufacturing Engineers, Dearborn, MI.)

Table 5.2 Hydraulic Press Calculations for the Determination of Press Action Requirements

The number of strokes per minute for a hydraulic press is determined by calculating a separate time for each phase of the ram stroke. The rapid advance time is calculated, then the pressing time, (the work stroke); then, if there is no dwell time, the rapid return.

The basic formula for determining the length of time in seconds for each phase of the stroke:

$$\frac{D \times 60}{IPM} = T$$

T = time in seconds
D = distance of a particular phase of ram travel in inches
IPM = ram speed of a press expressed in inches per minute

Example: A hydraulic press with a 600 IPM rapid advance, 60 IPM pressing speed, and 600 IPM rapid return. The work requires a 3" advance, 1" work stroke, and 4" rapid return.

Rapid close 3" × 60 ÷ 600 IPM	= .300
Press 1" × 60 ÷ 60 IPM	= 1.000
Rapid return 4" × 60 ÷ 600 IPM	= .399
Allowance for valve shift time	.500*
	2.199

60 ÷ 2.199 = 27 cycles per minute

*Electrical actuation and valve shift time varies depending on the type of hydraulic circuit. One half second is a reasonable average figure.

(Courtesy of Greenerd Press & Machine Company, Inc., Nashua, New Hampshire.)

Savings in hours are then converted to dollars by using the appropriate machine operators' rate of pay. Indirect savings occur in all facets of total machining costs. Included would be maintenance, sometimes tooling, programming (which does not occur on conventional machines), scrap, floor space and material handling.

The Manufacturing engineer must constantly be on the alert for new and improved machines and equipment which could give his company a competitive advantage. In recent years, for example, the state of the art in machine tool technology has moved ahead rapidly, after years of no real advance in machine tools. In many instances, it is far more economical to dispose of an otherwise good piece of equipment and replace it with new equipment that offers significant improvements in production rate and quality. Improvements come from the use of robotics, from automatic assembly, from automatic versus manual testing, from an automated material-handling system, or through the use of lasers rather than more conventional welding and cutting methods. Manufacturing technology in today's world is changing dynamically. Computers alone are making changes not anticipated several years ago. The manufacturing engineer must take the lead in bringing these new technologies and methods to his company.

Refurbishment

There are many instances where a major machine tool or piece of equipment has given good service over several years, but is now showing signs of wear and requiring excessive maintenance and downtime. New machines or equipment on the market offer little in the way of technological improvement over the present item. Also, the cost of replacing the machine or equipment at present may put a severe strain on company resources. In a situation such as this, the manufacturing engineer should investigate the possibility of having the machine or equipment rebuilt or otherwise repaired to extend its useful life for several more years at a fraction of the cost of a new machine. Often this refurbishment can be done by the original manufacturer. In addition to the refurbishment cost, there would be the cost of shipping the machine to the manufacturer's plant and of returning the rebuilt machine to your plant. In most cases the rebuilt machine will carry a limited warranty and will perform like a new machine.

Worn Beyond Repair

Equipment that is worn beyond economical repair should be replaced. It is the responsibility of the manufacturing engineer to know the condition of all the major machines and equipment in his plant. With this knowledge, he must develop a replacement plan for this equipment which will become part of the plant capital equipment budget for the current fiscal year and of the 5-year capital equipment plan. In the case of production equipment and machine tools, downtime logs should be maintained on all items, showing time lost for maintenance and repair. In addition, there should be periodic checks of machine tool accuracy, including checks for spindle run-out, table travel, and condition of the ways, and a host of other checks which indicate the overall condition and whether or not the machine can hold the tolerances it was designed to meet.

One of the worst possible situations is to have a worn-out machine which cannot be used for production, sitting idle in the middle of the machine shop, occupying valuable floor space which could and should be used for production. Worn-out and surplus equipment should be disposed of quickly. Usually it is just a matter of calling a surplus equipment dealer and having him come out, appraise the equipment, pay for it on the spot, and haul it away the same day.

5.2 PROGRAMMING TO MEET REQUIREMENTS

The plan that finally evolves that includes all of the known and approved requirements is called the capital budget plan. This plan identifies in a time-phased manner what funds are needed, when the funds are needed, and what the funds are needed for. It provides top management with a road map of capital funding plans for up to 5 years. Top management may decide to defer some of projects in the plan, or it may cancel them altogether. The plan, when finally approved by top management, with changes and revisions, then becomes the approved plan that manufacturing engineering operates with.

5.2.1 The Capital Budget Plan

Table 5.3 shows a summary 7-year plan for both capital equipment and facility floor space. This provides management with the funding requirements by year for both facility floor space and capital equipment. Table 5.4 illustrates the backup information that will answer the question of what is required and will be provided with the funding requested. As shown, this provides both near-term and long-term requirements in a specialized area of manufacturing known as braided structures.

5.2.2 Time-Phasing of Requirements

The need for facilities and equipment, the availability of needed funding, the time required for orders to be placed, the time for the equipment supplier to build or fabricate the equipment, the time

Table 5.3 Summary Capital Plan: Additional Capital Dollars and Floor Space Requirements by Year

Year	Capital	Floor space (ft^2)	Floor space (dollars)[a]
1984	$527,024	5,600	$ 252,000
1985	622,056	8,000	360,000
1986	—	—	—
1987	—	1,900	142,500
1988	80,000	9,500	712,500
1989	230,216	14,444	1,083,300
1990	380,432	—	—
1991	—	—	—

[a]Conventional floor space calculated at $45.00 per square foot.

Table 5.4 Equipment and Floor Space Dollar Requirements by Year

	1984	
2	Braider × 142,624.00	$285,248.00
1	(HACOBA) auto bobbin winder with creel	19,000.00
1	Heat tunnel	150,000.00
1	Cutoff saw	30,000.00
1	Auto resin mix	35,000.00
3	Sets, quick change carrier (3 × 2592.00)	7,776.00
		$527,024.00
Floor space 5600 ft^2 × $45.00		$252,000.00

required to contract for and construct new facilities—all of these must be considered in planning the time-phasing of the requirements. The approved capital plan for an upcoming fiscal year must be broken down into a detailed acquisition plan, showing dates during the year when individual items and facilities must be ordered if they are to be received when needed and placed in service. This plan should be implemented, monitored, and controlled by manufacturing engineering. It is especially important to take into consideration the time it requires to prepare the capital equipment or facilities package with all justifications, quotes, specifications, and so on, the time it takes to obtain necessary approvals, and the time to have Purchasing actually place the order.

5.3 EQUIPMENT AND FACILITIES ACQUISITION

The equipment acquisition function is one of the most important duties of the manufacturing engineer. Once a need has surfaced for a particular machine or piece of production equipment, efforts by the manufacturing engineer would include:

1. Gathering of information and vendor literature on the machine.
2. Visists to other plants or facilities that have the machine to determine its performance, reliability, and customer satisfaction for similar work or application.
3. Preparation of an equipment specification detailing the requirements the machine must meet, including warranty, service, and spare parts needs.
4. Solicitation of vendor quotes to the equipment specification and any other requirements. The vendor quote should address not only price and delivery, but each detailed requirement in the equipment specification.
5. Tentatively selecting a vendor for the machine, and requesting the names of other companies that have the machine and the names of the persons to contact.
6. Making follow-on contacts with the companies having the machine to ask detailed questions about possible problems. This can involve a personal visit to see the machine in operation or a telephone call to the key person in the organization who knows the most about the machine and any possible problems.
7. Preparing the equipment order package with all required quotes, justifications, cost trade studies, and the equipment specification as part of the purchase agreement.

Figure 5.3 shows a record of a typical conversation between the manufacturing engineer and a key person in another company which reveals potential problems that would have to be overcome to use a

Subject: Sundstrand Omni-Mill OM1A Machine

He claims that they are utilizing the machine for three shifts and do their preventive maintenance on Saturday. The key point of the machine is a memory pin that memorizes the location of the tool in the holding turret. This pin tends to come out of position and the tool is then not stowed correctly in the turret. They utilize a preventative maintenance for each shift to check on the location of the pin. He believes that this problem has been solved.

He compared the machine to the K&T and to the DeVlieg. He contends that the Sundstrand is much more rigid than the K&T. He contends that the DeVlieg is probably more accurate than the Sundstrand but does not have the rigidity of the Sundstrand. He also discussed the bifurcated drives for the three axes of the machine. He stated that they are able to hold better than 0.001 inch on the X and Y axis and 0.001 inch on the Z axis. However, he stated that the rotary table that has the 36,000 positions only has an accuracy of 30 arc seconds. They utilize a 360° rotary table that has an accuracy of 10 arc seconds. Of course, this table can be utilized only if the indexing is in increments of 1°. This is probably satisfactory for the boring operation but may not be satisfactory for the milling operation around the periphery of the housing.

We discussed the run out of the two ends of the housing. He stated that they had not done any boring up to 8 inches in diameter. I told him that we would probably utilize a mill cutter for these operations such as the threads of one end and the O.D. in the other end. He indicated that the run out is a factor of four over the eccentricity and warned that 0.004 TIR may be difficult to obtain with the 36,000 indexing head. He stated that he was able to hold 0.002 TIR utilizing a 6-inch boring bar.

We discussed his reliability problems. The main problem has been the attachment of the memory pin on the tool turret. The next problem has to do with the keyboard. Chips and dust tend to get in the keyboard and cause troubles. This is a micro switch keyboard and he keeps a spare on hand at all times. However, if a chip does get in the keyboard you cannot input the correct data.

It is his opinion that the Sundstrand is more rigid than the K&T or the DeVlieg. However, the DeVlieg has better positioning accuracy. He felt that we should be able to hold 0.001 in X and Y and probably better than 0.002 inch in Z axis. I gave him my phone number and he stated that he would call me back if he thought of anything else on the machine. His three machines are used continuously and he could not fit us in in this application.

FIG. 5.3 Record of telephone conversation.

proposed machining center if it were purchased. The point to remember is that the purchase price of the machining center is a major investment of company funds to acquire a capability that the company needs to remain competitive. The effort expended in acquiring as much information as possible is time well spent.

5.3.1 Equipment Specifications

Any time a major machine or item of capital production equipment is purchased, the manufacturing engineer must write an equipment specification. This spcification must spell out in detail all of the performance requirements that the machine or equipment must meet. It must clearly describe what the acceptance criteria will be and how these criteria will be met. There should be minimum design and construction criteria that must be met and requirements for the power and electrical requirements should be spelled out, as well as the maintenance, service, and lubrication requirements.

Electric motor and other power drives should have their requirements clearly described. For example, should electric motors be explosion proof? Do they require antifriction bearings? Should the motors be alternating or direct current, and what rated voltages are we to get? If D.C., do we want series, compound, or shunt motor windings? If A.C., do we want wound rotor or squirrel cage windings? Do we want temperature rise characteristics to meet the latest NEMA standards?

How should the machine or equipment be prepared for shipment, and how should it be shipped? The requirements for engineering drawings of the machine should be spelled out, as well as the number of service and spare parts manuals to be delivered with the machine. What color should the machine be painted? The specification should require that a list of recommended spare parts be provided by the manufacturer, as well as cost and delivery for these. What about bearings and the requirements for lubrication?

If a machine tool, will the machine manufacturer be expected to provide tooling? If numerically controlled, will he be expected to provide programming. In some instances, a machine is purchased as a part of a turnkey package; that is, the machine manufacturer provides a completely tooled, programmed, and fully set-up machine that will produce acceptable parts. The final acceptance criterion for the machine or equipment would be to produce a given number of good parts, after the machine was set up and checked out in your plant or production facility.

Another important consideration is training. If the machine or the equipment is numerically controlled, will the manufacturer be expected to train company personnel in the programming, operation, and maintenance of the machine. If this is an important consideration, then the requirements for this should be clearly spelled out in the basic specification, including the exact number of people to be trained and when the training should be scheduled.

An example of a general equipment specification is shown below. It covers and describes a specially designed system for applying a protective coating to a composite tube.

Specification for a Protective Coating System

Description and intended use: This specification describes the requirements for the design, build, and acceptance of a system to grit blast, zinc coat, and paint composite tubes. The system will use a conveyor to transport the tubes through the processing cycle at rates up to 75 tubes per hour.

Performance requirements: The system shall be capable of performing the following functions and sequences of operations:

The system must deliver tubes at rates up to 75 tubes per hour and perform a protective coating cycle on fiberglass tubes which includes grit blast, zinc coat, and spray paint.

Zinc coat thickness must not exceed 0.005 in. and enamel paint thickness must not exceed 0.003 in.

The entire system must fit into an area of 24 by 44 ft.

The system must have a dust collection system that meets Occupational Safety and Health Administration (OSHA) requirements for dust and noise, and local environmental regulation requirements.

The system must operate in ambient temperatures up to 110°F and 95% relative humidity.

Acceptance criteria: Tubes processed by the system will be subjected to the following tests:

Bending test consisting of applying a 200-kg load to the center of the tube. Any cracking of the coating will constitute a failure.

Adhesion test consisting of scribing an X with a sharp knife that will fall within a circle having a diameter of 0.25 in. penetrating the coating. A piece of 0.75-in.-wide pressure-sensitive tape will be applied over the scribed area and pressed in place to ensure firm contact. The tape will then be removed with a quick, continuous pull approximately 90° perpendicular to the tube. Any removal of the coating constitutes a failure.

All tubes processed by the system must pass these tests for the system to be accepted.

Design and construction: All controls such as buttons, switches, levers, gauges, and regulating devices shall be identified, grouped, and mounted in a location affording the operator clear and unrestricted view of the controls and working area of the system when in normal operation.

The electrical control and safety instrumentation shall have a fail-safe provision and be mounted on a separate instrument panel. All of the electrical equipment and its installation shall conform to the latest National Electrical Code.

Power shall be supplied from one lead (four wire), 440 volts, three phase, 60 Hz; the fourth wire shall be for grounding purposes. Electric motors shall be totally enclosed, fan-cooled, with antifriction bearings.

All steel surfaces to be painted shall be thoroughly cleaned and free of rust, grease, loose mill scale, dirt, and other foreign material that could cause improper bonding of the primer. All exterior surfaces shall be painted red earth, using the highest commercial quality paint and workmanship in accordance with paint manufacturer's instructions.

The system shall have suitable provision for manual lubrication of all moving parts, if not permanently lubricated.

An exhaust system shall be provided to remove fumes or vapors developed during the cycle. A fresh-air intake with suitable filter shall also be provided. Both the air intake and exhaust shall have dampers for airflow adjustment.

Shipping and installation: Authorization to ship the system upon completion will be given after acceptance criteria have been met, as approved by the manufacturing engineering representative on site at the plant of the system manufacturer.

Upon delivery of the system, the manufacturer shall install, set up, and debug the system. He will conduct final acceptance tests. If the tests do not conform to specified requirements, the maufacturer shall make necessary changes or adjustments to the system, and conduct further testing until the system meets all performance requirements.

Spare parts and documentation: The manufacturer will provide five sets of engineering drawings of the system and five maintenance and spare parts manuals at the time of system delivery.

The manufacturer will provide a recommended list of spare parts, and warrants that any spare parts order will be manufactured to the same standards and criteria as the original equipment.

Warranty: The manufacturer guarantees that the system will operate free of defects in material or workmanship for a period of 1 year from final acceptance of the system.

The equipment specification ensures that your company will get a machine or piece of production equipment that meets your specific requirements. Nothing is left to chance; everything is clearly spelled out. When the order is placed for the machine or equipment, the specification is made part of the purchase order. When bids and quotes are obtained for the equipment from the various manufacturers, they are asked to quote to the specification. This ensures that all are quoting to the same baseline, or set of requirements, and that there is a clear understanding of the requirements at the outset.

To summarize, whether you are buying a single piece of capital equipment or an entire system consisting of several pieces of major equipment possibly tied together by a central computer, you must have a specification. The equipment specification, as a minimum, should contain:

1. A description and statement of intended use of the equipment.
2. Performance requirements that the equipment must meet, clearly spelled out in as much detail as necessary to ensure understanding and leave no doubt as to exactly what the requirements are.
3. The acceptance criteria for the equipment. This should be tied down to quantitative requirements that leave little doubt about what the equipment must do to be accepted.
4. Design and construction features that are required to ensure a first-class piece of equipment.
5. Shipping and installation requirements that must be followed to eliminate the possibility of shipping damage and to ensure proper installation in your facility.
6. Spare parts and documentation requirements.
7. Warranty requirements that you must have as a minimum to ensure that the equipment will function as envisioned, and that if service is required in the first year it will be provided promptly, at no cost to you.

The equipment specification provides and defines the minimum total requirements that the manufacturer must meet to satisfy your company's needs. It remains for the manufacturing engineer, with the assistance of the purchasing department, to solicit bids and quotes from the various equipment manufacturers and to select the one that meets the minimum equipment specification and offers something more than the minimum requirements. This would be the equipment that has the best quality and that more than exceeds the minimum requirements.

5.3.2 Bids and Quotations

The importance of doing a thorough job in the solicitation of bids and quotations on capital equipment cannot be emphasized too strongly. The manufacturing engineer must play an active part in this process. He cannot leave this work to the buyer alone. The first step is the location and identification of qualified sources for the equipment. During the initial planning phases some of these qualified sources provided information and data to the manufacturing engineer for his research and preparation of the equipment specification. Certainly these firms should be solicited and given the opportunity to quote, as well as the other qualified sources. Even though the buyer may be experienced in buying capital equipment, he is usually not an engineer, and may not know all the qualified vendors for the equipment. The manufacturing engineer must assist in providing a list of the possible suppliers he would like to have quote on the equipment.

As quotations are received, the manufacturing engineer should review each quotation and evaluate the key points in it. Capital equipment should never be purchased on the criterion of cost alone. Cost is extremely important, but so is delivery in many cases, and then there is service, and training offered (a vital consideration in the case of numerically controlled equipment). If tooling and programming as well as the equipment itself are being purchased, and you are buying a turnkey package, then the planned approach to this would probably be a more important consideration than cost.

The manufacturing engineer should prepare a matrix of important features which highlights the key points in each manufacturer's quote that are important for the equipment acquisition. For example, in the coating system for tubes described earlier, a key consideration was the type of process used for dust collection. A wet process meant that zinc dust would be collected in a waterfall system and discharged into the local sewer system, which required a long and costly permitting process and negotiation with local authorities. The other approach was a dry process which involved construction of an outside dust collection shed, separate from the main building where the coating equipment was set up, which would collect the dust in bags for reuse in the process of coating; this required no special environmental permits, since no toxic materials would be discharged into either the air or the sewer drain system. The dry process for dust collection was initially more costly then the wet system, but in the long run was the least expensive choice.

The importance of contacting other companies that already have a certain manufacturer's equipment in operation in their plant was discussed earlier. This is the best way to find out about problems with the equipment that could become your problems if you purchase the item, or steps to ensure that they will not, before the equipment is purchased. The manufacturing engineer is the logical choice for this follow-up with other users of the equipment, as he, more than anyone else, would know what questions to ask and could judge the quality of the information received from the answers that were given.

5.3.3 Placing the Order

Once it has been decided which manufacturer has the best and most competitive quotation, the next step may be to have the buyer place the order. It is more likely that the next step will be a series of meetings and negotiations with the selected manufacturer, or possibly the two manufacturers that submitted the most responsive bids, to resolve minor differences, clarify points in their quotations that were not clear, and even negotiate final price and terms.

The manufacturing engineer must participate in these meetings and in the final negotiations. For example, in a recent acquisition of a three-axis, numerically controlled maching center for a special application by a small company, it was possible to get the manufacturer of the machining center to guarantee not only that the tooling, program, and machining center would produce good parts, but also that it would produce at the rate of one part every 35 min. This was critical to the small company, and represented a concession by the equipment manufacturer that involved a degree of risk, in order to make the sale.

5.3.4 Acceptance and Equipment Installation

No equipment manufacturer wants to have an unsuccessful installation. It can hurt him as well as you. Most manufacturers are most interested in seeing that the customer's people have the know-how to make the installation successful. Usually they have prepared manuals, organized training programs, and offered engineering assistance to go along with the equipment. The manufacturing engineer must make certain that these services are utilized.

On large machines or equipment, the manufacturer's field service people will supervise the actual physical installation and instruct your operator in getting started. Because of the manufacturer's own self-interest, he will do everything in his power to give you an important backup resource to make certain your installation works. Again, use this service and support to the fullest.

It is especially important that the manufacturing engineer know and understand the environment and utilities the new machine or equipment will require. In many instances special foundations, power requirements, compressed-air requirements, and other support utilities will be needed. These requirements should be in place before the machine or equipment arrives. For example, with a numerically controlled machine, its electronic control unit may be sensitive to certain shop environmental conditions. Temperature, humidity, and electrical interference are factors to be considered. Some electronic control units will malfunction if there are wide electrical voltage variations. For this reason, a voltage regulator may be required.

Equipment acceptance is contingent upon successful installation and shakedown. If the equipment acceptance is based on performance (and it should be), then not only the manufacturing engineer but also the quality control people must be involved in inspecting parts or units produced and performing any testing required. Often machines and equipment must be calibrated before they can be put in service and this equipment calibration must be renewed at periodic intervals. Finally, once the acceptance is complete, a maintenance schedule should be prepared to ensure that the equipment remains in a good state of repair and in first-class operating condition.

5.4 FUNDAMENTALS OF LAYOUT PLANNING AND MATERIAL HANDLING

Plant layout and material handling are engineering disciplines in which every manufacturing engineer should be proficient. The process of manufacture must dictate the plant layout and determine the requirements for material handling. A poorly planned plant layout can be so ineffective in arranging related work areas that materials and products cannot be produced on an economic basis. The plant layout must be planned with well-defined objectives in mind. The principal objectives should be as follows:

1. To facilitate the process of manufacture
2. To eliminate or minimize required materials handling
3. To ensure a high degree of turnover of work in process
4. To maintain flexibility of layout and operation
5. To utilize all floor space as economically as possible
6. To keep equipment investment to a minimum
7. To assist in the most effective utilization of manpower

5.4.1 Plant Layout

Plant layout embraces the physical arrangement of industrial facilities. This arrangement includes the spaces needed for material movement, storage, indirect personnel, and all other supporting activities or services, as well as for operating equipment and personnel. There are three classical types of plant layout:

1. *By permanent or fixed position*: This type of layout is used where all operations are performed with the material staying in one fixed location. All tools, machinery, manpower, and other parts and materials are brought to it. The complete job is done, or the product is manufactured with the major component staying in one location. One man or crew makes the complete assembly, bringing all parts to each assembly point. The work force may or may not move from one product assembly location to another. The advantages of this type of layout are as follows:

Handling of the major assembly unit is reduced (through increased parts handling to the assembly point).

Highly skilled operators are allowed to complete their work at one point, and responsibility for quality is fixed on one person or assembly crew.

Frequent changes in product design and in sequence of operations are possible.

The layout is adapted to a variety of products and to an intermittent demand.

It is more flexible in that it does not require highly organized or expensive layout engineering, production planning, or provisions against breaks in work continuity.

This type of layout is used in shipbuilding and frequently in the aircraft industry.

2. *By process or function*: This type of layout applies to situations where all of the operations or processes of the same type are performed in the same plant area or department, with similar machines or like assembly operations being grouped together as closely as possible, consistent with good material handling and processing practice. All machining is done in one area, all welding in another, all drilling in another, and all painting in a paint shop. This type of layout offers the following advantages:

Better machine utilization, allowing lower capital investment in machines and other production equipment.

It is adapted to a variety of products and to frequent changes in sequence of operation.

It is adapted to intermittent demand (variable production schedules).

It is easier to maintain continuity of production in the event of machine or equipment breakdown, shortages of material, or absent workers.

This type of layout is generally best for job shop or job order manufacturing, or for large-volume production of nonstandard products. It can also be used in high or mass production as a feeder division. The majority of manufacturing in the industrial world today is job order manufacturing and would have a layout by process or function.

3. *Line production or layout by product*: In layout by product or line production the machinery is laid out, machine after machine, regardless of machine type, with all necessary operations in their proper sequence to follow the flow of the product in its progress through the various steps of manufacture, including machining, finishing, assembly, and testing for the product involved. This is the most popular type of layout for mass production in a manufacturing plant. In this type of layout (unlike layout by fixed position) the material moves. This type of layout places one operation immediately adjacent to the next. The advantages offered by this type of plant layout include:

Reduced handling of material
Reduced quantities of parts and materials in process, allowing reduction in production time (time in process) and lower total investment in parts and material
More effective use of labor through greater job specialization, ease of training and use of less skilled labor
Easier control of production, control of workers, and much easier supervision.
Reduced congestion and floor space otherwise allotted to storage and aisle space.

In practice, most plant layouts are a combination of these classical layout types. They are made to utilize the advantages of all three types. In deciding which type of layout to use, the following considerations would apply:

Use layout by fixed position or fixed material location when:

1. Material forming or treating operations require only simple hand tools or simple machines.
2. You are manufacturing only one or a few items of product.
3. The cost of moving the major assembly or piece of material is high.
4. The skill of workmanship lies in the abilities and craftmanship of the workers, or it is desired to fix responsibility for the quality of the product with one worker.

Use layout by process or function when:

1. Machinery is expensive and difficult to move.
2. You are manufacturing a variety of products.
3. There are wide variations in the times required for different operations.
4. There is small or intermittent demand for the product.

Use line production or layout by product when:

1. There is a large quantity of parts or product to be manufactured.
2. The design of the product is frozen or standardized and will not change.
3. The demand for the product is steady and predictable.
4. Balanced operations and continuity of material flow can be maintained without difficulty.

In plant layout, as in any engineering discipline, there are certain guiding principles or fundamentals that must be followed to develop a fully engineered and successful layout. These are as follows:

1. Plan the whole and then the details. Start with the layout of the plant as a whole, or entire entity, and then work toward the details. Determine the general requirements in relation to the volume and to production anticipated. Next develop a generalized overall plant layout. The next step is to proceed with the detailed layout within each area. This involves the actual positioning of the machines, equipment, materials, operators, and supporting activities which constitute the detailed layout.

2. Plan the ideal and from it the practical. The initial layout concept should be a theoretically ideal plan, without regard for existing conditions and without considering cost. Later adjustments will incorporate the practical limitations of existing buildings and other considerations. This approach ensures that the best possible layout will at least be considered.

3. Follow the cycles of layout development and make the phases overlap. There are four cycles of layout development which must follow in sequence. First, you must determine where the layout will be, or where the facilities to be laid out are to be located. Second, an overall layout should be planned for the new production area. Next to follow would be the detailed layout plan, and then the installation cycle. In actuality, the four cycles will overlap.

4. Plan the process and machinery around the material requirements. Material is basic to the layout plan. Product design and the specifications dictate what process must be used. The quantities and rates of production must also be known to determine the processes to be used. The process and machinery must be designed around the material requirements.

5. Plan the layout around the process and the machinery. The space required and the location of the production processes, machinery, tools, and other equipment are the heart of the layout plan.

6. Plan the building around the layout. Where machines, service equipment, and the layout will be more permanent than the building, the building should be set around the most efficient layout. If the layout will be less permanent than the building, follow the same rule, but design a general-purpose building around several feasible layouts that may occupy the building. For an existing building, plan the building modifications around the layout.

7. Plan with the aid of clear visualization. The experienced analyst or manufacturing engineer realizes that aids to clear visualization are critical in his work. They help in gathering facts and in analysis. It is essential when he must discuss his plans with others and with management.

8. Plan with the help of others. Plant layout is a cooperative type of activity. The best layout will not be obtained without the cooperation of all those concerned. Their ideas should be requested and they should be made a part of the project. Shop supervisors

are especially important, because they have the detailed knowledge of the job and are the ones who will make the layout work.

9. Check the layout. When one phase of the layout plan is developed, it should be checked and approved before moving ahead with the next. This will save later headaches and ensure that the plan is on course and soundly designed.

10. Sell the layout plan. Often the most difficult part of layout work is getting others to agree. Every layout is a compromise, and this fact should be recognized at the outset. Time and skill expended in getting others to participate are critical to a truly successful layout plan.

As each plant layout project runs its course from the initially stated objective to the installed physical reality, it goes through the four phases of layout planning. These phases are:

1. *Location*: Here it must be decided where the area to be laid out will be. This is not necessarily a new site problem. More often than not, it is one of determining whether the new layout (or relayout) will be in the same place it is now, in a newly acquired building, or somewhere else.
2. *Overall layout*: This phase establishes the basic flow patterns for the area being laid out. It also indicates size, relationship, and the configuration of each major activity, department or area.
3. *Detailed layout*: This phase is concerned with the preparation of the detailed layout, and includes planning where each piece of machinery and equipment will be placed.
4. *Layout installation*: This phase is concerned with planning the installation and the actual making of the necessary moves.

These phases follow in sequence, and for best results they should overlap. The analytical part of planning the general overall layout begins with the study of the input data:

I. An analysis of the flow of materials:
   A. Operation process chart
   B. Flow process chart and flow diagram
   C. Routing sheets, methods sheets, process sheets, shop travelers
   D. Engineering drawings

II. A relationship analysis of activities and areas:
   A. Proximity analysis (Fig. 5.4) is an analysis of the importance of location of one function versus another. It answers the key question of how important it is for two related functions to be immediately adjacent to one another.
   B. From-to chart (Fig. 5.5) analyzes the number of unit loads moving from one department or machine to all of the other departments or machines.

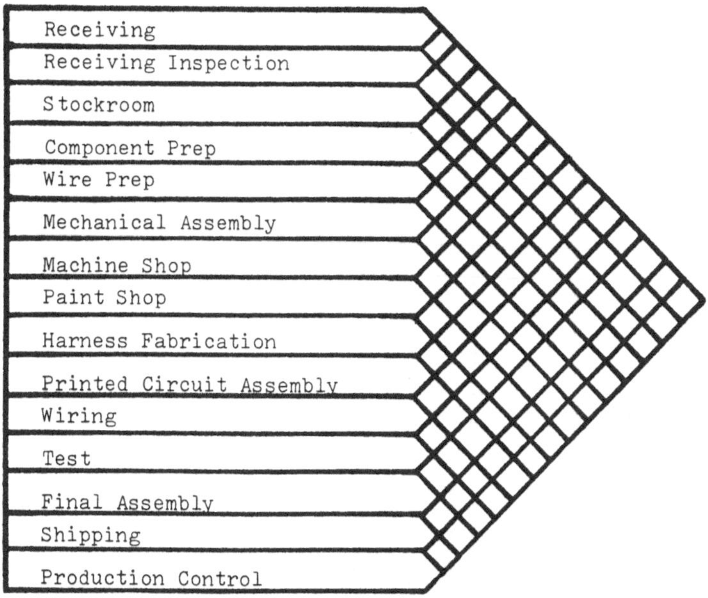

VALUE	CLOSENESS
A	Absolutely Necessary
B	Especially Important
I	Important
O	Ordinary Close
U	Unimportant
X	Undesirable

FIG. 5.4 Proximity analysis.

UNIT LOADS/DAY

FROM \ TO	Press	Weld	Machine	Assembly	Paint	Stockroom	REMARKS
Press		1	9				Varies with Rate
Weld					1		
Machine		3		15			Two Shifts
Assembly					3		
Paint				7	9		
Stockroom	12	5	1				Raw Material

FIG. 5.5 From-to chart.

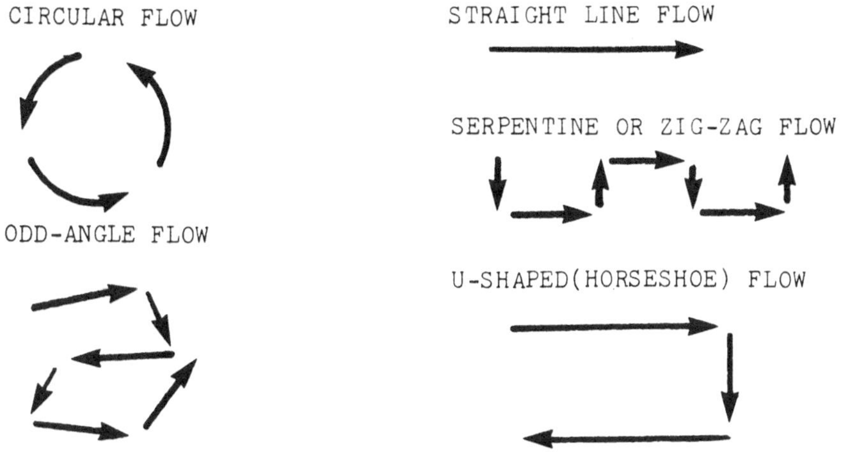

FIG. 5.6 Conceptual flow patterns.

C. Conceptual flow patterns determine the kind of flow the plant layout will provide (Fig. 5.6). These basic flow patterns are:
  1. Straight-line flow
  2. Serpentine or zigzag flow
  3. U-shaped (horseshoe) flow
  4. Circular flow
  5. Odd-angle flow

III. Space requirements are determined next. There are four methods for determining the size of a given area in the plant layout:

A. Calculation. This is the most accurate method for determining space requirements. The space required for each and every machine is tabulated; areas for workers are added; and then the floor space needed for aisles, material setdown, work in process, and so in determined, which in turn defines the size required for the area or department.

B. Conversion. In this method, the amount of space required is determined for each machine and operator, based on what it is now. This is then adjusted to what it should be, and then converted by some factor or multiplier to determine what will be needed for new requirements.

C. Space standards. This method is more appropriate for offices, where, for example, the standard is determined by management (a manager's office would be 200 ft^2, a director's office 250 ft^2, etc.). Another standard that is frequently used is 50 ft^2 per operator for hand assembly operations.

D. Ratio trend and projection. This method determines future requirements by projecting current requirements against time and future growth in the production operations.

Detailed layout planning comes next and begins with a clear understanding of the plan on the part of the manufacturing engineer. He must visualize how the layout will look and how it is going to work. He must have a clear picture or reproduction of his layout to discuss with others. The others must have something that clearly shows the layout plan being proposed. The three common means of layout visualization are:

1. Drawings and diagrams
2. Templates and layout board
3. Three-dimensional models

Of the three, drawings and diagrams are the most basic. They are readily made, easily altered, and inexpensive.

In actually putting together a reproduction of a detailed layout, the templates are the most valuable. There are many kinds and they can be used in many ways. Templates permit reproduction of as many different layout proposals as desired, simply by rearranging the templates on the master.

To prove and check the layout plan, or to help others visualize what is planned, the three-dimensional model comes into play. For these purposes the model is supreme. Models themselves are not a substitute for layout planning. They glamorize the job, help sell the proposed layout, develop interest, help in training workers, foremen, and staff people, and are a check on the planning of the layout engineer.

The most common pitfall into which the inexperienced engineer falls is jumping into the use of templates and models too early in the project, before he has the necessary facts to evaluate the various layout plans his visualization may create. He should not start right in with sketches and templates, and merely cut and try, until by trial and error he comes up with a layout that looks as if it might work. The stages required to develop a detailed layout with templates and models include the steps below:

1. The engineer works up his proposed arrangement by analyses, diagrams, and templates on master layout board.
2. The engineer sets up scale models on the layout plan, according to the planned arrangement, or places them on master layout board. He checks his plan and explains his ideas. He invites comments from those concerned. Suggested improvements are tried out by moving models or otherwise evaluating the ideas.

3. When the final arrangement is decided upon, the layout board is rearranged to agree, or the models are removed and templates fastened securely to the floor plan in their place, and the board is made ready for photographing or reproduction.

How do we go about evaluating the various layout alternatives? The best layout is a compromise of the various factors, considerations, layout objectives, and types of layout. The most frequently used methods of layout evaluation are the following:

1. List the advantages and disadvantages of each layout being evaluated. This is the easiest way to evaluate alternatives. It is surprising how often such a listing quickly clarifies which alternatives should be selected.
2. Factor analysis. This method selects the factors or considerations on which the decision will be made. Each factor is given a weight value according to its importance (10, 9, 8, etc.). All the alternatives are then rated on one factor at a time. The rating is then converted to a number and multiplied by the weight value. The weighted ratings are totaled for each alternative and a numerical comparison is made. This method of layout evaluation increases the objectiveness of what can be a very subjective decision-making process.
3. Cost comparison. In important projects, cost will nearly always become a basis for selecting the best alternative. This means the cost of everything that goes into layout installation and operation. Costs would include: Investment costs (for initial installation of new facilities, accessory costs, as well as depreciation and obsolescence costs) and operating costs (material, labor and overhead costs).

Layout installation is the final phase in layout planning. Information needed for layout installation includes:

A list of all new machinery and equipment to be installed, or existing equipment to be moved or changed in location
A layout print, drawing, or photograph explaining details of new locations
A schedule of moves
A specification sheet to show how each machine should be disconnected, moved, and hooked up.

In any relayout problem, care must be taken to ensure that all of the moves are planned so that the movers do not try to move something into space already occupied. This problem, together with that of keeping up production during the change, is frequently a major obstacle limiting the design of the layout itself.

The decision as to when to install the new layout is a difficult one at best. Although the time to install the layout is important, it is generally a matter of selecting the least inconvenient time rather than finding a time that is fully acceptable to everyone. Frequently preferred times include the time of annual changes in product design, the time of plant shutdown for vacation, the slack or slow season, and weekends or an extended weekend.

In planning the installation of the new layout, there are seven key stages (plan, provide, prepare, move, install, start-up, and clean-up):

1. *Plan*: Start planning early. Sound planning ahead saves time during the fast action stages. Determine the sequences of moves, considering practical operating problems. Make an inventory of everything to be relocated. Get a disposition for the balance. Schedule the moves in detail. Set up a timetable with specific data and times. Assign a move number to each item. Mark it on the inventory sheet, and check it against the machine tag.
2. *Provide*: Consider the use of outside moving and installation contractors, and get bids from several of them. Have adequate help. Set up key men in each department or area involved. Obtain ample moving equipment. Consider renting equipment to assist in moving or keep the operations running during installation. Provide good communications. Have telephones and lead-men at both ends of the move.
3. *Prepare*: Prepare new location. Foundations, partitions, cleaning, painting, and auxilliary service lines should be ready. Broadcast the plans. Let everyone know what is going on, and take advantage of new ideas or suggestions. Tag every item to be moved. Use color coding, and mark identification, moving date, and destination. Notify employees what to do and when, where, and how to do it. Get the equipment ready to move, and check it out before releasing it to movers.
4. *Move*: Keep the move on schedule. Post accomplishments each day. Move equipment intact. Try to keep equipment together to reduce the reassembling time before it can operate. Move as close to the installation point as practical to reduce handling time by skilled installation crews. Keep movers coordinated by frequent briefing and notification.
5. *Install*: Expect to have last-minute changes. Do not be upset if the plan does not work perfectly. Make use of temporary hookups, with permanent service connections later. Flag equipment ready for installation inspection, and have installation crews post their accomplishments daily.

6. *Start-up*: Check the installation. Be sure the placement and hookup are correct. Release equipment for tryout and acceptance.
7. *Cleanup*: Inspect the installation and note any inconsistencies. Set a deadline for cleanup; otherwise the installation remains a temporary one and production output reflects the same attitude.

The manufacturing engineer should be on hand, ready to make any changes in the layout plan deemed necessary during the installation phase. No matter how good the layout engineering has been, some adjustments will be needed at installation time.

5.4.2 Material Handling

There are several basic operations involved in any manufacturing or production activity, including processing, assembling, inspecting, and material handling. What is not generally recognized is that the last of these, material handling, more often than not accounts for the largest portion of the total production cost. Material-handling activities include:

1. Transportation from vendors
2. Unloading
3. Receiving
4. Stores
5. Issue to production
6. In-process handling
7. In-process storage
8. Workplace handling
9. Packaging
10. Finished goods warehousing
11. Loading
12. Shipping
13. Transportation

Inefficient material handing can be the cause of many problems and losses in profit. The first step toward overcoming problems involving material handling is properly identifying them as such, then implementing the necessary solutions. The following material-handling checklist is useful in identifying problems in material handling:

Is the material-handling equipment more than 10 years old?

Do you use a wide variety of makes and models of material-handling equipment, which requires a high spare parts inventory?

Are equipment breakdowns the result of poor preventive maintenance?

Do the lift trucks go too far for servicing?
Are there excessive employee accidents due to manual handling of materials?
Are materials weighing more than 50 lb handled manually?
Are there many handling tasks that require two or more employees?
Are skilled employees wasting time handling materials?
Does material become congested at any point?
Is production work delayed due to poorly scheduled delivery and removal of materials?
Is high storage space being wasted?
Is material being damaged during handling?
Do shop trucks operate empty more than 20% of the time?
Does the plant have an excessive number of rehandling points?
Is powered equipment used on jobs that could be handled by gravity?
Are many handling operations unnecessary?
Are single pieces being handled where unit loads could be used?
Are floors and ramps dirty and in need of repair?
Is handling equipment being overloaded?
Is there unnecessary transfer of material from one container to another?
Are inadequate storage areas hampering efficient scheduling of movement?*

There are some 20 basic principles of material handling which generally apply to any material-handling problem. These include the following:

1. Planning principle: Plan all material-handling and storage activities to obtain maximum overall operating efficiency.
2. Systems principle: Integrate as many handling activities as is practical into a coordinated system of operations, covering vendor, receiving, storage, production, inspection, packaging, warehousing, shipping, transportation, and customer.
3. Material flow principle: Provide an operation sequence and an equipment layout optimizing material flow.
4. Simplification principle: Simplify handling by reducing, eliminating, or combining unnecessary movements and/or equipment.
5. Gravity principle: Utilize gravity to move material wherever it is practical.
6. Space utilization principle: Make optimum utilization of building cube.

---

*Checklist courtesy of the Material Handling Institute (from copyrighted materials).

7. Unit size principle: Increase the quantity, size, or weight of unit loads or flow rate.
8. Mechanization principle: Mechanize handling operations.
9. Automation principle: Provide automation to include production, handling, and storage functions.
10. Equipment selection principle: In selecting handling equipment consider all aspects of the material handled—the movement and the method to be used.
11. Standardization principle: Standardize handling methods as well as types and sizes of handling equipment.
12. Adaptability principle: Use methods and equipment that can best perform a variety of tasks and applications where special-purpose equipment is not justified.
13. Dead weight principle: Reduce ratio of dead weight of mobile handling equipment to load carried.
14. Utilization principle: Plan for optimum utilization of handling equipment and manpower.
15. Maintenance principle: Plan for preventive maintenance and for scheduled repairs of all handling equipment.
16. Obsolescence principle: Replace obsolete handling methods and equipment when more efficient methods or equipment will improve operations.
17. Control principle: Use material-handling equipment and activities to improve control of production, inventory, and order handling.
18. Capacity principle: Use handling equipment to help achieve desired production capacity.
19. Performance principle: Determine effectiveness of handling and performance in terms of cost per unit handled.
20. Safety principle: Provide suitable methods and equipment for safe handling.*

Often a material-handling problem cannot be fully defined until all of the relevant data are collected and analyzed. This information-gathering effort is critical and should be done by trained engineers. Often many of the graphical techniques used in plant layout and in preproduction planning are used to generate this information. For instance, in a real manufacturing situation, a preliminary idea of material flow can be obtained by constructing an operation process chart. This graphical tool shows the component makeup of the product, the relationship between the various parts, and the sequence in which components are assembled. It defines material flow patterns,

*List courtesy of the Material Handling Institute (from copyrighted material).

including the sequences of production and assembly operations. It also gives an indication of the relative space requirements.

From-and-to charts, as well as activity relationship diagrams, provide visibility of material movement between related activities, and the flows between departments. These data, plus the information in the operation process chart, provide the information needed by the manufacturing engineer to form a concept of an idealized material-handling system. Once the problem has been defined and quantified, the solution is usually obvious.

It is impossible to process materials without handling operations. However, surveys reveal that a large portion of handling is unnecessary. The number of pickups, setdowns, and transports which many materials experience is far out of proportion to the actual processing time. These are the movements that can be eliminated, simplified, mechanized, or automated. In making equipment layouts, and in designing the material-handling system, the manufacturing engineer should consider the following material-handling axioms:

1. Provide aisles wide enough to accommodate safely the latest types of mobile material-handling equipment. This should take into consideration both the largest loads anticipated and the pedestrian traffic.
2. Provide space to set down unit loads of work in process in such a prepositioned manner as to eliminate the need to rehandle them.
3. Keep work at a convenient working level.
4. Never set materials directly on the floor unless absolutely necessary. It requires manual labor (as a rule) to unload and to reload each time.
5. Eliminate isolated, fenced-in storerooms whenever possible. Storerooms of this kind require additional handling to receive and dispense materials, and often additional paperwork.
6. Plan first operations as close to the point of receiving as possible. If possible, move material directly to the first point of receiving inspection.
7. Determine whether the material can be received in containers, or in a manner which does not require its removal or rehandling between receipt and first use.
8. In processing materials with high scrap ratios, determine whether the largest amount of scrap can be removed early in the processing cycle to eliminate much of the handling of the material that will eventually become scrap.
9. Provide an adequate scrap removal system. Waste materials must be removed constantly and without delay or production

cannot be carried on.

10. Whenever possible, keep materials flowing from one workstation to the next without intermediate setdown. When this can be done, mechanized handling is warranted.
11. To conserve floor space, utilize overhead means of conveying or storage.
12. Handle materials in bulk or unit loads when they must be intermittently handled and stored.
13. Plan inspection stations in the flow of work (if possible) to avoid lateral movements of materials.
14. Plan packaging operations as an integral part of the process. Avoid repacking and rehandling at another location.
15. Ship direct from the packing floor if loads are repetitive to one destination and quantities warrant. It is less expensive to handle paper than materials.
16. In planning receiving and shipping areas, provide for recessed docks to permit entry into trucks and railroad cars with material-handling equipment. Receive and ship in unit loads wherever possible.
17. Use yard storage when materials do not require protection from the weather.
18. Materials which require protection, and which must be stored for long periods of time, can often be stored outdoors with temporary protection.
19. Consider mechanical aids when workmen must lift more than 75 lb, must handle the same kinds of material more than 30 min each day, must move materials more than 50 ft, or are exposed to unusual safety hazards.
20. Consider using the following types of mobile handling equipment where skids or pallets can be utilized:

    Movement within a group (hand-powered lift jack).
    Movement within a section or department, not over 150 ft (hand steer lift truck).
    Movement and stacking within a section or department, not over 150 ft (motorized, hand steer, elevating fork truck).
    Movement between departments and buildings, about 300 ft (power-operated, rider-type fork truck or elevating platform truck).
    Movement between departments and buildings of large loads for distances over 300 ft (tractor-trailer trains).

    An operator with a hand truck costs considerably less than an operator with a fork truck, but performs about 10% as much work.
21. Wherever possible, use power-operated fork trucks to unload

trailer trains and move and stack loads at workstations.
22. Where materials follow a fixed route repetitively, even for appreciable distances (up to 1000 ft), consider the use of conveyors. In some instances, longer distances can be effectively provided with conveyors.
23. Where large quantities of material are used constantly but are purchased, handled, and stored in small containers, study the possibility of mass purchase and storage.

After the plan for solving the handling problem has been developed, the final step is to make the solution happen. The approaches for doing this have been covered earlier in this chapter. Material handling is as much a part of preproduction planning as are plant layout and process planning. It must be treated accordingly.

QUESTIONS

1. What are some of the basic requirements in preparing a meaningful long-range facilities plan?
2. When a new item of production equipment has been identified as needed for new production requirements, what is the first thing the manufacturing engineer should do in order to satisfy the requirement?
3. When is it more economical and prudent to refurbish a major item of production equipment rather than replace it with a new machine?
4. What are the duties and responsibilities of the manufacturing engineer in the acquisition of new production equipment and facilities?
5. What are the three classical types of plant layout and some of the advantages of each?
6. In plant layout, what does it mean to plan the ideal and from it the practical?
7. What function does the priximity analysis serve in plant layout work?
8. What are the five basic flow patterns of work through a plant or facility used in plant layout?

9. What are the three most frequently used methods for evaluating plant layout alternatives and what are the advantages of each?
10. What is the first step in overcoming problems in material handling and movement?
11. How can the movement of material and work in process, as well as material staging and storage be improved without expanding an existing building or facility?

6

Special Processes

6.1 INTRODUCTION

In modern industry there are almost unlimited numbers of special or unusual processes that require specialized process controls, equipment that is unusual or specialized, and, in many cases, special environmental or pollution controls. Production operators involved in these processes require specialized training, and often the entire process is controlled and monitored by computer. Equipment maintenance is extremely critical, and maintenance routines and schedules are very carefully planned and strictly adhered to. Often a process that is special in one plant is routine in another. Often, because of the large investment in special-process facilities and capital equipment, special processes are run 24 hr per day, 7 days per week.

6.1.1 Definition

A special process can be defined as any process or operation that is based on a highly technical or specialized technology, requiring that operators have special training and that equipment and facilities be either specially designed or specially ordered from specialized manufacturers or suppliers. These processes require special documentation and work instructions, careful monitoring of process parameters to ensure process control, and may require special instrumentation.

6.1.2 Examples

Some examples of special processes are given in the following list, which is by no means total or comprehensive:

Electroplating
Electrostatic painting
Heat treating
Dip brazing
Electron beam welding
Conformal coating

Laser cutting and welding	Chemical milling or etching
Wave soldering	Fabrication of composite structures

Many of the examples of special processes listed above may be further subdivided into one or more sub-tier special processes. An example of this subdivision would be in electroplating, where we would have:

Chromate conversion coating of aluminum
Passivation of stainless steel
Electroless nickel plating

6.2 SPECIAL-PROCESS PLANNING

The manufacturing engineer should be fully knowledgeable about any special processes in his plant. The planning and analysis that precede the installation of any new special process are highly technical in nature and require the professional engineering knowledge and experience that only the manufacturing engineer has. He must plan the process, the facilities, and the equipment required, write the specifications, order the tooling, make certain that all safety and environmental requirements are met, prepare the necessary work instructions and process control documentation, assist in the training of operators, direct the start-up, shakedown and validation of the new process, and monitor the day-to-day operation once the special process is fully on-line and fully operational.

6.2.1 Facilities and Equipment

The basics of planning facility and equipment requirements were covered in Chapter 5. However, when special processes are involved there are additional planning steps that must be taken. For example, if a new metal finishing line is being planned, the first step in planning the new line would be to determine the total metal surface area to be treated by each of the sub-tier finishing processes to find the total surface area to be treated in square feet per hour. Figure 6.1 shows how this can be determined and laid out in tabular form. For each finishing process, the surface area is calculated for each metal part to be finished, and this is multiplied by the number of parts per hour. This is the basic information needed for determining the size of the plating and rinsing tanks required, as well as for the analysis of the vapor and effluent discharges shown in Figs. 6.2 through 6.6.

The size of process tanks also depends on the size of the parts to be processed through the metal finishing line. Parts must be totally immersed in the tanks. The smallest tank must be big enough to accommodate the largest part to be processed. Adequate rinse tanks must

Assembly Part No.	Vapor degreasing	Alkaline cleaning	Deoxidizing	Anodize	Alodine	Passivate
	Surface area, ft² per assembly					
10698027	110	55	70	20	50	—
10697901	15	20	15	15	5	—
10697902	150	150	55	55	55	—
10758000	40	60	40	40	20	—
10758003	165	40	40	40	—	—
10758005	290	225	100	100	40	Minor
10758007	15	15	15	15	3	2
10117775	10	4	4	2	2	—
10117776	5	—	—	—	—	5
10117777	5	5	5	2	4	—
Total	805	574	344	298	179	7
	Surface area, ft²/hr					
15 sets of parts/4 weeks (15 days/week, 8 hr/day)	75	54	32	27	17	0.7

FIG. 6.1 Metal surface area to be treated.

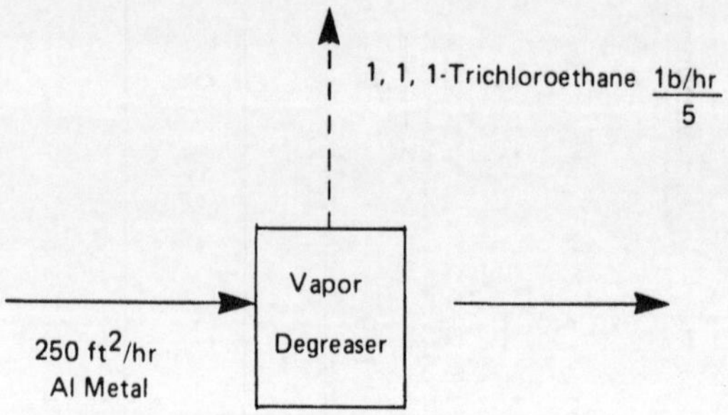

Notes:

1. Per EPA Report, EPA 450/2-77-022 the average loss from an open-top vapor degreaser is 0.5 lb/hr-ft^2 and most of this is lost by diffusion out of the top. Also in this report tests on degreasers with cover, correct freeboard ratios, and operation, etc., on 1,1,1-tri have emissions of 0.1 lb/hr-ft^2. Using this "controlled" emission factor for a 40" W by 15' L degreaser is 5 lb/hr. For an 8 hr/day, 5 days/week operation, this is 40 lb/day, 5.2 tons/yr.
2. 1,1,1-trichloroethane is an exempt solvent.
3. Sludge cleanout and solvent recovery should be infrequent on an as-needed basis.
4. Treatment rate is the estimated capacity for the anodizing step.

FIG. 6.2 Vapor degreasing. (Courtesy of IT Corporation, Knoxville, Tenn.)

be provided for each finishing process in the line. Typically, rinse tanks are constructed so that fresh water enters from the bottom and leaves from the top. When two or more rinse tanks are cascaded (in series), the work should flow counter to the flow of water, with baffles placed to provide agitation of the water and flow from bottom to top. Time of contact between the work and the rinse must be adequate to ensure a good mix (some form of agitation must be provided so that the adhering film of drag-out mixes well with the rinse water). Some form of mechanical agitation, either by propeller, pump, ultrasonics, or air, is an absolute necessity. The importance of agitation in metal finishing increases as multiple rinse tanks are employed, since the flow of water is less and less effective as water usage is reduced.

Unless the work is to proceed to further processing solutions, the final step in metal finishing is drying. If the last rinse is a hot-water immersion, evaporation of the water is easier and quicker, and if the work is very large, its retained heat may evaporate the water film without further effort. Otherwise, the work should be dried by centrifuging or by use of ovens.

In metal finishing, as in many special processes, consideration must be given to the amount of chemical waste that must be treated or otherwise disposed of. Table 6.1 summarizes wastewater discharges associated with the metal finishing line discussed above and described in Figs. 6.1 through 6.6. In planning the facilities for this line, provision must be made for treating the wastewater in the quantities indicated. Local and state regulations now normally require a permit to handle and dispose of these kinds of chemical wastes. This would especially include rinse water discharges, which in prior years could be disposed of in the municipal sewer system or by dumping into ponds for evaporation. The amounts of the contaminants are measured or calculated in parts per million (PPM), and in the example cited would include chromium and copper.

6.2.2 Tooling

Like other tooling, tooling for special processing must be carefully thought out during the preproduction planning phase. Tooling that is special or peculiar to a product or part should be considered in light of length of production run, the potential of the product or part for further engineering or design changes which would affect the tooling, and the potential for additional or repeat business. The cost of special-purpose tooling adds to the cost of the product significantly. For example, special tooling that costs $1000.00 over a production run of 1000 units adds $1.00 to the cost of each unit.

Special processes can be very expensive operations which, even though necessary, turn out to be significant cost drivers. Tooling can be a major part of this cost unless action is taken during the preproduction planning phase to hold tooling costs down. One way that this can be done is through the use of standard tooling to the maximum extent possible. Standard tooling can be used over and over again on a variety of products and parts, and has an unlimited life that is not controlled by product configuration or length of the production run.

In the case of a metal finishing or processing line, tooling includes the racks and fixtures used to hold or retain the parts during the finishing cycle. If the parts to be treated or finished are very small, there is a good possibility that they could be barrel plated (placed in a revolving barrel or basket that holds the parts during the entire processing cycle), thus eliminating the need for any other fixtures or racks. Larger parts would be hung on standard plating racks or fixtures; they would be moved by hand from tank to tank through the processing cycle, or the racks would be hung on a conveyor over the finishing line and moved through the process cycle automatically, with the conveyor programmed to dip the parts for the required period of time in each tank as it moves through the process line.

If the special process was wave soldering of printed circuit card assemblies, standard wave solder fixtures which can be adjusted for the

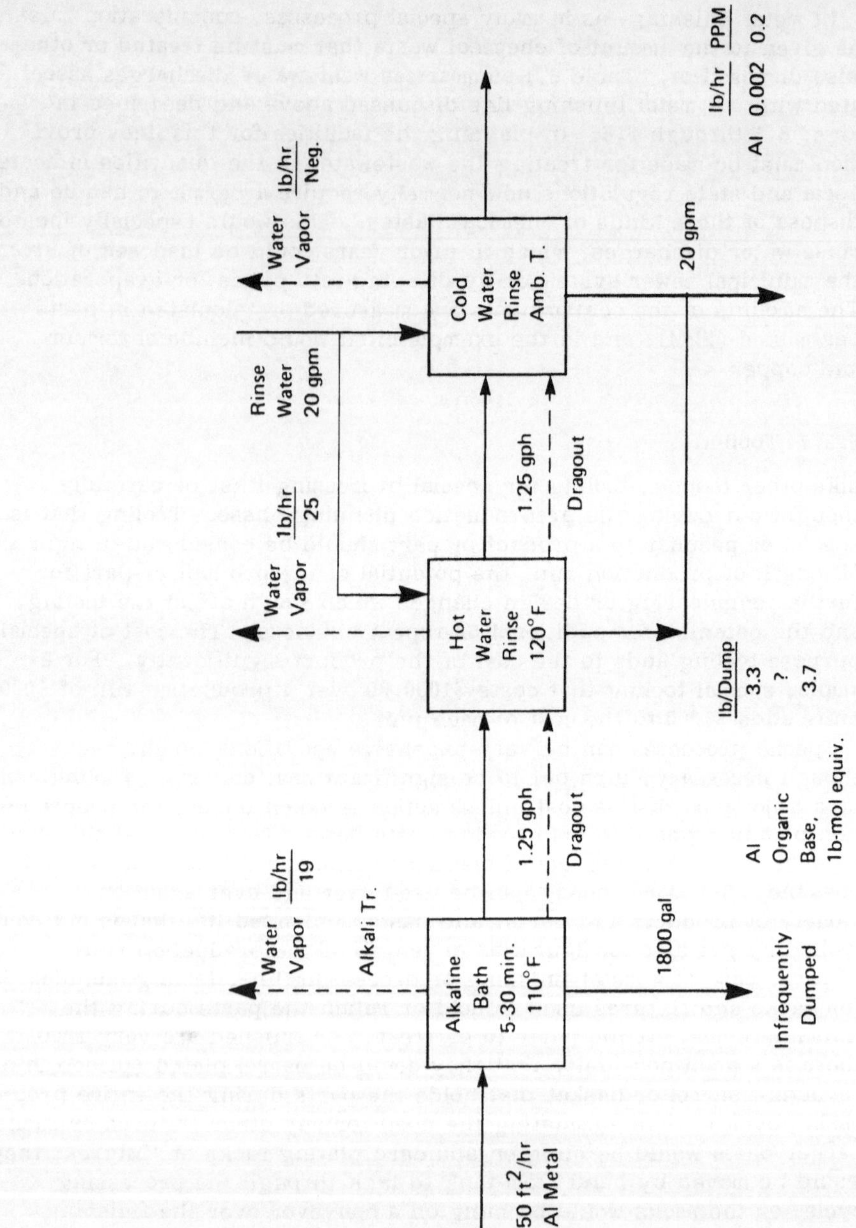
250 ft²/hr
Al Metal
Alkaline
Bath
5-30 min.
110°F
Water
Vapor
Alkali Tr.
lb/hr
19
1800 gal
Infrequently
Dumped
1.25 gph
Dragout
Hot
Water
Rinse
120°F
Water
Vapor
lb/hr
25
lb/Dump
Al 3.3
Organic ?
Base, lb-mol equiv. 3.7
Rinse
Water
20 gpm
1.25 gph
Dragout
Cold
Water
Rinse
Amb. T.
Water
Vapor
lb/hr
Neg.
20 gpm
lb/hr PPM
Al 0.002 0.2

Notes:

1. Bath additive is Turco 4215LT-NC, which is "nonchromate." It does not contain caustic, therefore, it will not etch aluminum.
2. Dragout is based on an estimated rate of 0.005 gal/ft^2.
3. Rinse water will be slightly alkaline since dragout is estimated to be ~0.25 N.
4. Alkaline bath is discarded when parts cannot be cleaned satisfactorily or sediment is excessive. Metals content estimated from above dragout. Base estimated from ~6 oz/gal Turco 4215LT-NC additive being ~0.25 N. The organic is glycol ethers and the like that is part of the additive.
5. Evaporation is based on a rate of 0.045 gal/hr-ft^2 at 110°F and 0.06 gal/hr-ft^2 at 120°F. (These rates are for ambient conditions of 75°F, 75% relative humidity, and 95% mole fraction water.) If the alkaline bath is air agitated, the evaporation rate will be higher.
6. Treatment rate is the estimated capacity for the anodizing step.

FIG. 6.3 Alkaline cleaning, type I. (Courtesy of IT Corporation, Knoxville, Tenn.)

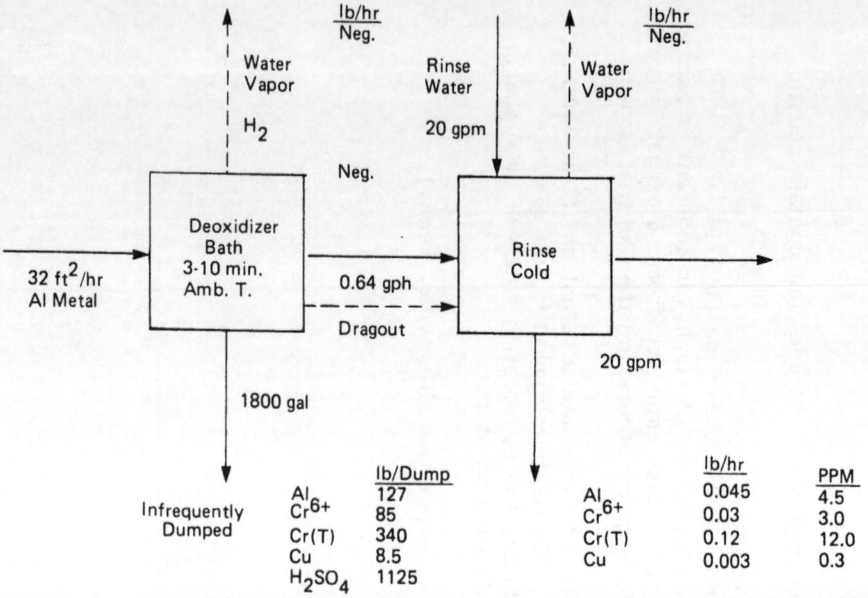

Notes:

1. Deoxidizer bath has an original concentration of 2.2 oz/gal of potassium dichromate (3 oz/gal Amchem 7) and 5% H_2SO_4 by volume. Active chromic acid content is kept at approximately this level by deoxidizer addition based on titration. When the spent chromate to active ratio is ~4.7 to 1, it is recommended the bath be discarded. Sulfuric acid content is maintained at 8—12 oz/gal.
2. Deoxidizer treatment is designed to remove ~0.0001 in. of metal. With a 32 ft^2/hr rate and a 93.4% Al, 6.3% Cu, 0.3% Mn alloy, this is 0.045 lb/hr Al and 0.003 lb/hr Cu.
3. Dragout is based on an estimated rate of 0.02 gal/ft^2.
4. Chromic acid consumption is estimated to be equivalent to 2 lb Cr/l lb Al removed.
5. With the above dragout rate and the Al and Cr reaction quantities assumed, the deoxidizer bath theoretically would not reach a condition where it would need to be discarded. The given quantities in the rinse water and the deoxidizer bath are based on the system at equilibrium. It is estimated that it would take >5000 operating hours to reach this condition with an 1800-gal bath size and 32 ft^2/hr of aluminum treated.
6. It has been assumed that if any H_2 is formed by the reaction of H_2SO_4 with Al, it is readily oxidized in the bath.

FIG. 6.4 Deoxidize with sulfuric acid. (Courtesy of IT Corporation, Knoxville, Tenn.)

TABLE 6.1 Wastewater Discharge Summary

	Cleaning	Deoxidize	Anodize	Alodine	Total (gpm)	Total (lb/hr)	Total (ppm)	30-day average Titusville (ppm)	30-day average EPA (ppm)
Option A—All rinse flows at 20 gpm									
Flow, gpm	20	20	40	20	100				
Al, lb/hr	0.002	0.045	0.045			0.092	1.8	1.5	
$Cr6^{+}$, lb/hr		0.03	0.08	0.011		0.121	2.4	0.5	
CrT, lb/hr	0.003	0.12	0.08	0.011		0.214	4.3	1.0	0.80
Cu, lb/hr		0.003				0.003	0.06	0.015	1.09
Option B—Rinse flows reduced[a]									
Flow, gpm	16	12	20	1	49				
Al, lb/hr	0.002	0.045	0.045			0.092	3.7	1.5	
$Cr6^{+}$, lb/hr		0.03	0.08	0.011		0.121	4.9	0.5	
CrT, lb/hr	0.003	0.12	0.08	0.011		0.214	8.8	1.0	0.80
Cu, lb/hr		0.003				0.003	0.1	0.015	1.09

[a]Rinse waters reduced to quantity to give 20 ppm Cr in those steps where Cr is in the treating bath. Other rinses reduced to give same drag-out/rinse ratio as the Cr steps.

Source: IT Corporation, Knoxville, Tenn.

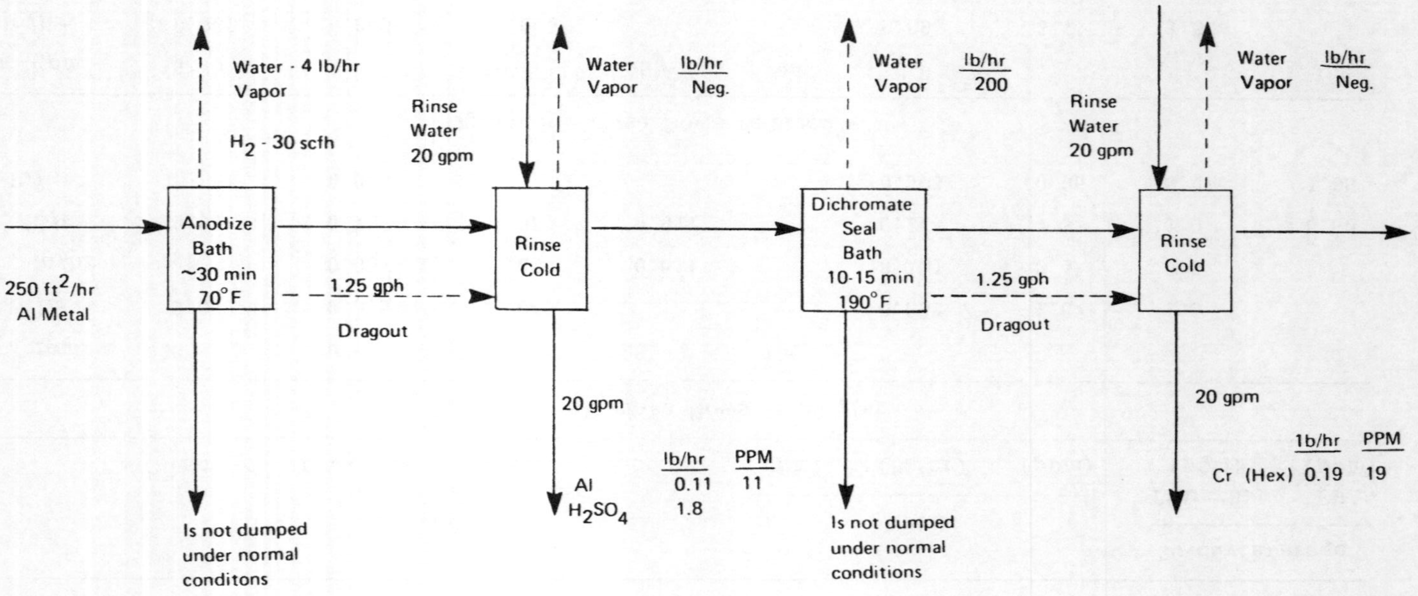
Water - 4 lb/hr
Vapor
H_2 - 30 scfh
Anodize
Bath
~30 min
70°F
250 ft^2/hr
Al Metal
1.25 gph
Dragout
Is not dumped
under normal
conditons
Rinse
Water
20 gpm
Water
Vapor
lb/hr
Neg.
Rinse
Cold
20 gpm
Al
H_2SO_4
lb/hr
0.11
1.8
PPM
11
Water
Vapor
lb/hr
200
Dichromate
Seal
Bath
10-15 min
190°F
1.25 gph
Dragout
Is not dumped
under normal
conditions
Rinse
Water
20 gpm
Water
Vapor
lb/hr
Neg.
Rinse
Cold
20 gpm
Cr (Hex)
1b/hr
0.19
PPM
19

Notes:

1. Anodize bath is maintained at 15–18 wt% H_2SO_4. Maximum alumina concentration allowed is 20 g/liter. It was assumed that at equilibrium with dragout the concentration would only reach 10 g/liter; this is equivalent to 0.00003 in. of metal being lost.
2. H_2 evolution was estimated by assuming that a 0.0004-in. film of Al is reacted with O_2 and that the equivalent H_2 is evolved when O_2 is produced by water electrolysis.
3. Dragout is based on an estimated rate of 0.005 gal/ft^2.
4. Dichromate seal bath concentration is 4–6 wt% $K_2Cr_2O_7$.
5. Evaporation is based on a rate of 0.5 gal/hr-ft^2 at 190°F.
6. Treatment rate is the estimated capacity for the this step.

FIG. 6.5 Anodizing, type II. (Courtesy of IT Corporation, Knoxville, Tenn.)

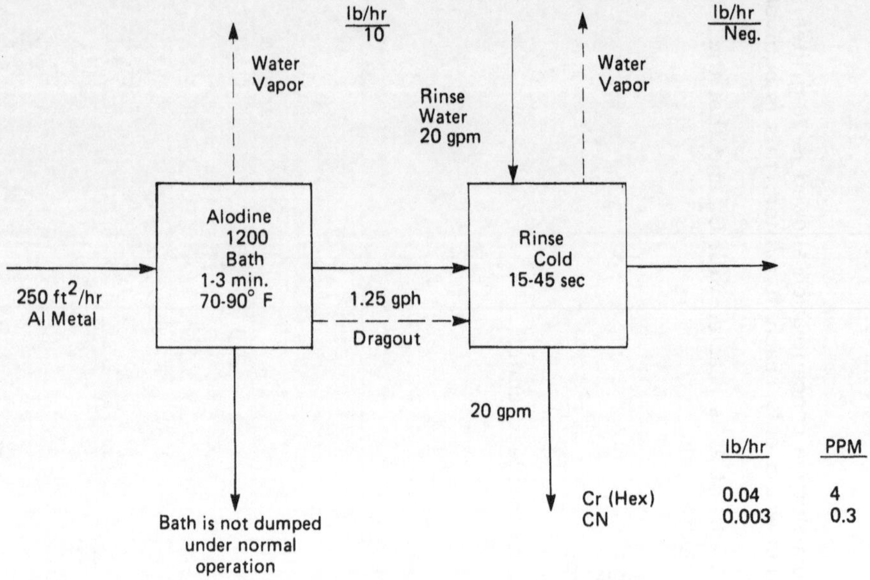

Notes:

1. Alodine 1200 bath is kept at ~1 oz/gal of additive, which is estimated to be ~90% chromic acid.
2. Dragout is based on an estimated rate of 0.005 gal/ft^2.
3. Evaporation is based on a rate of 0.025 gal/hr-ft^2 at 90°F.
4. CN bath concentration has been measured at 260 ppm.
5. Treatment rate is the estimated capacity for the anodizing step.

FIG. 6.6 Alodine coating. (Courtesy of IT Corporation, Knoxville, Tenn.)

various board sizes would be less expensive than special fixtures designed to hold only one board size or configuration. The same would hold true if the special process was conformal coating of the printed circuit card assemblies after wave soldering and cleaning.

Often, in very short or limited production runs, the product or part can be built on "setup." This means without tooling, and in the case of a fabricated part, it would be carefully laid out and cut to size by the sheet metal mechanic or operator. This is the least expensive way since only a few units or pieces are being fabricated and no tooling can be justified. This is seldom the case with special processes. If we look at painting as an example, considerable labor is usually involved in the masking of parts to be painted and the demasking after the parts are painted. In order to hold the cost of masking and demasking to a minimum, standard tooling is required in the form of plugs to place in holes and precut plastic tubing to place over appendages that would otherwise require hand cutting and fitting of masking tape and subsequent hand removal after painting and curing. Some special processes such as painting and heat treating would require only the simplest, most inexpensive standard tooling, such as paint dollies and wire hooks to hold the parts during painting, or a standard rack or shelf to hold the parts in the heat-treat oven.

In special processes such as the encapsulation and molding of ribbon wire cables, expensive, specially designed tooling in the form of cable molds is an absolute requirement regardless of the number of units to be built or the probability of repeat business. Figure 6.7 shows such a molded cable for a 556-ft television antenna. Even in this instance significant labor savings on each cable produced can be realized by use of support tooling in the form of standard resin mixers to eliminate hand weighing and mixing of molding compound, and standard high-pressure injection guns to replace manual hand-crank injection guns. The savings will be realized regardless of the special mold cost, or the number of cables to be fabricated, or the number of different cable configurations.

In planning for special processes, tooling and how it is used can make a significant impact on the efficiency of production and the cost of the product.

6.2.3 Process Procedures and Methods

Process control is contingent upon process control procedures and methods. The manufacturing engineer must clearly define the process and all its steps. This then becomes the basic documentation that defines the process and how it is to be done. From these basic data process routings and work instructions are prepared. Many times we find these procedures prepared by engineering rather than manufacturing engineering. The end result is the same: process definition and process requirements. As a minimum, process procedures should contain the following information:

1. Title—Describing briefly what process procedure is being explained
2. Scope—Describing exactly what is covered by the process procedure
3. Equipment and tools—A comprehensive listing of equipment and tools needed to perform the process procedure
4. Material—A listing of the process materials needed, both by drawing part number and commercial trade name
5. Procedure—The step-by-step description of how the materials are prepared, mixed, weighed, applied, cured, or otherwise processed
6. Safety provisions—Cautionary information covering safe handling, processing, and toxicity

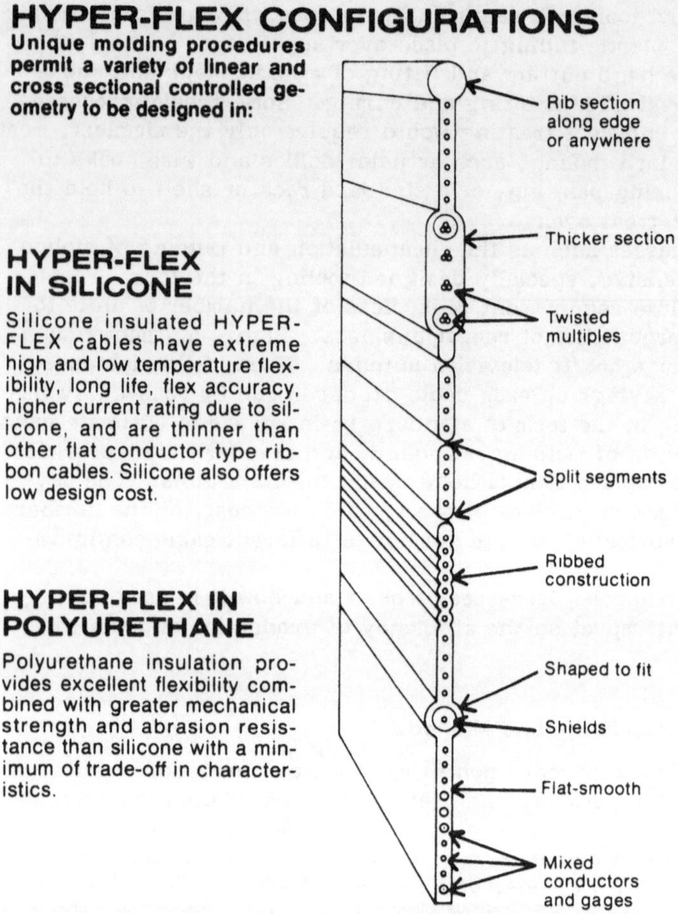

FIG. 6.7 Antenna cable cross-section configuration.

7. Quality assurance provisions—Acceptance criteria for the processed material

Figure 6.8 is a process procedure for the preparation of RTV silicone 655 used in the potting and molding of electrical connectors. In many instances illustrations or diagrams should be used in the process procedure to help clarify instructions or steps in the procedure. It is also absolutely essential that manufacturing engineering maintain control of all process documentation through revision letter control and control of document issue and distribution. Obsolete documentation should be immediately removed from the production floor when superseded.

MANUFACTURING PROCESS PROCEDURE	APPROVED:	PAGE 1 OF 4
	CONCUR:	REVISION: ORIGINAL
	PREPARED BY: J. P. TANNER	DATE RELEASED: 12-15-81
		MPP NO. 801

TITLE - PREPARATION OF RTV SILICONE 655

SCOPE:

This procedure details the approved method for preparing and applying RTV Silicone 655.

EQUIPMENT AND TOOLS:

- Clean container which is approximately four times the volume of the base material being weighed
- Power mixer or clean spatula
- Clean scale and weights
- Tongue depressors and acid brushes
- Bell jar and vacuum pump (evacuator)
- Curing oven

MATERIAL:

- Silicone RTV - 655A compound *
- RTV - 655B curing agent *

* **NOTE:** Limited shelf life material. Do not use after expiration date.

- Primer SS-4120

FIG. 6.8 Manufacturing process procedure.

PREPARATION OF RTV SILICONE 655	Page 2 of 4
	Revision: original
	MPP No. 801

PROCEDURE:

1. **MIXING** - Silicone 655 has a mixing ratio of 10 to 1. It requires (10) parts of Silicone 655A and (1) part of Catalyst 655B.

- Place empty container on left scale platform and adjust balance weight until the indicator reads exactly zero.
- Take a weight corresponding to the required quantity and place on right scale platform. For example, a requirement for five ounces requires a weight of five ounces on the right scale platform.
- Pour Silicone 655A into container until indicator reads exactly zero for the selected weight.
- See table below for required weight of catalyst for selected weight of Silicone 655A:

SILICONE 655A	CATALYST 655B
10 oz.	1 oz.
5 oz.	1/2 oz.
3 3/4 oz.	3/8 oz.
2 1/2 oz.	1/4 oz.
1 1/4 oz.	1/8 oz.
5/8 oz.	1/16 oz.

- A weight corresponding to the weight of the catalyst to be added should be placed on the right scale platform with the weight for the Silicone 655A before adding Catalyst 655B.
- Pour Catalyst 655B into the container until scale indicator reads exactly zero.
- Take tongue depressor and stir mixture vigorously for 3 or 4 minutes. Mixture will become white and foamy. Scrape bottom, sides and corners of container repeatedly making certain you get a thorough mix.

FIG. 6.8 (Continued)

PREPARATION OF RTV SILICONE 655	Page 3 of 4
	Revision: Original
	MPP No. 801

1. **MIXING** (Continued) - When using a power mixer, avoid high mixer speeds for prolonged periods, which could entrap large amounts of air or cause heating of the material resulting in reduced pot life.

2. **EVACUATION** - Place mixture in evacuator (bell jar and vacuum pump) for 8 to 10 minutes until all bubbles have vanished. At the start the mixture will bubble up or froth to approximately four times its original volume, crest, and recede approximately to its original level.

3. **APPLICATION** - After mixing and evacuation, the RTV Silicone 655 may be applied by pouring, brushing or dipping.

   - Pouring should be done carefully to avoid air entrapment.
   - For complex assemblies, pouring under vacuum or evacuation after pouring may be necessary to prevent voids in the cured material.
   - Pot or mix life of catalyzed RTV-655 is four (4) hours at 77°F. Pot life may be extended by refrigeration. (Pot life is defined as the length of time between catalyst addition and cessation of easy pourability).

4. **BONDING** - With the use of Primer SS-4120, RTV-655 can be bonded to metals, most plastics and glass.

   - The surface to be bonded must be clean and dry.
   - Surface roughening will improve the bond to metals.
   - Apply primer by brushing, spraying or dipping. Film thickness should be more than 0.1 mil but less than 0.5 mils.
   - The primer should be allowed to air-dry for one hour before RTV-655 is applied.

FIG. 6.8 (Continued)

PREPARATION OF RTV SILICONE 655	Page 4 of 4
	Revision: Original
	MPP No. 801

5. **CURING** - RTV-655 should be oven-cured at:

150°F. for four (4) hours

212°F. for one (1) hour

- Consideration should be given to time to bring large or heavy assemblies up to cure temperature.

SAFETY PROVISIONS:

- RTV-655B Catalyst can generate flammable gas on contact with acidic, basic or oxidizing materials and such contact should be avoided.
- Primer SS-4120 contains methanol and should be used in a well ventilated area. Care should be exercised to prevent contact with the skin or eyes. Avoid prolonged or repeated breathing of the vapor.

QUALITY ASSURANCE PROVISIONS:

- The RTV-655 should be free of foreign particles.
- Bubbles of maximum diameter of 1/32 inch are acceptable provided there are no bubbles or voids in areas adjacent to wires or connectors.
- Certification samples are not required.

FIG. 6.8 (Continued)

6.2.4 Process Instructions and Controls

If the special process involves a series of separate and distinct processing steps, as in a metal finishing line, there must be process documentation addressing the makeup and control of each tank in the line, as well as the sequence of steps or operations that must follow in the line itself. Tank makeup procedures should contain as a minimum the following information:

1. Cover sheet—Showing issue or revision letter of each page in the procedure
2. Tank usage—Such as hot-water rinse, cold-water rinse, passivation of stainless steel, and so on
3. Tank capacity—Tank size, working depth, capacity in gallons, and other physical features
4. Temperature requirements—Operating temperature; how the tank is heated, and so on
5. Tank construction and ventilation—Whether the tank is plastic-lined steel, plastic, or other material, and what ventilation is needed if any
6. Tank makeup and disposal procedures—The exact quantity of chemicals needed to charge the tank, and how they will be disposed of
7. Initial bath analysis—How the analysis is to be made, what the limits are, and instructions for making adjustments
8. Solution control—What chemicals to use, limits if any, and frequency of monitoring

Figure 6.9 shows such a tank makeup procedure for a tap-water rinse tank used in a stainless steel passivation line. It clearly explains what is required in terms that the operator can understand.

In addition to the tank makeup procedures, there must be a standard manufacturing process dealing with the total process line itself. Figure 6.10 is an example for the special passivation of stainless steel. As a minimum, such a standard procedure should contain the following information:

1. Cover sheet—Showing issue or revision letter of each page in the process
2. Scope—Describing the extent and application of the process
3. Equipment and tools—A listing of tools and equipment that are required to carry out the process
4. Materials—A listing of the process materials and the chemicals required
5. Solution makeup and disposal—Describing how the chemicals and materials must be mixed and disposed of after use

Tank Makeup

Tank No.: 3, 7, 11, & 1D. *Use*: Tap-water rinse.

Tank Size: 10" W × 18" L × 18" D.

Capacity: Full 14 gal. *Working*: Depth, 16". Volume, 12 gal.

Temperature: Water main.

Heaters: None.

Type of Tank: Steel—plastic lined.

Ventilation: None

Disposal Procedure

The platers are authorized and instructed to drain and refill this tank at any sign of contamination. The appearance of color, particles, stray metal, piece parts, etc., are to be considered signs of contamination. This tank need be kept running only when in use.

Makeup Procedure

Add to the tank.

Description	*Quantity*	*Instructions*
Tap water	To overflowing	Constant medium flow when in use

Initial bath analysis: not required.

Solution control: not required.

FIG. 6.9 Tank makeup procedure.

6. Procedure—The step-by-step sequence of operations that are required to execute the process by the plating operator
7. Safety precautions—Measures that must be taken by the plating operator to prevent accident or injury, and what to do if an accident does occur
8. Process cycle—The steps or sequence of finishing operations in block diagram form, giving tank numbers and dwell times in each tank

Such special-process documentation or work instructions provide the minimum essential information needed to ensure process control. The individual part or assembly routing will list the operations needed in the correct sequence for each part or assembly configuration and, using the example in Fig. 6.10, will call out to "Passivate per SMP 1275."

6.2.5 Safety and Environmental Considerations

Special processes are often associated with chemical stresses that affect the human operator. Every manufacturing engineer who is involved in the design and development of special processes should be knowledgeable about the industrial hygiene and safety fundamentals involved in the special process. Chemical stresses are created by liquids, gases, dusts, fumes, mists, and vapors in the form of air contaminants and skin irritants. The following definitions are those of the American Standards Institute:

1. Fume—A substance composed of solid particles formed by condensation from a gaseous state. These particles are microscopically small (odorous gases and vapors are not fumes).
2. Gas—A substance that will diffuse to evenly occupy the space in which it is enclosed. A gas does not appear in the solid state or liquid state at standard temperature and pressure.
3. Vapor—A gaseous form of a substance that is normally a liquid or a solid.
4. Mist—A suspension in air of very small drops, usually formed by mechanical means (atomization) or by condensation from the gaseous state.
5. Dust—A substance consisting of solid particles that have been reduced to a small size by some mechanical process.

Physical stresses on the human operator are created by electromagnetic and ionizing radiation, noise, vibration, and extremes of temperature and pressure.

Exposure is defined as the entering into the body of a health hazard. Such a hazard can affect the nervous system by entering through the eyes, ears, and breathing area of the mouth or nose, or it can be absorbed into the system through the skin on contact and possibly find its way into the digestive tract or other organs of the body. Except for skin diseases, most occupational diseases are acquired by inhalation.

Certain types of processes should arouse suspicion of a health hazard unless it has been specifically established that they are properly controlled. Some of the more common processes of this nature are outlined in the following list (taken from *Occupational Safety and Health Management*, by Thomas J. Anton, McGraw–Hill, New York, 1979, p. 146-147).

SPECIAL PASSIVATION OF STAINLESS STEEL

Scope:

This cycle will cover the following passivation callout per print. Passivate per:

Applicable SMPs
MIL-F-14072
MIL-STD 171

Applicable Documents:

See scope above.

Equipment & Tools

Racks
Hooks

Materials:

As specified per documents called out and the applicable SMPs specified.

Solution Make-up and Disposal:

Per applicable SMPs referenced.

Procedure:

The plater will:

Check the Process Card

Any error found on the Process Card must be called to the attention of the responsible Process Engineer and corrected before proceeding with the lot.

Allow no deviation from the assigned cycle without written instructions from the responsible Industrial Engineer.

The plater will:

Follow the cycle assigned by the Plating Instructor.

Report to the responsible Engineer any unusual reactions on the parts being processed.

Be responsible for his own safety insofar as wearing protective clothing while performing the above operations.

Safety Precautions:

HF Operation: This solution can cause severe burns. Persons working with the HF are cautioned to wear eye protection (goggles), sleevelets, smock, apron, and rubber (Neoprene) gloves.

Magnesium oxide paste is provided near the HF booth. In the event of a burn, wash out immediately with cold tapwater, apply some of the magnesium oxide paste generously to the affected area, and get the victim to the Medical Department. Water only should be used for the eyes—*do not* apply magnesium oxide to the eyes.

Passivation solution: wash the area affected with cold tapwater and take the victim immediately to the Medical Department.

Note that the shower and eyewash are to be used to help bring cold tapwater to the victim.

Personnel are cautioned to note any rash or other allergy symptoms they may get while working with these solutions. These should be brought to the Medical Department's attention and the responsible Industrial Engineer should also be advised.

FIG. 6.10 Standard manufacturing process.

Passivation Cycle:

Step #1
Vapor Degrease: Use hooks, racks, baskets, as convenient. Spray to keep dripping Trichlorethylene.
Time: 30–45 sec

↓

Step #2
Rack parts

↓

Step #3
Immerse in HF tank #1, SMP 1070
$2\frac{1}{2}$ min

↓

Step #4
Rinse in standing tap-water battery jar
Tank #2, SMP 1078
5–10 sec

↓

Step #5
Immerse in running tap water
Tank #3, SMP 1074
5 min

↓

Step #6
Hook parts on other racks to fit subsequent tanks if necessary

→

Step #7
Soak in caustic tank #4, #5, or #6, SMP 1073
30 min

↓

Step #8
Rinse in running tap water
Tank #7, SMP 1074
10 sec

↓

Step #9
Soak in hot tap water
Tank #8, SMP 1077
3 min

↓

Step #10
Passivate
Tank #9 or #10, SMP 1075–1072
1 hr

↓

Step #11
Rinse in running tap water
Tank #11, SMP 1074
10 sec

→

Step #12
Soak in hot tap water
Tank #12, SMP 1077
3 min

↓

Step #13
Soak clean
Caustic #4, #5, or #6, SMP 1073
20 min

↓

Step #14
Rinse in running tap water
Tank #7, SMP 1074
10 sec

↓

Step #15
Soak in hot tap water
Tank #8 or #12, SMP 1077
3 min

↓

Step #16
Blow parts only with filtered compressed air

FIG. 6.10 (Continued)

1. Any process involving combustion should be inspected to determine what by-products of the combustion may be released to the outside environment and, possibly, how high the burner noise level is.
2. Any process involving high temperature with or without high combustion should be examined to determine whether workers are exposed to excessive heat and noise.
3. Any process involving induction heating, including microwave heating, should be examined with regard to effects of heat on workers, and also the level of heat, if it is suspected to be in a high range.
4. Any process involving the melting of metal should be studied to determine the toxicity of the metal fume and possibly of dust, if any is produced in the process.
5. Any process involving an electrical discharge in the air should be studied to determine whether ozones and oxides of nitrogen are produced.
6. Grinding, crushing, or combining materials involves the hazards of dust.
7. Conveying, sifting, sieving, screening, or bolting of any dry material presents a dust hazard.
8. Mixing of dry materials presents a dust hazard.
9. Mixing of wet materials presents possible hazards from solvent vapors, mists, dust, and noise.
10. Dry grinding operations, including milling and sandblasting should be examined for both dust and noise hazards.
11. Cold blending, forming, and cutting of metals or nonmetals should be examined for hazards arising from contact with a lubricant, inhalation of lubricant mist, and excessive noise.
12. Handling of small parts can present hazards from repeated motion and mechanical shock.
13. Coating operations, which generally involve solvent degreasing, can be hazardous because the solvents can act as irritants if they are ingested or come into contact with the skin.
14. Electroplating (metal finishing) should be examined to determine the toxicity of various metallic salts, acids, and alkalis and the hazards they represent when skin contact or inhalation occurs.
15. Hot bending, forming, or cutting of metals or nonmetals may create hazards from lubricant mist, decomposition products of the lubricant, contact with the lubricant, heat, noise, and dust.
16. Painting operations should be examined for the possibility of hazards from the inhalation of or contact with toxic and often irritating solvents, and the inhalation of toxic pigments.
17. Ceramic coating presents the same hazards from toxic pigments as painting, plus hazards of heat from the furnaces and from the hot ware.
18. Mechanical coating with metals presents hazards from dusts and fumes of metals and fluxes in addition to heat and radiation.

19. Explosive processing produces gases (from the explosion itself), mainly carbon monoxide and oxides of nitrogen, and dust from the materials being processed.

The most widespread and serious of all hazardous conditions in the plant or at the workplace are the airborne contaminants which exist as gases, dusts, fumes, mists, or vapors. Threshold limit values (TLVs) expressed in parts per million (ppm)—that is, parts of vapor or gas per million parts of air by volume—provide allowable time-weighted concentrations for a 7- or 8-hr workday and a 40-hr work week for up to 350 different toxic substances. The TLV tells how many parts per million of vapor or gas can be present in the air of a room for a normal 8-hr workday without endangering worker health. TLVs should be used as guides in the control of health hazards, not as fine lines between safe and dangerous concentrations.

Many special processes present hazardous waste problems which must be considered by the manufacturing engineer. Often, a new technology will create such a problem. Every new discovery, breakthrough, or innovation creates the potential for new types of pollution, whether through the creation of new forms of waste or through new pathways of release for familiar pollutants.

Iron and copper, for example, have been industrial waste problems since the turn of the century, first as mining and foundry tailings, and more recently as contaminants in the waste streams of high-technology printed circuit manufacturers. Among newly created wastes, the National Institute for Occupational Safety and Health (NIOSH) has listed tens of thousands of new combinants, derivatives, and synthetics about which little is known with regard to exposure effects. Some 2500 of the listed substances have been identified as suspected carcinogens.

When new pollutants or new pathways of release are created, a number of unknowns are presented. How will each new process or waste material react to external factors in the air, ground, and water? What is the potential for damage? With every new special process these questions must be answered. Once the presence of hazardous waste materials is determined and the pathways of release have been identified, control systems must be established. In establishing these control systems, the following questions should be considered:

1. Are employees who handle hazardous material adequately trained?
2. Are all hazardous materials identified and categorized?
3. Have approved procedures and contingencies been established?
4. Are storage inventories, process volumes, and disposal levels under constant monitoring?
5. Do disposal procedures comply with local, state, and federal requirements?
6. Have all potential release routes to air, ground, sewer, and water been identified, and are these monitored?

QUESTIONS

1. What are the responsibilities of the manufacturing engineer when it comes to special processes in his production line or facility?
2. What is the first step in planning for a new metal finishing line?
3. In addition to the basic calculations of metal surface area and number of parts per hour, what other key factor must be considered in sizing plating-and-rinse tanks in a metal finishing line?
4. What is the primary advantage of having a final hot water rinse at the end of the metal finishing process cycle?
5. In a metal finishing line, tooling consists primarily of what kinds or types of tools?
6. In painting of parts and assemlies, what is the most expensive and time consuming part of the operation?
7. What forms of standard tooling are available to eliminate much of the hand cutting and fitting or masking tape for parts and assemblies to be masked and demasked for painting?
8. In a special process such as a metal finishing line which involves a series of sepeaate and distinct processing steps, what form should the process documentation take?
9. What is the difference between a gas and a vapor?
10. What are some of the health hazard considerations to be examined in a process involving the melting of metal?
11. What health hazards are usually present in painting operations, and what can be done to control these hazards?
12. What is a threshold limit value and what does it tell us?

7

Tooling

7.1 INTRODUCTION

Product tooling can range from inexpensive hand tools to highly sophisticated automatic assembly machines, from motion economy devices for simple parts assembly to highly complex milling fixtures for precision machining. Product tooling can be molds for die casting, or dies for extrusions. Tooling requirements for any manufacturing operation depend basically on the process requirements, the sizes and types of products, and the quantity of products manufactured. The idea and principles discussed in this chapter apply to virtually any type of process or product.

When production levels are high over a period of several years, extensive and sophisticated tooling often is easily justified, but when production levels are low, or limited to a single run, the tooling must still perform its intended functions, but it is difficult to economically justify anything but the lowest costs. As mentioned in Chapter 6, a $2000.00 assembly fixture amortized over a production run of 200,000 units adds only 1 cent to the cost of each unit of product, but a simple holding fixture, costing $200.00 and needed for the assembly of a production run of 200 units, adds $1.00 to the cost of each and every unit. The factory costs of low-volume or short-run products therefore are dependent not only on the direct labor and material costs of production, but also to a very large degree on the costs of tooling.

Tooling costs and hence product costs can be kept as low as possible by following the rules outlined below:

1. Utilize standard tools and tooling components whenever possible.
2. When special tooling is required, utilize the design concepts of previously well-engineered tools and fixtures.
3. When special tools and fixtures are required, use low-cost tooling materials in place of tool steel wherever possible.

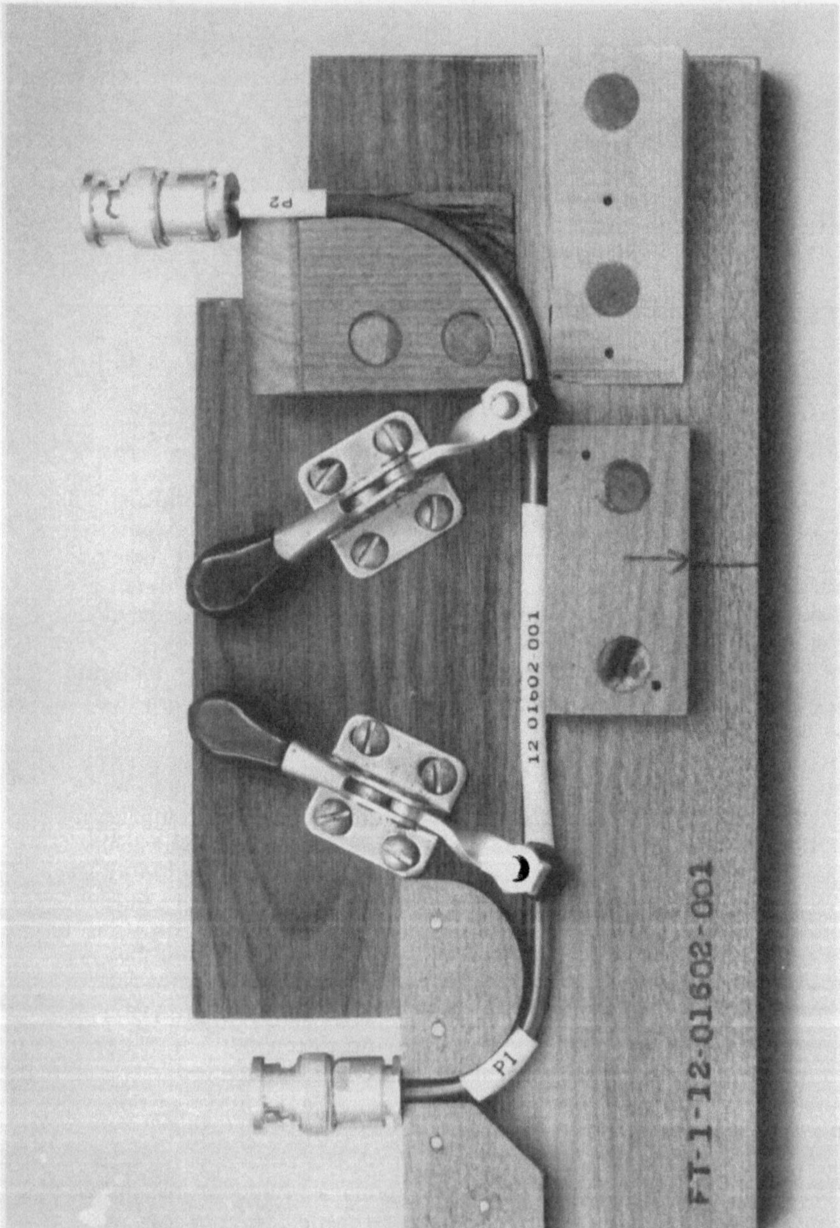

FIG. 7.1 Cable form tool. (From *Assembly Engineering* Magazine, October 1971, by permission of the publisher, © Hitchcock Publishing Company, all rights reserved.)

FIG. 7.2 Assembly fixture. (From *Assembly Engineering* Magazine, October 1971, by permission of the publisher, © Hitchcock Publishing Company, all rights reserved.)

4. Maintain close communication between the manufacturing engineer, the tool designer, and the toolmaker through all phases of the tooling program.

Figure 7.1 shows a wooden fixture with standard clamps and shaped, wood form blocks which bends a semirigid cable assembly to accurate tolerance. The total cost of the fixture was $98.00, while a similar fixture constructed of tool steel would cost $250.00. Figure 7.2 shows a standard fixture modified with standard clamping devices to make an assembly fixture for computer relay panels. The total cost of the reusable fixture is $73.00.

7.2 BASIC TOOLING PHILOSOPHY

During the preproduction planning phase for the production of any new product, or the continued, economical production of existing lines of products, it must be decided what the approach will be for tooling. In many instances, the process itself dictates tooling requirements, leaving little room for deviation. This is the case for foundry and investment casting operations, as well as die casting, injection molding, and similar processes. In the production of automobiles, tooling is the key to successful production of a quality product, and little is spared in ensuring that the tooling is right and will do the job.

Unfortunately, most companies spend far more on tooling than is really necessary. In the metal fabrication and forming industry, for example, an experienced manufacturing engineer can utilize standard shop tooling on many occasions to accomplish the same job that is often tooled with special-purpose designed tooling. It cannot be emphasized too strongly how important it is to review each drawing and ask yourself, if the tooling looks apparent, what other method exists, or whether hand tooling can be modified or improvised to avoid this expense.

In the metal fabrication industry a new material was developed some years ago, called chromolly. It is nickel steel with part chromium. It is normally purchased in 3/32 thick sheet stock. This material, even though in sheet form, is harder than tool steel in performing its intended function. It can be spot-welded, pierced, filed, or sawed. It needs no heat treatment. To make a rectangular blanking tool, for example, requires four pieces of this material measuring 6 by 6 in. Two of the four pieces will act as a base for spot welding the punch and die portions, and for mounting the two guide pins. One other, which will become the punch portion, is cut to the exact size of the hole it will punch. It is then spot-welded to a 6 by 6-in. base. The remaining piece is now sawed or punched, leaving a cutout in the piece of the desired size plus 0.007 in. This becomes the die and the 0.007 in. is for die clearance. This piece is now mounted to the other base, again by spot-welding. Two guide pins are inserted in opposite corners, or in the center of the cutout in the female portion of the base. Two holes of the same diameter as the pins plus 0.003 in. for clearance are added opposite the pins in the other base, which is the punch. Slide one onto the other, using the pins for alignment, and the blanking tool is complete (see Fig. 7.3).

This tool is capable of piercing 800 large holes per sharpening. It can be sharpened twice before being discarded. It takes 6 hr to make this economy tool. For comparison, a tool with the same function made in a tool shop would cost $750.00 or more, plus a number of weeks for delivery. This tool has great value, in that this or similar tools can be made in any sheet metal shop by a journeyman mechanic. The tolerances to be expected from this form of economy tooling are identical to those expected from a permanent tool. Under normal conditions, this would be plus or minus 0.005 in.

On many press-brake bending jobs, solid blocks of urethane and other elastomers can be used to replace conventional machined dies. The

FIG. 7.3 Economy punch and die.

savings, of course, are great. The value of the urethanes lies in their ability to transmit kinetic energy. Urethane, unlike rubber, is virtually incompressible. If a blank is forced into a block of the material by a punch, the energy of the punch is transmitted almost evenly in all directions. If the material is confined by a retainer, the energy is reflected back by the retainer walls since it has no place else to go. It will force the blank against the punch with uniform pressure. The advantages of urethane tooling can be summarized as follows:

1. Low tool cost. Urethane is usually used in block form with minor relief to direct metal flow. It can be machined with ordinary shop equipment.
2. Minimum setup time.
3. Extreme flexibility.
4. Nonmarring of surface finishes.
5. Excellent part definition when properly used.
6. Production of sharper bends or smaller radii than with metal dies.
7. Sharp reduction of springback and wrinkling.
8. No compensation needed for variation in stock thickness.

Urethane is now accepted as a standard die material, and its use is standard practice in most progressive sheet metal shops.

The tooling philosophy should reflect a low-cost yet fully adequate tooling approach. In an increasingly more competitive world, the company that is successful is the one that goes a step beyond the standard tooling approach, and finds ways to do things better and cheaper. This requires the knowledge, experience, and resourcefulness of the manufacturing engineer. The tooling preplanning should determine how the low-cost tooling philosophy will be carried out.

7.3 TOOLING PREPLANNING

The first and most important phase of a new tool or fixture design is the preplanning accomplishment by the manufacturing engineer. He must review the entire fabrication or assembly operation, define the tool or fixture requirements, and consider such factors as:

How the part or assembly will be held or clamped in the tool or fixture
What orientation and accessibility the tool must provide the machine or operator so that the work may be performed most efficiently
What relationship the tool will have to the workplace (if an assembly tool), parts bins, hand and power tools, and other equipment required for the operation
The spatial relationship between the product, part, or assembly and the operator to provide an efficient and comfortable working position for that operator (assembly tool)

To really understand the importance of tooling preplanning, it is necessary that we explore in more detail the tooling requirements, how the tool is conceptualized, critical tool features, tool design standards, and the importance of maintaining a standard tool inventory.

7.3.1 Defining the Requirements

This is the first step in developing the tooling program for any product. The requirements of the tool must be thoroughly understood by the manufacturing engineer. In the injection molding industry this need for understanding the requirements is fundamental for the product to meet quality needs. A thorough understanding of all the conditions that prevail during the molding operation should ensure that necessary precautionary measures are incorporated in the mold design to safeguard the expected features in the product.

The function of an injection mold type of tool is to receive molten plastic material ranging in temperature from 350° to 750°F at pressures between 5,000 and 20,000 psi. In the injection process, the plastic comes from a heated nozzle and passes through a sprue bushing into feed lines, or runners, and then through a gate into a cavity. The cavities are maintained at temperatures ranging from 30° to 350°F at which solidification takes place. Means are provided for control of the temperature. The range of temperatures and pressures depends on the type of plastic material. The plastic is held in the cavity of the mold for a prescribed time until full solidification takes place. At this point, the mold opens, exposing the part to the ejection or removal operation.

From the general concept of the molding operation, it is recognized that it is important to design a mold that will safely absorb the forces of clamping, ejection, and injection. Furthermore, the flow conditions of the plastic path must be adequately proportioned in order to obtain, in cycle after cycle, uniformity of product quality. Finally, effective heat absorption from the plastic by the mold has to be incorporated for a controlled rate of solidification prior to removal from the mold. The tool designer must incorporate all details that are conducive to good moldability.

Understanding the tooling requirements is basic to all else that now follows. The manufacturing engineer must thoroughly understand why a tool is needed, what function it will perform, and what process-type parameters it must meet.

7.3.2 Conceptualizing the Tool

Because investment castings are made in small lots and very large lots, and range from the simplest design to extremely intricate configurations, many different methods of tooling are possible. If the requirements are small or moderate, a single-cavity tool may be sufficient (see Fig. 7.4).

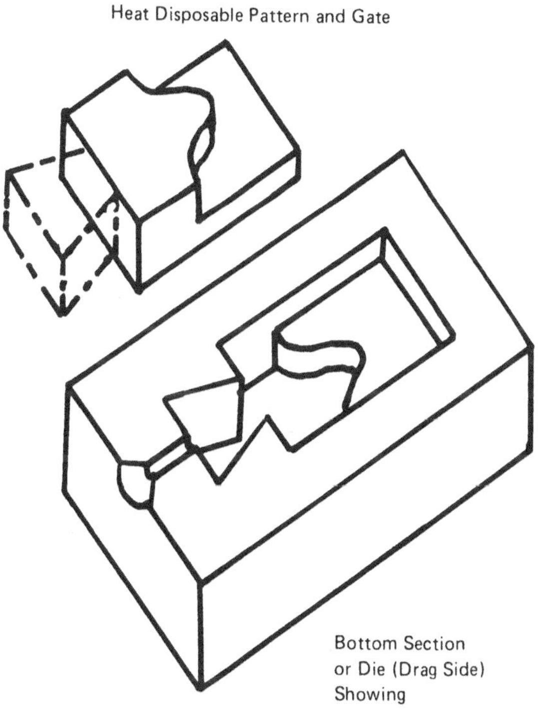

FIG. 7.4 Single-cavity tool. (Courtesy of the Investment Casting Institute, Dallas, Tex.)

The production rate per hour from a single-cavity tool can vary considerably. The pattern rate from a single-cavity, simple die can vary from 20 to 50 parts per hour. A highly automated single-cavity die with hydraulic openers and compressed-air pattern removal can produce as many as 250 patterns per hour.

Automated dies machined from steel or aluminum are used where thousands of patterns monthly are necessary. Once this tooling is ready for production, it is expensive to make changes. For this reason, the engineer must be sure of the product design before the manufacturing engineer proceeds with development of what amounts to high-production tooling. If an area of possible design change is known, the manufacturing engineer should be advised so that provisions for such changes can be incorporated in design of the die at the original concept stage.

These are the kinds of considerations that must be addressed by the manufacturing engineer when he develops his tooling concepts. A well thought out tooling concept for a special tool to open the insulation track in aluminum window sash bars is shown in Fig. 7.5. This was an especially troublesome problem in the manufacture of aluminum

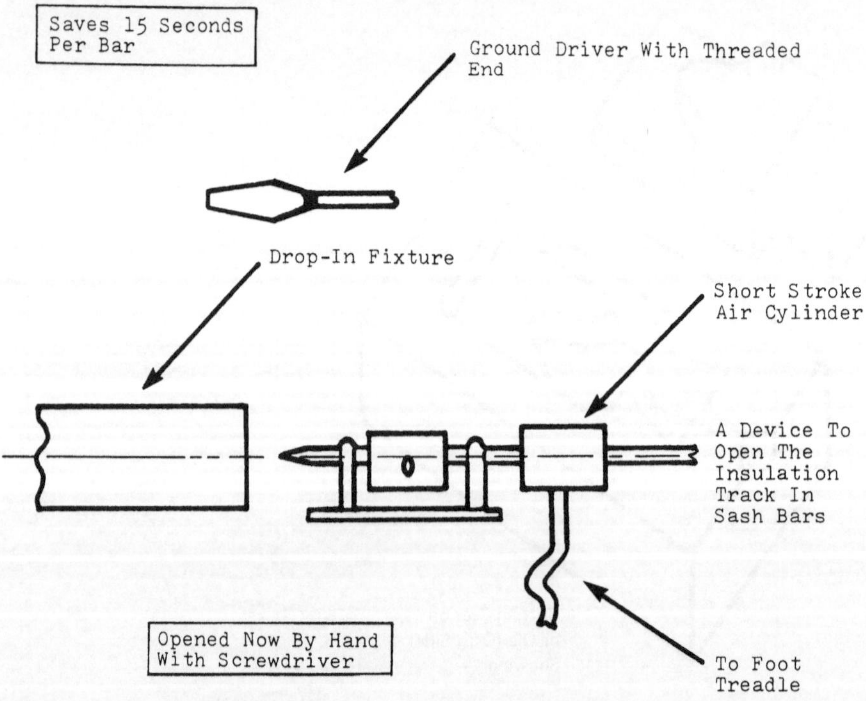

FIG. 7.5 Tooling concept for air-operated driver to open sash.

window frames, as the extruded shape of the sash bar was such that the insulation track was necked-down at the sash bar cutoff operation, requiring an extra manual operation with a screwdriver to open the track. The new tool cut the time for this operation in half. The concept sketch tells the tool designer exactly what kind of tool to design and what the tool is expected to do.

Figure 7.6 shows a concept for a lazy Susan fixture to hold and turn large window frames while caulking and insulation strips are applied to hold the glass panels in place. Previously this required two operators to physically pick up the frame and turn it 180° on the workbench. Now one operator can turn the window frame by depressing a foot pedal, without leaving his workplace.

An excellent way to assist the manufacturing engineer in tool preplanning and conceptualizing and to avoid having each engineer "reinvent the wheel" would be to provide a manual that catalogs, illustrates, and describes tooling that was successful in the past. Such a manual would show applications of standard tools and tooling components that the company has in its tooling inventory. These would include such items as small vises, stops, quick-acting clamps, ball-joint or swivel fixtures rotary fixtures, air cylinders, small electric motors, punch and die

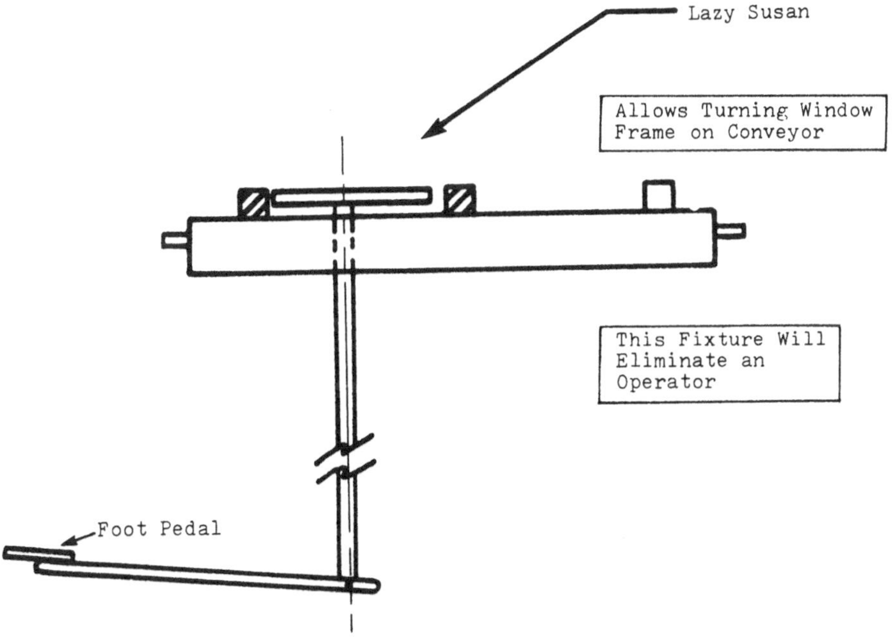

FIG 7.6 Tooling concept for lazy susan fixture.

sets, and so on. Reference to such a manual would provide the manufacturing engineer with ideas that might help with his current tooling requirements, and could uncover a design or application of a previously used tool that might be used with minor modification for the new tooling application.

It is one thing for the manufacturing engineer to know exactly the tool he requires for the fabrication or assembly operation, but often another thing for him to adequately communicate his concepts to the tool designer. Concept sketches such as those shown in Figs. 7.4 and 7.5 go a long way toward conveying this information, as does the written tool order used in many companies, but nothing takes the place of direct verbal communication between the manufacturing engineer and the tool designer. The tool designer must understand how the operation is to be performed and must discuss with the manufacturing engineer precisely what is required of the tool. The tool designer must learn critical dimensions and features, as well as desired tool locating points and surfaces, and stops and clamps required. The expected life of the tool in terms of anticipated total production of the product should be discussed, as this is the basis for the degree of durability required of the tool. For example, tooling made of wood or plastic components in place of tool steel can, in some cases, provide the required durability for low-volume, short-run production and reduce the total cost of tooling by about 40%.

The tool order is the primary document used by the manufacturing engineer in most companies to formally authorize the design and fabrication of tools. Figure 7.7 shows a tool order format for a small to medium-sized company which contains all of the necessary information blocks to convey information to the tool designer for the design of the tool. In addition to conveying information, it is a formal work-authorizing document. It is used by accounting to identify and collect costs by individual tool for a given job. All tool order forms should be similar to this form. They should be printed as an 8.5 by 5.5 in. snap-out form, no carbon required, in four separate snap-out copies, color coded as follows:

Original—white
First copy—pink
Second copy—blue
Third copy—yellow

The system or procedure for processing the tool order is shown in Fig. 7.8 and clearly outlines the necessary steps in preparing and processing tool orders. In larger companies, much of this may be done by computer, but the steps would be essentially the same.

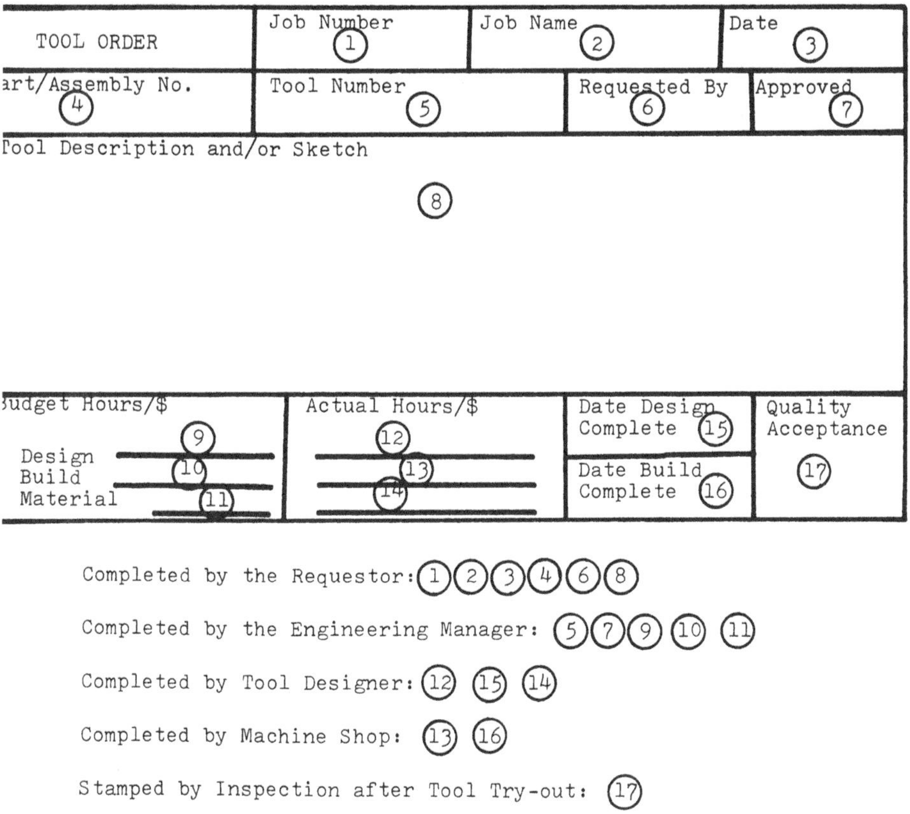

FIG. 7.7 Tool order format.

7.3.3 Critical Tool Features

Once the necessary concept and descriptive information have been transmitted to the tool designer via the tool order, concept sketch, and verbal communication from the manufacturing engineer, it becomes the tool designer's responsibility to adequately communicate with the toolmaker. The primary medium of that communication is the formal tool drawing. Beyond following standard drafting practices, tool designers should observe the following practices to assist in communications and reduce tooling costs:

1. Tool drawings should describe the tool completely and also specify optional construction wherever possible.
2. Tool tolerances should be specified as loosely as possible, with every attempt made to avoid extremely tight tolerances.

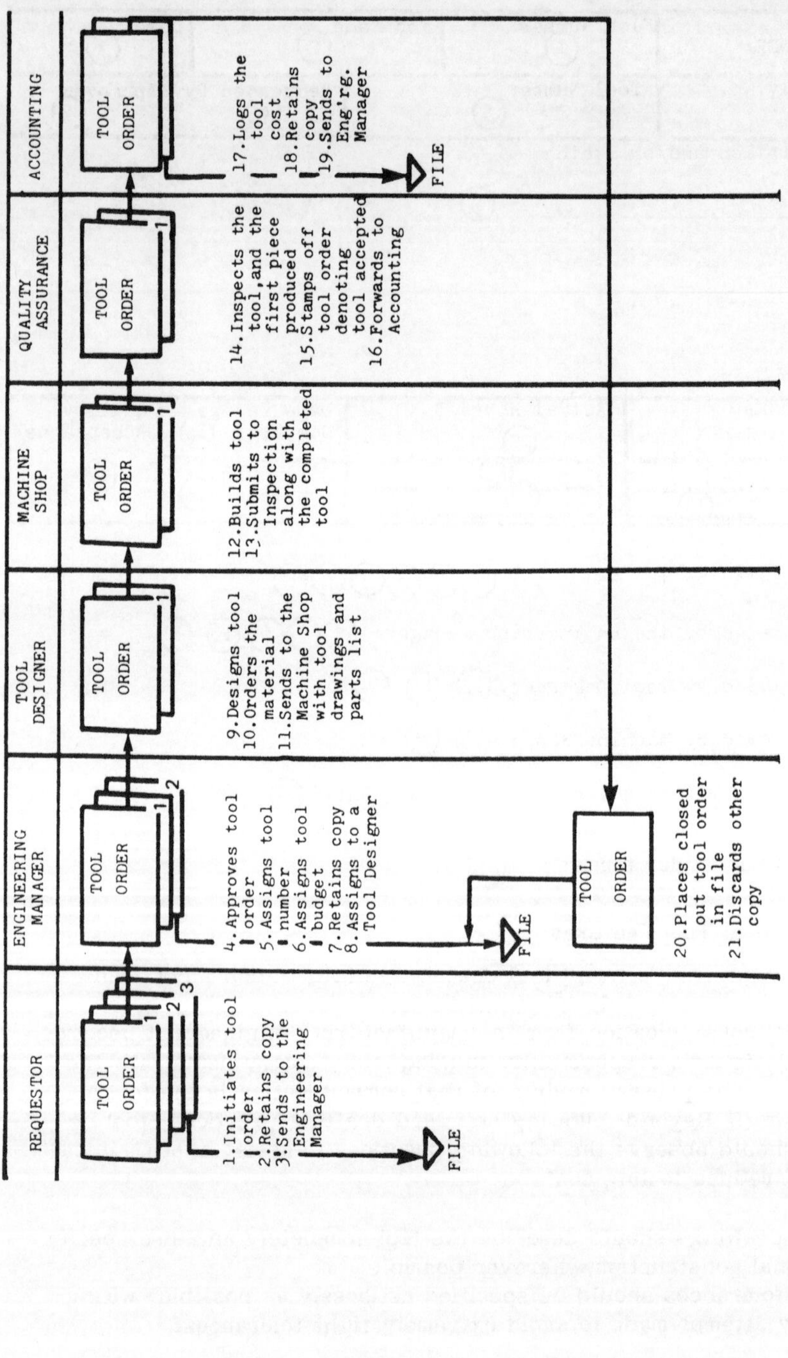

FIG. 7.8 Tool order procedure.

3. Tool drawings should be dimensioned completely so that scaling is not required by toolmakers.
4. Noncritical screw and dowel locations should not be dimensioned; instead, a note such as "locate approximately as shown" should be indicated.
5. A complete material list should be included on the first sheet of the tool drawing.
6. A list of standard sizes and shapes of materials stocked in the toolroom should be available to each tool designer, and materials should be specified from this list whenever possible.
7. Clamps, stops, locating pins, rest buttons, and holddown screws should be located so that the part or assembly will not be distorted or have its surface damaged.
8. The tool drawing should specify marking the tool number, part number, and any other numbers to allow positive identification of that tool in the future.
9. The current revision letter or number of the part or assembly should be noted on the tool drawing and the tool itself, anytime the tool must be updated because of a revision in the part or assembly.

When the tool drawing is completed and checked, the manufacturing engineer should review the drawing and sign off denoting approval and acceptance of the design concept. At that time, and before signing off on the tool drawing, the manufacturing engineer should estimate the cost of building the tool and review the economics of the tool and the manufacturing operation to determine whether a more or less elaborate tool would be more economically feasible.

Although the tool drawing should be so complete that no other communication with the toolmaker is needed, a few minutes of explaining the function of the tool is usually time well spent. If the toolmaker knows what is expected of the tool, he can normally do a better job of building that tool, and can often offer suggestions for improvements or cost reductions.

Upon completion of the tool, both the tool designer and the manufacturing engineer should observe the tool tryout or, if shop conditions permit, try out the tool themselves by performing one or more complete operations with production parts. Any required debugging or modification can be accomplished before starting manufacturing operations. All changes and modifications made during tool build and tool tryout should be recorded so that the tool master drawing can be updated to agree with the final configuration of the tool.

7.3.4 Tool Design Standards

Designed tools should be identified and marked with a tool number. Tool identification should consist of a prefix indicating the type of tool and the number of tools made for the part or assembly, followed by the engineering drawing number of the part or assembly. For example, in HF-11-5000294-001, the HF prefix indicates holding fixture and 11 is the number of tools made for part 5000294-001. Recommended tool prefix identification letters are as follows:

AL—aligning fixture
AD—adapter
AR—arbor
AT—assembly tool
AN—anvil
BD—press brake die
BF—brazing fixture
BT—bending tool
CB—cable board
CM—cam
CT—cutting tool
DA—disassembly tool
DE—extrusion die
DH—die holder
DI—die
DJ—drill jig
DT—drill template
ES—emulsion screen
FT—form tool
FP—face plate fixture
FD—form die
GA—gauging fixture
GR—grinding tool
GT—inspection gauge
HF—holding fixture
HT—heat-treat tool
IT—insertion tool
LF—locating fixture
MA—masking tool
ML—milling fixture
MO—mold or pattern
OC—operational chuck
PF—plating fixture
PP—punch press tool
PS—peg stamp
PT—pierce template
PU—punch and die
PO—potting tool
RA—rack
RM—reamer
SC—spinning chuck
SF—setting fixture
SK—staking tool
SO—soldering fixture
SP—spinning tool
SR—stripping tool
SS—silk screen
ST—stamp fixture
SW—spot-weld fixture
TE—test equipment
VD—vendor tool
VJ—vise jaws
WD—welding fixture
WR—wrench
WT—Wiedemann template

There should be a tool record card on file in tool design for each designed tool. The card should contain the tool number and name, the date designed, date built, date the tool passed tool tryout, and, if it is a vendor tool, the name of the vendor. Duplicates of the same tool will be carried on the same tool record card. Tool status (active, inactive, or obsolete) should also be shown, as well as the revisions or changes that have been made in the tool.

There should be a manual of standard practices that all tool designers must use when designing tools. An example of the kind of information that should be found in this tool design manual would be the use of pins for fixturing. In most progressive companies, tool designers have discovered the basic error in attempting to use two full round pins as locators. That approach attempted to provide full location at two positions simultaneously. Analysis indicated that a part fixtured on one full round pin is constrained in all respects except that it can rotate about that pin. The second pin is needed only to prevent rotation, and to accomplish this requires a full diameter on only a limited part of its surface. The balance of the full-diameter surface contributes nothing to part location except interference.

Tool designers soon learned that to provide full constraint, only one of the pins needed to be full round, and to avoid interference, one pin must be relieved. The diamond-shaped locating pin was soon developed to provide that relief. Problems in using pin fixtures, such as hole damage, pin breakage, and loading and unloading difficulty, can be traced to pin designs that do not provide sufficient relief. Fig. 7.9 shows both a diamond-shaped locator pin and the new oval locator pin, which can provide even more relief than the diamond pin. Figure 7.10 shows some ways in which relieved locator pins can be used. The tool design standard practices manual should contain this and similar kinds of information.

7.3.5 Standard Tool Inventory

As mentioned earlier, an inventory of commonly used standard tools and tooling components should be built up specifically to meet the requirements peculiar to your shop and your plant operations. This inventory would include standard cutters, end mills, broaches, reamers,

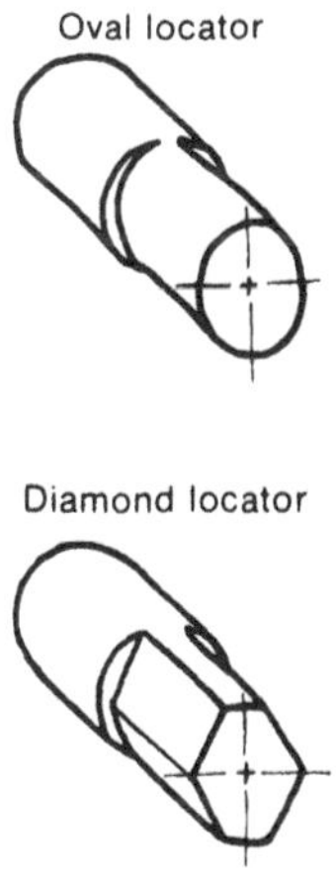

FIG. 7.9 Relieved locator pins.

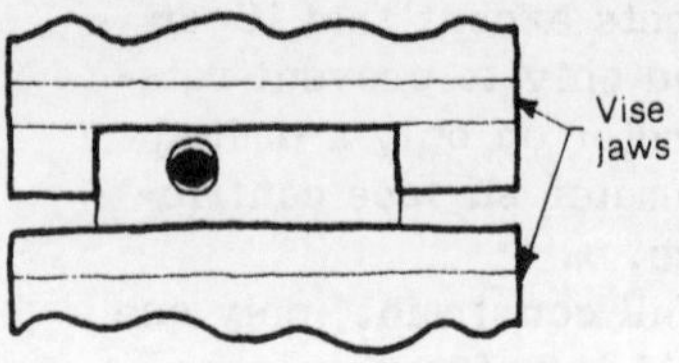

Relieved pin allows edges of vise jaws to establish location in one direction.

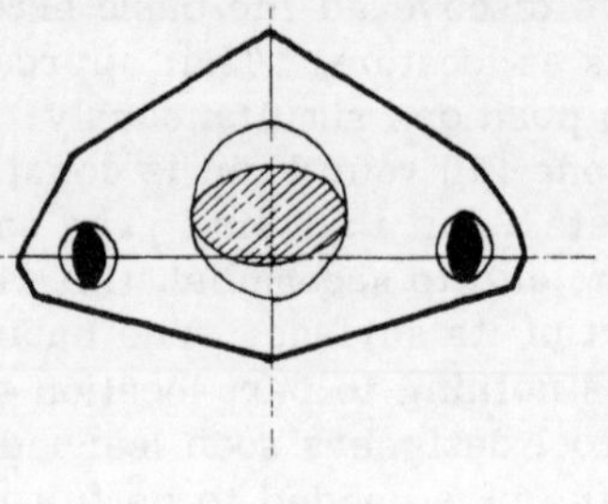

Three relieved locators provide full constraint without interfering with loading and unloading.

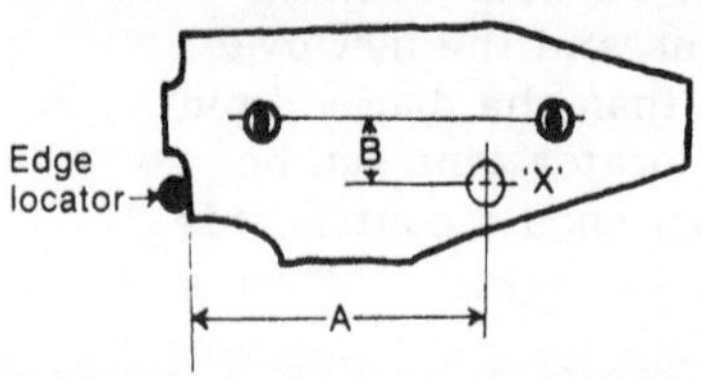

Hole at X is machined to close tolerances relative to two planes—one established by simple edge locator.

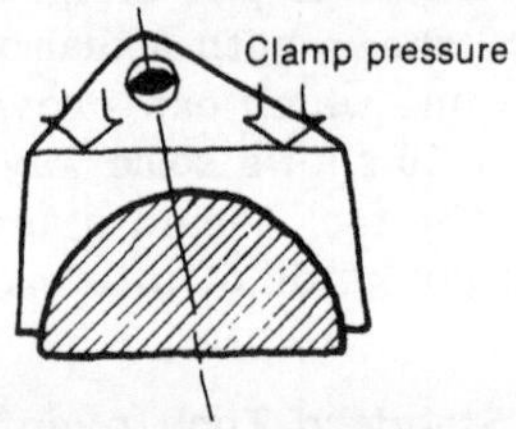

Clamp pressure holds part against large, half-round locator. Addition of secondary relieved locator solved problems caused by cocking of parts and clamping distortion.

FIG. 7.10 Applications of relieved locator pins. (Courtesy of *Machine Design* Magazine, a Penton Publication.)

drills, taps, and similar metal-cutting tools if your manufacturing operation was machining. Also included in an inventory for a machine shop would be a variety of different chucks, collets, hand tools such as files, scribes, deburring tools, and vernier calipers, and micrometers.

Tooling components such as drill bushings, quick release clamps, tool balancers, and small vises would also be included. The manufacturing engineer should determine what the standard tooling inventory should contain. He should specify the items to be stocked, from whom they should be ordered, and how many of each to order. Expendable tools and cutters should be placed on "max-min," which means that when the actual inventory of an item reaches a certain minimum quantity, an order is immediately placed for a predetermined number of the items to bring the inventory back up to a certain maximum level.

The standard tool inventory is usually stocked and issued from a central tool crib. This tool crib can be a part of the manufacturing engineering department, but more than likely is under production control department jurisdiction. The tool crib would also stock items such as shop rags, work gloves, lubricating and cutting oils, C-clamps, machinist's scales, safety glasses, and shop aprons.

If the plant operations include structural and mechanical assembly, the standard tool inventory would include welding rod, standard clamps and holding fixtures, ball peen hammers, power nut runners, combination wrench sets, socket wrench sets, screwdrivers, adjustable wrenches, pliers, clamping pliers, rubber mallets, power screwdrivers, rivet squeezers, and welding and brazing equipment. If operations include electronic assembly and wiring, the standard tooling inventory would include soldering irons and soldering iron tips, solder pots, temperature-controlled soldering irons, soldering aids, heat sinks, spools of cored solder, side cutting pliers, needle nose pliers, long nose pliers, chain nose pliers, assorted small brushes, and plastic squeeze bottles. Included also would be holding fixtures, power arms, and typewriter erasers.

7.4 TOOL DESIGN

The tool design section or group is an integral part of the overall manufacturing engineering department. It is staffed by competent professionals with background, experience, and training in tool design and tool engineering. Small shops and plants will not have a tool design department and may not even have a tool designer on the staff. The manufacturing engineer will work directly with the toolmaker or other shop personnel to make whatever tooling is required from a sketch or direct verbal communication, or both. This is an acceptable practice in small companies, where more formal procedures cannot be economically justified. The quality of the end-item product will determine whether this mode of operation is adequate.

As mentioned earlier, there are a number of proven, recommended practices that should be observed in designing tools, and these practices should be included in the company tool design manual. Some of these practices were highlighted earlier in this chapter. Then the knowledge and experience that have been accumulated over the years of proven techniques and methods are available to the newest and most inexperienced tool designer. This kind of information cannot be duplicated, except by trial and error in relearning the past.

7.4.1 Recommended Practices

A master tool drawing is prepared for each designed tool as a means of communicating tool dimensions to toolmakers and other interested activities. The master copy of a tool drawing is produced by the tool designer on vellum tracing paper or mylar sheets, as appropriate.

The master drawing is referred to as the original vellum or mylar. It functions only as a permanent record of the tool drawing, and serves as the master document to reflect changes, additions, or deletions to the tool configuration. The original is used for the specific purpose of producing additional printed copies of the tool drawing (blueprints, redlines, or Xerox vellums) needed for tool construction and required distribution. It is never issued or used as a working copy of the drawing. It is maintained in a permanent file location in the tool design department.

Vellum is a premium quality transparent tracing paper which is available in a variety of sizes, tints, weights, thicknesses, and surfaces. The quality ensures the strength, durability, and transparency needed for long-lasting, permanent drawings subject to frequent and often rough handling. Mylar is a transparent polyethylene sheet material that has maximum durability and reproduction quality for permanent drawings. Changes or additions can be made easily by eraser, thus preserving costly originals for an indefinite period.

The use of scale in making tool drawings is of the utmost importance. Drawing to scale reduces the possibility of costly errors during layout and detailing, provides a means of physically checking any and all dimensions, and is an aid to the toolmaker in accurately judging size. The primary points to be considered in the selection of a scale for a tool drawing should be clarity, legibility, practicability with regard to cost, and the convenience of the scale to the tool designer.

Drawings should be full size whenever practical; however, drawings to other scales are acceptable and are recommended in instances where full scale would make the drawing unreasonably large, or too small to be legible. The following are the only recommended scales for tool drawings. The scale block in the title block should be filled out accordingly.

1/10 size	2 times size
1/8 size	4 times size
1/4 size	8 times size
1/2	10 times size

Where it is impossible to make a drawing to scale due to tabulations, be sure to state "not to scale" on the face of the drawing, or in the scale block of the title block, whichever is appropriate. Out-of-scale dimensions that have been regarded as impractical or too costly to correct should be labeled as out of scale by underlining the dimension with a wavy line.

When more than one scale is used on a tool drawing, the predominating scale or the scale of the assembly, together with the words "and noted," should be placed in the scale block of the title block, and any view, section, or detail in which a scale is used other than the predominant scale must carry a note stating the particular scale used.

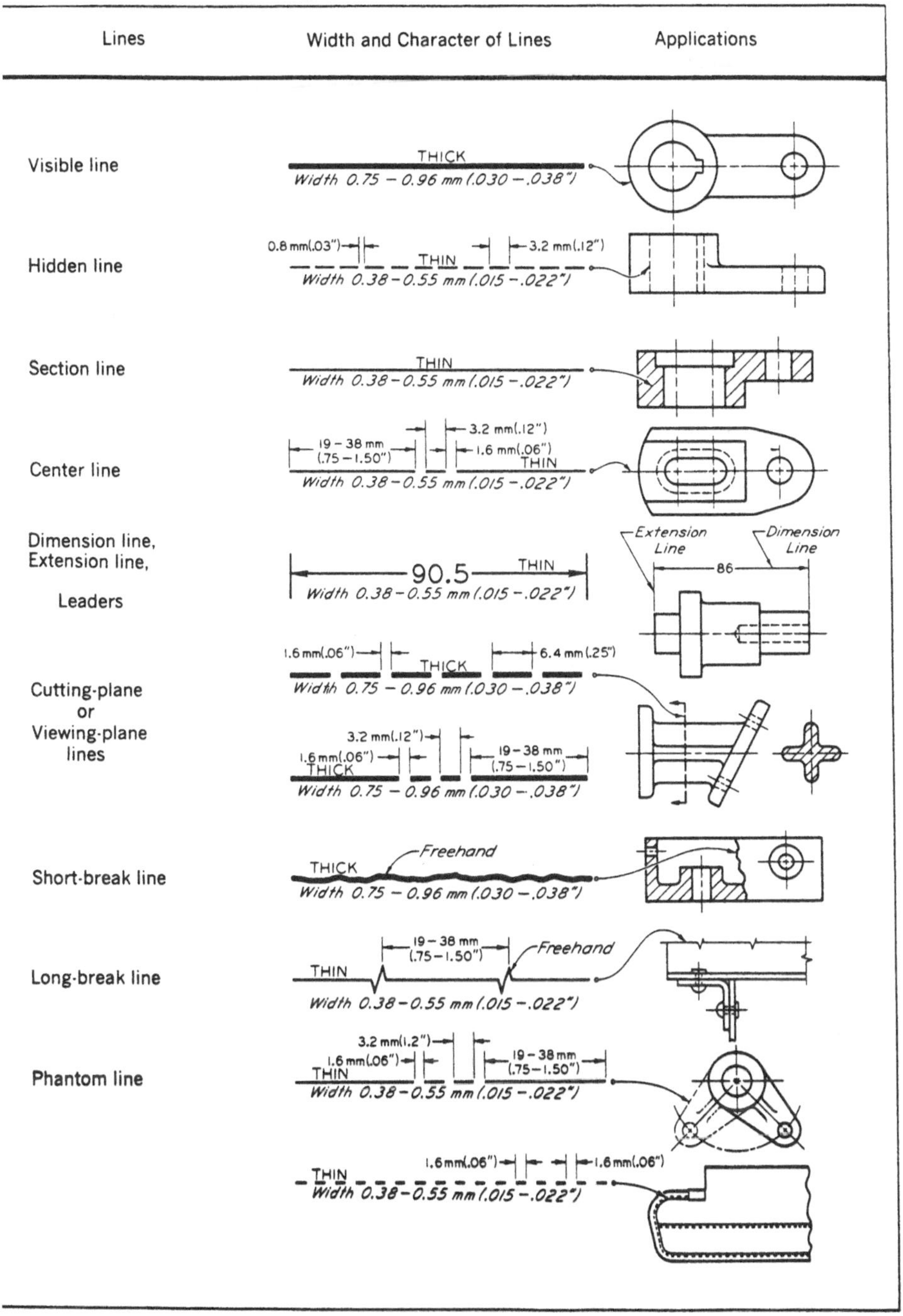

FIG. 7.11 Code of lines for tool drawings.

The basis of all drawing is lines arranged so as to impart, to anyone reading them, the thoughts and ideas of the person who made them. Tool drawings are no different. To obtain maximum legibility, clarity, and uniformity in all tool drawings, a code of lines has been agreed upon. All lines shall be of sufficient density to produce legible prints or reproductions. Figure 7.11 shows this code of lines. Production parts:

1. Shall be shown in relation to the tool by a phantom line
2. Shall be shown in all related assembly views
3. Shall not be overdone

Reference tool lines of green color shall be used to show the position of reference tools (master gauges, templates, cutters, arbors, or any other related tool) with relation to the tool being designed.

Leader arrows in tool drawings are as necessary as dimension lines. They are a means of relating noted information to a particular location on an assembly or detail, or in the detail identification, with either a firm or reference bubble callout. There are two different types of leader arrows: those used with local notes and those used with bubble callouts. The leader arrow used in conjuction with local notes is made by using straight lines only, with one end at the note and the other at the point where the information note is applicable (see Fig. 7.12). The leader arrow used in conjuction with bubble callouts is made with a a curved line to make it readily distinguishable from any other leader arrow or line. One end is attached to the bubble callout and the other end to the edge or surface of the detail to be identified (see Fig. 7.13).

It is desirable that leader arrows should never cross dimension or extension lines; however, this is not always possible. In such cases

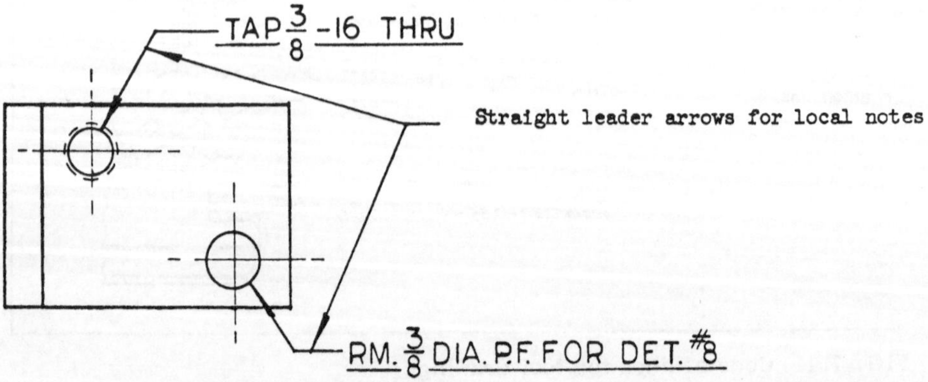

FIG. 7.12 Straight leader arrows.

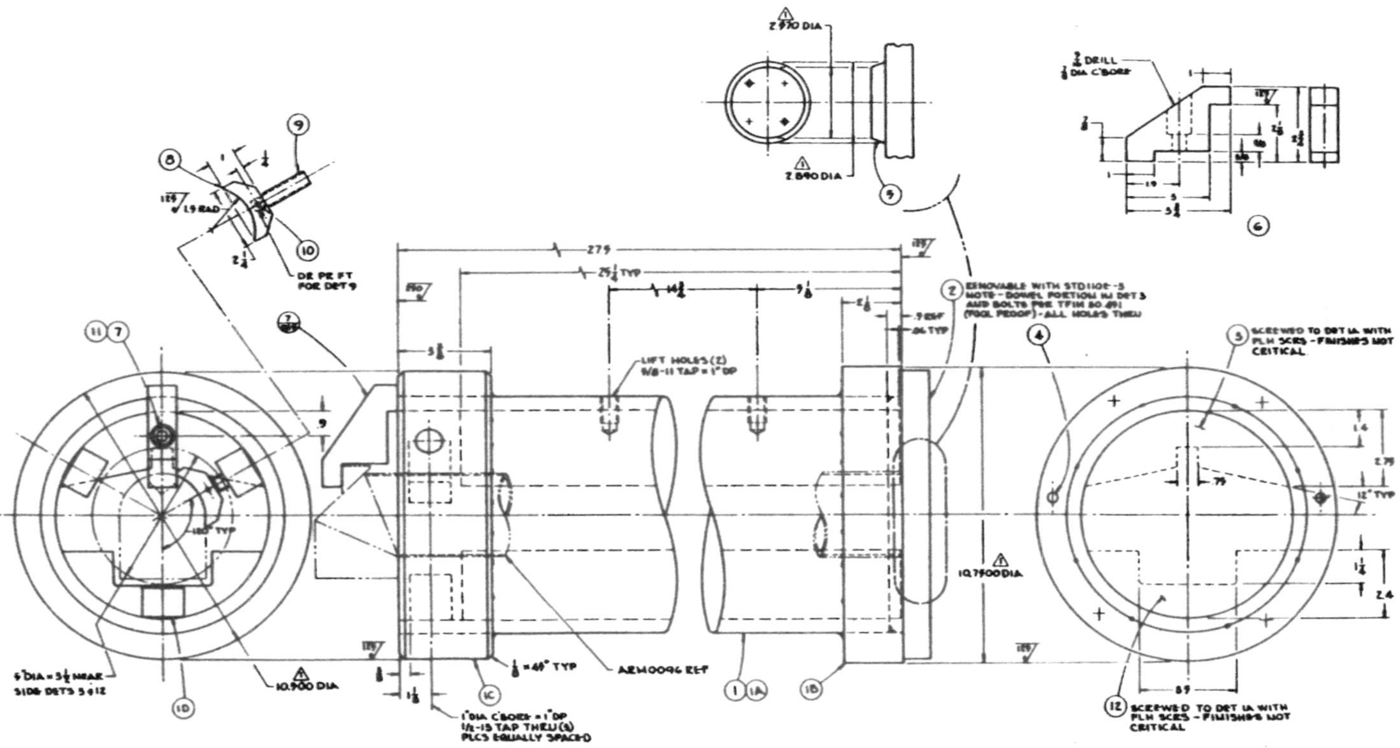

FIG. 7.13 Curved leader arrows.

the leader arrows should be broken on each side of the dimension and/ or extension line. Line weights for leader arrows should be the same as those used for dimension lines.

A recommended practice that should be in most tool design practice manuals involves bushings. Bushings have many and varied uses. Six main types of standard bushings are used:

Press fit	Fixed	Renewable
Slip renewable	Potted	Clinch

Figure 7.14 shows bushing nomenclature, definitions, and elemental data. Press fit bushings are made in two styles: headless and headed. These bushings are permanently pressed in place and used where replacement due to wear is not required during the life of the tool. Headed bushings should be used where the head is required for

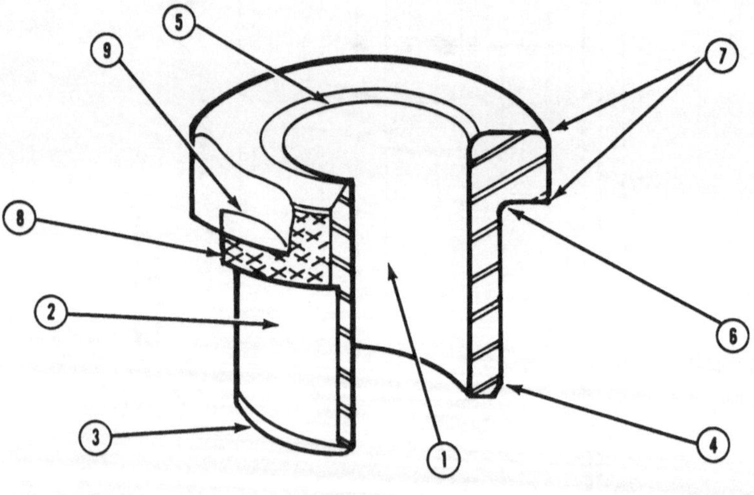

FIG. 7.14 Bushing nomenclature. 1, Inside diameter (ID), internally ground diameter, TD standards indicates specifications for counterbored bushings; 2, outside diameter (OD), ground OD bushings used; 3, Concentric ground lead necessary for proper initial bushing alignment and entry into tool; 4, 45° chamfer lead; 5, radius on ID to allow proper tool entry alignment—assists in preventing wear and breakage of tools; 6, two-way undercut allows bushings to sit flush on tool; 7, head chamfered, headed bushings; 8, knurled head on slip renewable bushings; 9, recessed area for lock retainers—slip renewable bushings only.

additional drill guide length or where an anticipated heavy load would force a headless bushing through the support plate. Headless bushings are recommended for most applications.

Renewable bushings are made in two styles: fixed and slip fit. These bushings are generally used in conjuction with a liner bushing that is pressed in place. Fixed renewable bushings are used where the drilling action is severe and the bushings are apt to wear out. Slip renewable bushings are used where more than one operation is to be performed, such as drill and ream, or drill and tap or counterbore. Renewable bushings must be retained in place during the drilling operation or they will rotate and withdraw with the drill. Use the standard lockscrew provided for this purpose. A typical installation of a renewable bushing, liner, and lockscrew is shown in Figs. 7.15 and 7.16.

Potting bushings is the process of locating the bushing by its inside diameter in an oversize hole and casting plastic between the outside wall of the bushing and the side of the hole in the metal plate. The advantage of the potted bushing is the accurate coordination and close tolerances between bushings that can be held. Potted bushings may be used to guide another bushing such as a slip bushing or a special type of bushing such as is used in pressure foot drilling. Potted bushings must never be used to guide a drill directly because the heat generated may soften the plastic.

Clinch or template-type bushings are recommended for metal tooling applications where the stock thickness of the tool body will not adequately support a press fit bushing. An example is shown in Fig. 7.17.

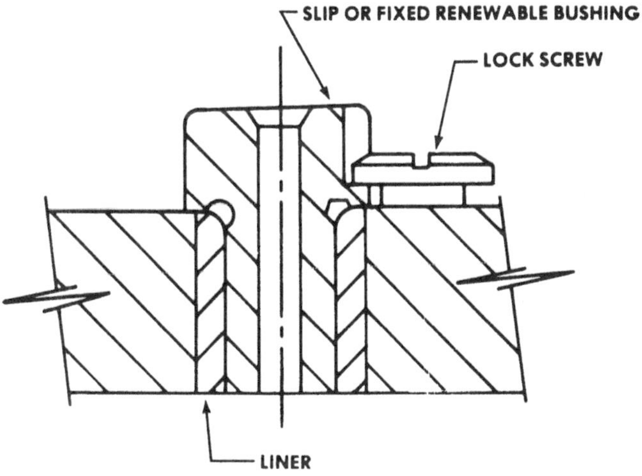

FIG. 7.15 Renewable bushing.

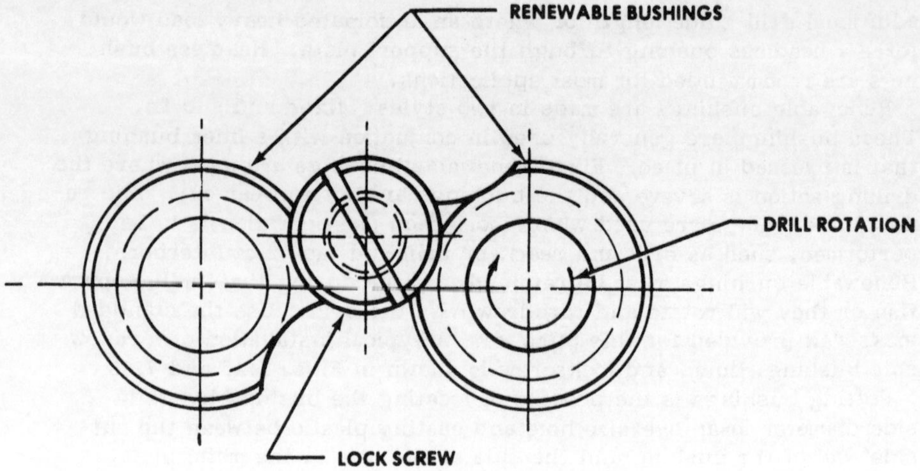

FIG. 7.16 Single lock screw for two bushings.

Bearing length limits for press fit bushings or liners shall be as follows:

Minimum—Material thickness equal to the outside diameter
Maximum—Material thickness equal to three times the outside diameter

Too short a press fit area will not adequately hold or align the bushing. Too long a press fit length adds to possible damage during installation.

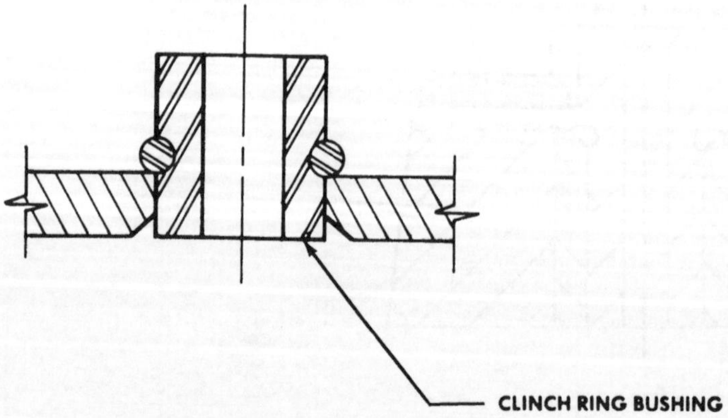

FIG. 7.17 Clinch-type bushing.

Clearance between the bottom of the bushing and the workpiece must be sufficient to allow chip removal and yet be close enough to properly guide the drill bit. Recommended clearance for most materials is to allow one-half to one drill diameter between the bushing and workpiece. Exceptions to this rule would be for reaming, for irregular workpieces, and for short-run precision jobs where the bushing should lightly contact or just clear the workpiece. The bottom of the bushing should be flush with the support detail to eliminate chip pockets (see Fig. 7.18).

Multiple hole patterns that are closely spaced present special problems in drill plates where press fit bushings are used. Warp of the plate carrying the pattern can be caused by the stresses introduced during the press fit operation. Pressures required to adequately retain bushings cause the surrounding metal to stretch, as the amount of stretched or compressed material builds up. The stresses tend to find relief by warping the drill plate. In addition to plate warp, bushing hole closure can result. The accumulated stresses in the plate can close the inside diameter of the bushing sufficiently to cause the tools to bind, resulting in breakage. When a multiple hole pattern must be bushed in a steel or aluminum plate, either pot the bushings in place or call for installation using a bonding agent.

The above represents the kind of information that should be found in a good tool design manual. This manual should contain information on approved practices in:

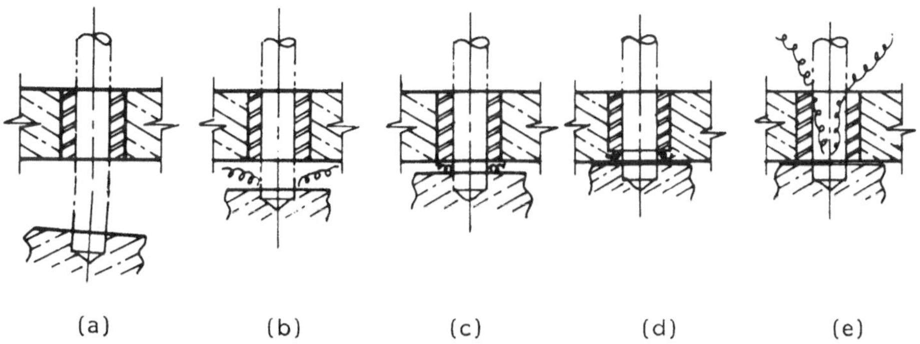

FIG. 7.18 Clearance between the bushing and workpiece. (a) Workpiece too far away; drill will wander. (b) Best condition, one-half to one drill diameter. (c) Workpiece too close; chips pack (less than one-half drill diameter). (d) Chip pocket not acceptable. (e) Permissible. Chips go through bushing (part flush with tool).

1. Tool design drafting and information callouts
2. Methods and techniques of designing tools to minimize problems and hold down tooling costs
3. Standard tooling components and parts that should be specified by the tool designer
4. Tool design practices that have caused problems in the past and should be avoided
5. Approved systems and procedures for preparing and processing tool design paperwork such as tool orders and purchase requisitions for tooling material

7.4.2 Tooling Economics

The importance of minimizing the cost of tooling cannot be given enough emphasis. At the same time, the tooling must perform its intended job. Some good examples of this can be found in the design of drill jigs for low volume production. The design and construction of these tools involves:

1. Simplification of tool design
2. Elimination of some unnecessary components
3. Use of inexpensive materials
4. Composite construction with reusable components
5. Use of temporary or emergency tools or of ordinary vises or chucks in tool fabrication

Examples of these tools are shown in Fig. 7.19 below:

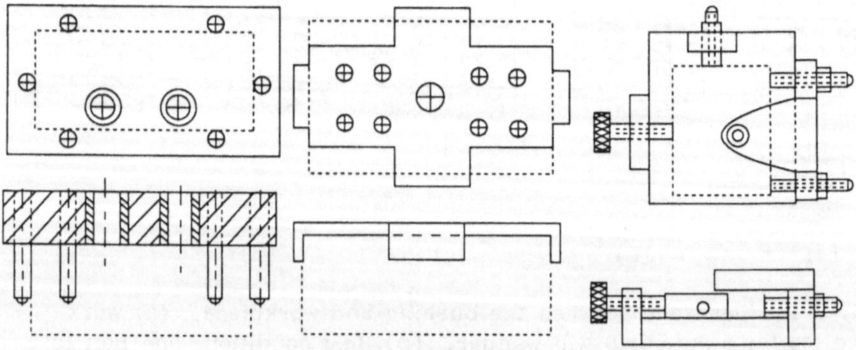

FIG. 7.19 Economical low volume production drill jigs which can be fabricated from inexpensive materials.

Since only a few parts are to be produced, the manufacturing engineer can sacrifice some of the correct principles of tool design. He need not bother with speed or ease of operation, easy maintenance, long life, replaceability of worn parts, or such features as automatic ejection of finished parts. Also, any surface that does not directly affect the job need not be finish-machined.

The examples shown in Fig. 7.19, show machined shoulders and bent tabs and dowel pins to provide stops and locate the workpiece. When a drill jig is supplied with stops, clamping force may be supplied by the operator in light duty work.

The primary purpose of any drill jig is to reduce manufacturing cost, but it must also provide workpiece interchangeability, accuracy, minimal operator adjustments and elimination of mistakes or rejects. The drill jigs in Fig. 7.19 provide these features for short run work, and they are economical to design and build.

The responsibility for good tooling economics falls primarily on manufacturing engineers. We often find careful analysis given to the selection of a numerically controlled (N/C) machining center and to the development of programming procedures, policies, and standards. Design and selection of the proper cutting tools are often given little or no study and analysis. The result is that a $300,000 machining center, backed up by a costly computer but equipped with an inadequate $50 cutter, represents $500,000 in headaches.

Before an N/C machine tool becomes operational, the manufacturing engineer should ensure that the following ground rules are carefully and rigidly adhered to:

1. Establish a set of cutting tool dimensional standards and make certain that all cutting tools ar purchased to these standards and are reground to the next N/C standard.
2. Use the shortest and most rigid cutting tool that can be applied to the job. It should be noted that more loss of accuracy on N/C machines comes from twist drills that walk because of improperly ground points, from long, weak drills that bend and wander, and from long end mills that flex and subsequently machine out of tolerance, than from any other source.
3. Use carbide and especially carbide indexable-insert type tools wherever possible in preference to high-speed steel. The potentially higher productivity of N/C machines and the investment in such equipment demand the longer and more constant wear life of carbide.

It has been determined that the average N/C machining center job is made up of 20% milling, 10% boring, and 70% hole making, in machine cycle time. Drills are the most common cutting tools used in numerical control. High-speed steel twist drills can cause many problems if the user does not insist on the best (precision drills versus standard drills). A so-called standard drill can vary significantly in lip height, flute spacing, web thickness, straightness, and contouring of the drill point. Even if its diameter is within tolerance, it will walk if its drill point is off center. Precision drills are a must, and the sharpening of worn drills must be to the same high standards.

7.4.3 Design Approval

Design tools can be approved only by the manufacturing engineer who originally ordered the tool. As simple as this seems, this is often not the practice, particularly in certain large and very large organizations. Accountability and responsibility for tools must rest with the manufacturing engineer. He must conceptualize the tool, approve its design, follow the build or fabrication of the tool, be present when the tool is tried out, assist in making any needed changes or revisions, and ensure the proper functioning and maintenance of the tool in production.

In a small company, where there is no formal tool design or tooling function, responsibility for tools used in production belongs to the manufacturing engineer. He may work directly with the toolmaker or machinist to develop needed special tools, without benefit of formal design or even concept sketches, but the responsibility for the tools and how well they do the job is still his, and his alone.

7.5 TOOL FABRICATION

Tool fabrication may be the function of a large toolroom department, or it may be the collateral duty of production operators or machinists, as a normal part of the job. As indicated earlier, the tooling may be highly sophisticated, precision fabrication or assembly tooling, or the most rudimentary shop-aid type tooling.

However tools are built or fabricated, it is important that there be direct communication between the toolmaker, the manufacturing engineer, and the tool designer. In some small shops, whatever tooling is required and built is left to the discretion of the operator. This practice is highly questionable and is not recommended. Often, the most economical fabrication method is to send the tools outside to be built. This offers several advantages, especially to the small shop, which may not have the latest in toolroom equipment such as jig borers, N/C milling machines, cutter grinders, and toolroom lathes.

Another consideration in deciding to have tools fabricated by outside sources is cost and delivery. If it is less expensive to go to outside vendors and the finished tools can be delivered when needed, then it makes sense to utilize this outside capability. It is extremely important to have reliable outside sources for tool build that can be called upon when your own shop is overloaded or that can fabricate needed tools on short notice. The problem with using outside sources for tool fabrication is the requirement for good tool drawings and other documentation. Also, it is more difficult to make changes or to communicate informally with an outside shop than it would be if the tool were being fabricated in-house.

7.6 TOOL TRYOUT AND ACCEPTANCE

There should be a formal procedure for tool tryout and acceptance, even in small companies. Such a procedure should include the following:

1. Inspection of the completed tool with the tool drawing
2. Use of the tool to produce a sample part or assembly
3. Verification that the part or assembly produced with the tool is a good part or assembly that meets drawing and quality requirements
4. A tool tryout tag which is affixed to all new tools, indicating tryout status
5. Removal and filing of tags from tools that have successfully completed tool tryout

The manufacturing engineer and the tool designer should be present when tool tryout is performed. This enables immediate action to be taken to make adjustments, corrections, or changes if the tool does not function as envisioned. Tool tryout and acceptance should be administered and controlled by the inspection or quality control department. A tool tryout and acceptance procedure also implies a tool control system which prevents the use of unauthorized and "jury-rig" tools in the shop or factory. Tool control is essential if product quality and tooling costs are to be controlled.

Figure 7.20 shows a sample tool tryout tag used by a small company. Normally, standard, nondesign jigs, fixtures, and motion economy devices which have as their primary purpose the holding or positioning of work at the workplace, but do not control critical dimensions or part configuration, are excluded from tool tryout requirements. Also, in some companies it has been found to be advantageous to leave the tool tryout tag on the tool in production rather than to remove it and file it. If this procedure is followed the tag should be of heavy card stock and wired to the tool to prevent tearing or removal during production use.

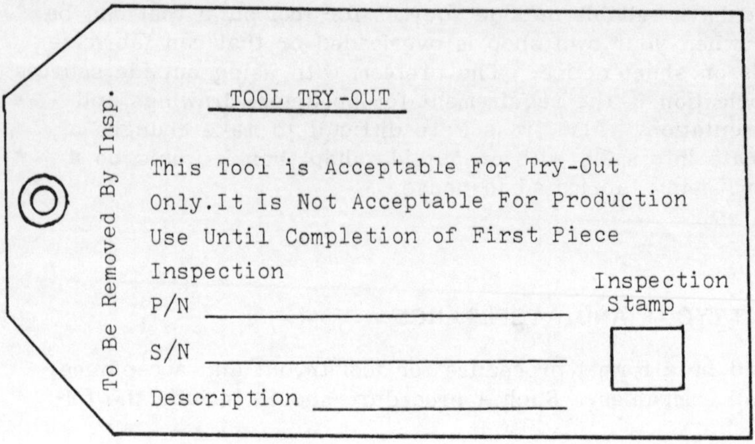

FIG. 7.20 Tool tryout tag.

7.7 INSPECTION TOOLS, GAUGES, AND TEST EQUIPMENT

Inspection tooling and gauges would normally be designed by the tool design department, where special design is required. Tool and gauge fabrication would normally be accomplished in the tool shop where production tools are fabricated, unless the tolerance or degree of precision required is beyond the capability of the tool shop. In many instances it is possible and desirable to have production tools serve as acceptance tools, thus performing a dual function and reducing the cost of tooling. Some examples would include alignment fixtures, assembly jigs, master layout templates, and optical tooling masters.

Mechanical and electrical test equipment is normally purchased either as capital equipment or expense equipment, and fixtured for the intended product application. Tool design would design test adapters and fixtures as it does any other design tool. Tool tryout procedures would apply to this special kind of nondesign tooling. Often we find items of commercially available mechanical and/or electrical test equipment used as building blocks for a special test console or station. The individual pieces of test equipment are rack- or cabinet-mounted and tied with interconnecting cables or lines to a central power source and to each other. The product to be tested would mate to a special test fixture or test adapter.

In many instances the requirements for the testing of electronic and electrical products can be satisfied by utilizing the product itself as a test bed for the various subassemblies that must be tested.

An example of this would be in the functional testing of printed circuit card assemblies by plugging them into the product top assembly; if the unit works, the cards are performing in a satisfactory functional manner.

For every design tool, whether production, inspection, test adapter, or other, there should be a tool record card on file in manufacturing engineering. The tool record card should contain the tool number and description, the number of duplicates of the tool, the tool revision letter, when the tool was designed, when it was built, when tool tryout was completed, and the present location of the tool. It should also indicate whether the tool is active, inactive, or obsolete.

7.8 TOOLING MAINTENANCE

Once a tool is in production, there should be an ongoing effort by the manufacturing engineer and tool designer to ensure that the tool continues to perform in a satisfactory manner. Tool wear can also be a problem especially with standard and special cutters and twist drills. Table 7.1 provides a checklist for diagnosing problems in production with carbide cutting tools. It is important that manufacturing engineers have knowledge and experience in troubleshooting tooling problems.

For example, twist drill breakage can be caused by lateral loading if the work surface is not normal to the drill axis and if the workpiece is not rigidly clamped to the drilling machine. A surface inclined with respect to the drill axis will result in the presence of lateral forces as the chisel edge of the drill point attempts to indent the work surface. If these lateral forces become high enough, the drill point will tend to slide down the inclined surface, causing lateral bending of the drill. If the flexibility of the drilling system is great enough, the drill will never indent the work, and the resulting stresses in the drill due to excessive bending may be enough to break the drill.

Such breakage is rare in production, where drilling is done through accurately aligned drill bushings in production drill jigs. Breakage does occur in production from excessive dulling. When excessive dulling is the cause, the resulting failure is nearly always due to torsional overloading.

The importance of good tool maintenance cannot be overemphasized. As tools wear they lose their dimensional accuracy. This applies to all tools, not just cutters. Tools that break are often "repaired" by shop people, but their function is impaired in some way which may show up as an out-of-tolerance part or series of parts, or as an assembly that will not mate with another assembly. A proper tool maintenance program prevents this from happening.

TABLE 7.1 Carbide Tool Troubleshooting Checklist

Problem	Probable Cause	Recommended Solution
Crater	Negative rake angles	Use positive rake
	Chip breaker	Increase width
	Coolant	Use coolant with good lubricating qualities
	Speed	Increase speed
	Feed	Decrease feed
	Carbide grade	Use more crater resistant grade
	Carbide grade	Use harder grade
Wear	Lead angle	Decrease lead angle
	Relief & clearance angle	Increase one or both
	Center height	Check to be sure tool is not above center
	Machine	Eliminate vibration, if any
	Coolant	Use copiously
	Speed	Decrease speed
	Feed	Increase feed
	Material	Check hardness
	Carbide grade	Use harder grade
	Nose radius	Too large—decrease
Built-up Edge	Coolant	Use coolant with good lubricating properties
	Speed	Increase speed
	Rake angle	Use positive rake
Poor Finish	Nose radius	Increase
	Lead angle	Increase lead angle
	Center height	Tool not on center
	Machine	Check for looseness
	Speed	Increase
	Feed	Decrease
Chipping	Nose radius	Increase nose radius
Cracking	Lead angle	Increase if possible
Breakage	Rake angle	Use negative rake
	Relief & clearance angle	Decrease
	Chipbreaker	Increase width
	Shank size	Increase if possible
	Tool overhand	Decrease if possible
	Center height	Check to be sure not above center
	Machine	Check for looseness

TABLE 7.1 (Continued)

Problem	Probable Cause	Recommended Solution
	Coolant	Be sure coolant is on cutting point of tool
	Speed	Increase speed
	Feed	Decrease feed
	Material	Surface scale (hone cutting edge)
	Workpiece	Check rigidity
	Carbide grade	Use softer or tougher grade
	Carbide grade	Too sharp a cutting edge (hone)
Chatter	Nose radius	Decrease nose radius
	Lead angle	Use if possible
	Rake angle	Decrease
	Relief & clearance angle	Increase
	Shank size	Increase
	Tool overhand	Decrease
	Center height	Check to be sure tool is not above center
	Machine	Check horsepower, spindles and slides
	Speed	Decrease
	Feed	Increase
	Workpiece	Check rigidity

Source: John D. Agnew, presented at Fifth Annual Carbide Cutting and Forming Seminar, June 25–27, 1974, at Purdue University, Lafayette, Ind.

QUESTIONS

1. What are some examples of low cost tooling materials, and of standard tooling components that could be used to keep the cost of special design tools for production as low as possible?
2. What can be done by the manufacturing engineer at the tool concept phase to ensure low tooling costs?
3. What are some advantages of urethane tooling in sheet metal work?
4. What are the important factors that the manufacturing engineer must consider when preplanning for a new production tool?

5. What is the fundamental premise in all tool preplanning?
6. What is the primary purpose of the tool order?
7. As a minimum, what should the completed tool design drawing convey in the way of information to the toolmaker to assist in communication of tool requirements and to hold down tooling costs?
8. What should the manufacturing engineer do after the tool drawing is completed, checked, and presented to him for sign-off and approval?
9. What should be done by the manufacturing engineer and tool designer after tool tryout, if it becomes necessary to change or modify the tool to make it work as intended?
10. What information should be contained on the tool record card?
11. Why is it important to have a master tool design drawing for each and every designed tool?
12. What are some of the tool design practices the Tool Design Manual should contain?
13. What actions should the manufacturing engineer take to ensure that N/C machines have the proper cutting tools?
14. Why is good tool maintenance so important in production?
15. What is one economical approach in the design of inspection tools and gauges?

8

Methods and Detailed Work Planning

8.1 PROCESSING AND OPERATIONS PLANNING

The manufacturing process and the sequence of steps or operations required to accomplish the process constitute perhaps the most important function of manufacturing engineering. Engineers who work in this important area must have technical engineering knowledge, a detailed knowledge of the processes of manufacture, and a high degree of skill and experience in analyzing engineering drawings and determining the most efficient process for manufacturing a product, piece part, or assembly.

During the preproduction planning phase (see Chapter 3), the process of manufacture, the tooling, and the equipment requirements have been determined, leaving the detailed step-by-step sequence of total operations to be determined at the processing and operations planning phase. This work will validate the initial preproduction, overview planning and will result in the development and release of the final process routings, work instructions, visual aids, and methods needed to actually accomplish the work. Major tools required will have been identified and ordered in the preproduction planning phase, but other tools and tooling requirements may surface in this final detailed planning phase.

8.1.1 Fabricated Piece Part Processing

Fabricated piece parts are the building blocks for subassemblies and ultimately the final assembly of the product. The principal processes involved include:

1. Casting and molding
2. Machining
3. Forming
4. Finishing

These principal processes plus sub-tier manufacturing processes are listed in detail in Table 3.1. In piece-part fabrication, as the material is received from its basic process, it becomes associated with one or more of the principal secondary process operations. This is shown in Fig. 8.1 in more detail. If, for example, the material is received as a forging or casting, it must be completed by a cutting-type operation. Cutting thus becomes the principal secondary process to follow. If the material from the basic process operation is in the form of pig iron, then casting becomes the principal secondary process of manufacture.

Part Print Analysis

The key to determining a sound manufacturing sequence lies in the selection of suitable terminal points for processing. Of necessity this must begin and end with the part print. In the initial study of the part print, the manufacturing engineer is aided in determining what is wanted in the final product. This must be known before any processing plan can be formulated. At the termination of the manufacturing plan or sequence, the manufacturing engineer must again make reference to the part print to determine by comparison whether or not the final piece part meets all design specifications. It is readily apparent that a great amount of detail must be completed between the terminal points in order to round out a complete manufacturing sequence.

The diagram in Fig. 8.1 is designed to show cutting as the principal secondary process operation. It should be noted that as the sequence of cutting process operations is developed, it is supplemented by various auxiliary and supporting operations. The manufacturing engineer first plans the major process operations. However, the terminal operations must be established before the remaining sequences can be planned. The part print or drawing depicts the part in its final condition, or rather desired final condition, thus establishing the ending point of the process. The material specification determines how the material will be received, thus determining initial machining or fabrication operations and how workpiece control will be established and maintained.

Determining Processes and Sequences

The critical areas on any workpiece are the areas used to provide workpiece location for subsequent machining or fabrication operations, and those areas identified as requiring demanding functional tolerances or shapes or finishes. It is essential in planning the process sequences to get these critical areas first to ensure that the important dimensional relationships of the workpiece can be maintained during the balance of the process sequence, and then to address those critical areas that do not control location, but are important functionally.

Critical areas may become critical due to a close surface finish or tolerance which cannot be maintained with standard shop machines.

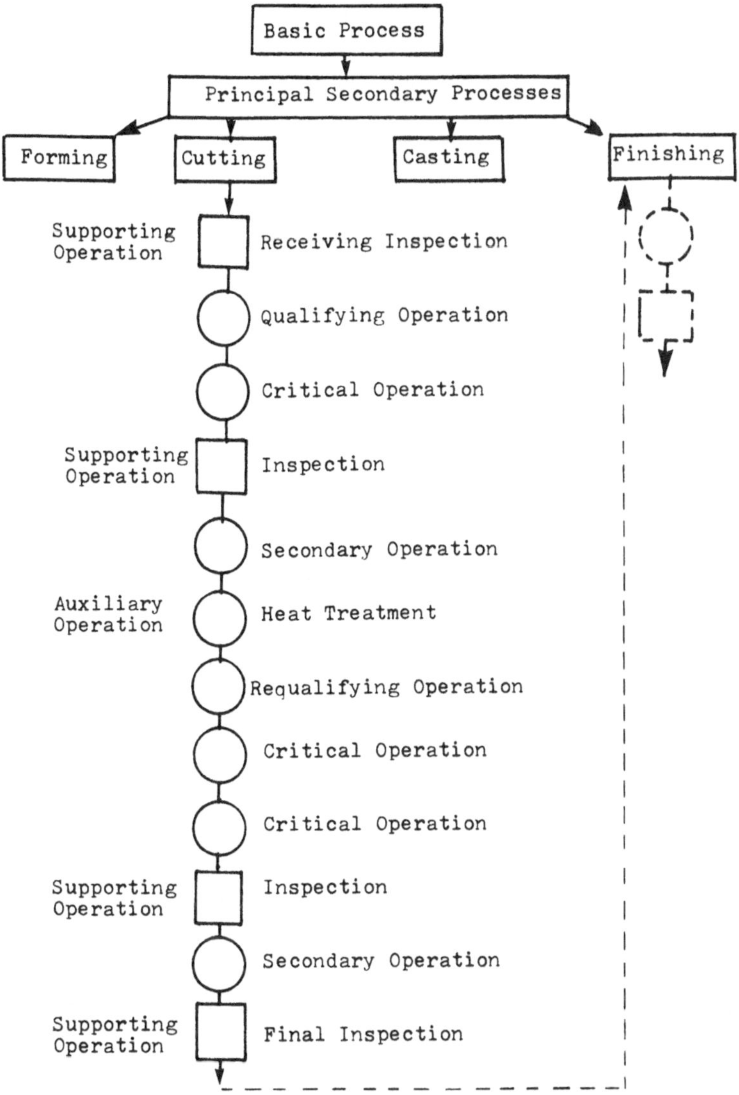

FIG. 8.1 Operation classification versus manufacturing sequence.

Operations in these areas will require special processes and special equipment to accomplish. The position of these operations within the process sequence are determined by the following conditions:

1. Surface finishes in critical areas may be such that in order to prevent damage, they must be accomplished at a point in the process when the surface can best be protected from damage or mutilation.
2. The sequence of machining operations that must take place before the operation in question can be performed must be considered. As an example, a hole must be drilled before it can be tapped.
3. The degree of accuracy that must be maintained between related surfaces may require that several less critical operations be performed before the final operation on the surface can take place.
4. The introduction of auxiliary operations into the major operation sequence may determine when the work can be performed. For example, if a given surface must be hardened, it may be necessary to grind the surface to obtain the required finish.

Qualifying and requalifying operations are those that precede the critical operations, and are required in preparation for the critical operations. Some examples would include stress relief and straightening operations after extensive machining but before the next critical operation is performed. Auxiliary operations such as welding and case hardening can cause this kind of distortion, depending on uneven heating and cooling, and so on.

Secondary operations are those that can be placed at any convenient point in the process, and should be placed to best facilitate efficient methods, handling, and movement of the work. Logic also plays a major role in determining when and where these operations occur. As an example, the sequence of drilling followed by reaming, turning by threading, or milling by polishing would be determined by the only practical order in which they can occur.

It follows, then, that the optimum sequence of manufacturing operations is dictated by the degree of control that can be maintained during the process cycle and by logical process order. These are controlled by the following considerations:

1. Part shape or geometry
2. Process limitations
3. Available process alternatives

Part geometry determines those surfaces which best control the part, and enable determination of the process sequence and number of operations that can be combined. It is easier to maintain part control from a large surface, and from a plane surface rather than one that is curved; in addition, part size is a major contributing factor. Also, when certain close dimensional relationships exist between part surfaces, care should be exercised to avoid tolerance stacking. This makes it more desirable to perform the operations which control close dimensions as early in the process sequence as possible.

The process itself imposes restrictions which result from a number of factors, the most important ones being:

1. Material limitations. The sequence for handling and machining a forging would differ markedly from that for machining the same part as a hogout from a solid block of material.
2. Machine flexibility limitations. Several different operation sequences may be possible on general-purpose machines, but a special-purpose machine that is designed to machine only one part to a given configuration has one fixed operation sequence.
3. Process capability limitations. Machines frequently vary in their capacity to machine accurately when heavy cuts are made, requiring more or fewer operations in the process sequence.
4. Tooling limitations. Grouping together of large-diameter cuts in the process sequence, rather than mixing them with cuts on smaller diameters which can be turned at higher speeds, reduces tool replacements and the changing of spindle speeds on turning operations.
5. Production volume limitations. Higher volume production can justify more efficient tooling. This may result in combining certain operations and also in shortening of the operation sequence.

When alternative ways of producing a part are available, the process sequence should be determined by first selecting the operation process sequence which will produce good parts. Several sequences may be tried until the optimum sequence is developed which will result in the greatest savings.

Perhaps the most important consideration in planning the process sequence is to guarantee the dimensional integrity of the piece part or workpiece through every operation that must be performed on it. This requires careful advance planning which addresses:

1. Selecting the best surfaces for location
2. Developing the best location system

The primary objective when selecting surfaces for location is to get the best possible part out of the process by controlling the variables inherent in the process and the workpiece. The location system matches the geometry generated by the process with that desired on the workpiece. This requires that each individual operation be planned separately to accomplish the intent of the part drawings and specifications.

In order to illustrate part processing, let us assume that we have a piece part made of 1050 grade steel, hot-rolled bar stock, which is to be completely turned on a general-purpose engine lathe. Also assume that calculations were made to determine the type of lathe required to produce the part most economically as a function of production rate requirements. Figure 8.2 shows the part as a flange adapter to be machined from 1050 hot rolled steel bars.

The first step in process planning for this part is to prepare a process sheet that includes the specific types of machine tools to be used for turning the part. The process sheet prepared for turning the part on an engine lathe is shown in Fig. 8.3. This is known as a summary process sheet, in that it lists and briefly describes each operation in the sequence in which it is to be performed. A part print or engineering drawing of the part must accompany this kind of outline or summary process. It lists the machines to be used for performing each operation, and would also list any tools required.

This is the kind of process documentation that would be provided to a skilled machinist, who knows how to set up his machine and who can read an engineering drawing. Along with the brief operation description, any pertinent information that should be brought to the attention of the machinist, and appears on the drawing, would be highlighted. See the reference callout in operation 70 of the process sheet (Fig. 8.3).

Process Routings

Another name for the summary process sheet described above is process routing. It is called a process routing because it tells production control where to take the part or batch of parts next, after the current operation is completed. The process routing sheet would then accompany the part or parts to the next operation, and so on. Often, provision is made on the process routing for inspection to stamp-off each of the operations as it is completed, if inspection is required.

A more detailed example of part processing is the following example of an automobile cylinder head coolant inlet which highlights the basic theories and reasoning that should be followed by the manufacturing engineer in planning a machining sequence.* Auxiliary and supporting operations are not considered; only the major process operations are. Our purpose is to produce a functional part with the highest degree of dimensional integrity. The engineering drawing of the cylinder head coolant inlet is shown in Fig. 8.4.

The drawing shows that the workpiece would be received as an aluminum sand casting. It indicates that four surfaces of the part need to be machined. These include the mounting face, the inside of the mounting face passage opening, three bolt holes, and three spotfaces on the bosses surrounding the bolt holes.

The mounting face was the most critical surface, and all other surfaces to be machined were located from this surface. Since this was a

---

*Following discussion adapted from *Process Engineering for Manufacturing* by Donald F. Eary and G. E. Johnson, Wiley, New York, 1968.

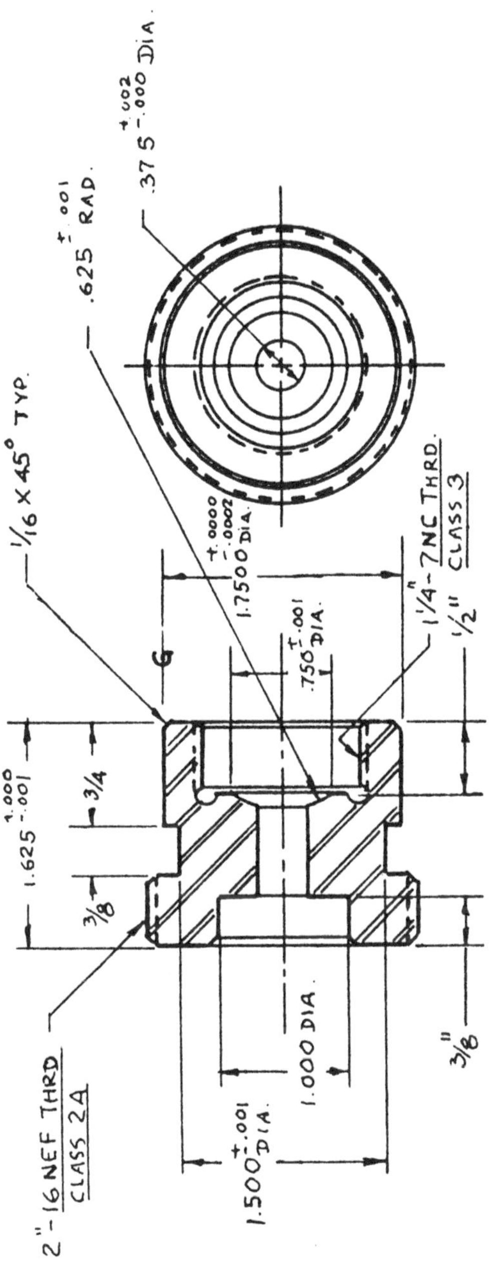

FIG. 8.2 Flange adapter.

PART NO. 10678-40 EFFECTIVE DATE 4-3-74 SHEET 1 of 1
PART NAME Flange Adapter PLANNER J. Watson

Dept. No.	Operation	Description	Machine Tool	Standard Time
68	20	Clean bars per Spec. NA 68-0756		
50	30	Inspect bars per QC-NA 12-0510		
77	40	Cut off to 1-3/4 length	Power Hack Saw	
3	50	Chuck O.D. (ref. 2"-16 thread)	Engine Lathe	
		Rough & finish O.D. face and O.D. groove (Note: leave grind stock)		______
	60	Drill, ream, counter bore, form radius, finish I.D. chamfer, and thread I.D.		______
3	70	Chuck O.D. (ref. 1.7500)	Engine Lathe	______
		Rough & finish O.D. & I.D. face end and thread		
5	80	Grind 1.7500 dia.	O.D. Cylindrical Grinder	______
8	90	Clean		______
68	100	Inspect		______

FIG. 8.3 Process routing.

casting, a means had to be developed to locate the casting in such a manner as to maintain the relationship between the centerline of the casting neck and the flanged mounting surface as it was being machined. Another important factor to be considered in planning this first operation was that the 7/16 finished dimension between the mounting surface and the spot-face on the three bosses had to be held to a tolerance of plus or minus 0.010 in. In performing this first operation, consideration had to be given to machining allowances in order to ensure that the three boses would finish up properly during the spotfacing operation.

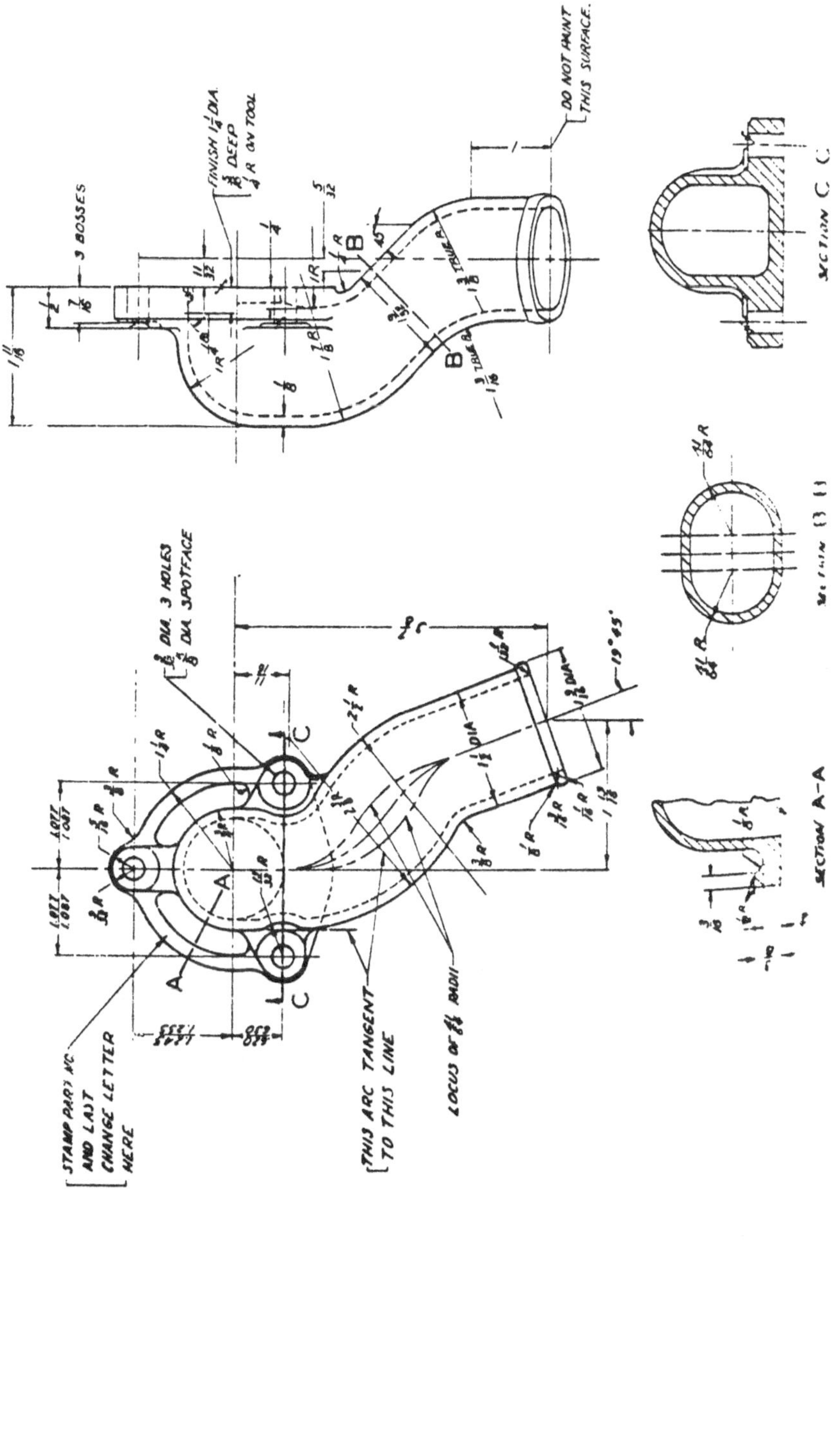

FIG. 8.4 Cylinder head coolant inlet. (Courtesy of General Motors Corporation, Flint, Mich.)

At least partial location of the workpiece in this first operation would have to take place on the bosses if adequate material allowance for spotfacing was to be ensured after completion of the first operation.

In order to accomplish this, locators were provided on two bosses, and a vee-type locator was placed on the side of the neck to control the position of its horizontal centerline. The positioning of the locators on the two bosses at section C–C was to provide the greatest geometric stability possible to the workpiece. This ensured the relationship between the centerline of the neck and the machined mounting surface. It also ensured sufficient stock allowance for machining the spotfaces on two of the three bosses. An adjustable support was placed under the third boss to provide mechanical stability during machining.

The second operation involved machining the passage opening. Examination of print dimensions indicated that the mounting holes were located from the centerlines of the passage opening, requiring that passage machining should be the next operation. In locating the piece part for this operation, consideration was given to:

1. Preserving the natural centerlines of the workpiece
2. Maintaining the specified wall thickness of the part
3. Blending the machined opening with the cored passage

The three initial points of location were established on the machined mounting surface. It was next necessary to fix the position of the machined opening in such a way as to maintain casting wall thickness. This can be done by placing locators on the outside wall itself. The locators were spaced at equal intervals around the outside wall to eliminate concentration of errors on one side of the casting. This not only reduced the total error, but minimized the effect of variation. A 1/4-in. radius was provided on the cutting tool to blend the machined opening with the cast passage. The end of the casting neck did not require critical location in this operation. It did require restraint, and for that reason one locator was placed against the side to control rotation about the axis of the holes during the machining of the piece.

The third operation was the drilling of the three mounting holes. For stability, the piece part was located on the machined surface of the flange. This was the only logical approach, as the drawing indicated that the three mounting holes were dimensioned from the centerlines of the machined passage opening for drilling inside this opening. To prevent rotational movement about the axis of the large hole, a locator was positioned to contact the neck of the casting at the extreme end from the mounting holes.

Normally, it is preferred practice to have drilling forces opposed by the locators rather than by the holding forces. In this case, just the opposite approach was selected because the shape of the casting would not allow sufficient space for properly positioning the drill bushings in the drill jig for the two side holes. Drilling from the machined surface caused the burrs to occur on the unfinished bosses, where they could be removed during the spotfacing operation. Also, drilling into the finished surface and breaking out through the cast surface of the bosses would force any surface inclusions from the cast surface out ahead of the drills. This results in longer drill life.

The fourth and final operation was spotfacing of the bosses. The locating points were identical to those of the third operation. Except that the cutting forces were exerted in the opposite direction. In the operations described above, combining tools was assumed. Drilling of the three mounting holes would be done simultaneously, as would the spotfacing of the three bosses. There is little problem in combining machining operations in this manner, since the location system would be the same for each individual hole as for all three. This kind of combining can reduce process time significantly and improve cost.

The key to good process planning is to develop an operation sequence in which the workpiece is handled as few times as possible, consistent with maintaining compatibility of the location system with the operation being performed. This calls for as many operations as possible being performed at one setup of the workpiece.

8.1.2 Assembly Processing

Assembly processing is another way of saying assembly methods. Assembly methods sheets, or work instructions, must clearly depict what is to be done, in what sequence, and with what tools and materials. Assembly methods sheets must minimize operator learning time. They must be economical to prepare, reproduce, distribute, and change. Assembly process planning must include an assembly process summary or process routing, detailed work instructions for each operation called out in the summary, an operations parts list for each operation, process sketches or visual aids, and a workplace layout for each operation. The work instructions should call out all tools necessary to perform the operation, and there should be a standard time on the process summary for each operation called out, broken down to the level of setup and run times.

In process planning for fabrication, whether for machining or forming, the skilled machinist or sheet metal mechanic could work to what amounts to an outline process routing supplemented by the engineering drawing of the piece part. As a skilled worker, he would set up his machine and perform the work with a minimum of written work instructions. Such is not the case with assembly operations. The work must be totally and carefully planned by the manufacturing engineer, and complete work instructions must be prepared. These are the two extremes. In most manufacturing plants today the process planning will fall somewhere in between.

If your plant is a high-volume producer of a single product line, then detailed assembly work instructions may be unnecessary. Once operators are trained to perform a short-cycle assembly operation, little else is needed except possibly some clear, concise visual aids showing the critical details of the operation in pictorial or exploded-view form. The manufacturing engineer must plan such production down to the most detailed level; he must prepare a layout of the assembly lines, show each and every workstation in plan view layout, plan the assembly tools required, write the complete description of the work performed at each station on the line, establish standard times, decide where visual aids would be needed and prepare them, and then "fine-tune" or balance the line, assist in training the operators, and finally shake down or debug the line.

All of the above documentation is necessary and required when assembly lines are initially established or set up, and to train the operators. Once the line is flowing smoothly and the operators have been trained, there would be less and less reliance on the written work instructions and even the visual aids. This initial planning documentation should always be available for ready reference and should be kept up to date by the manufacturing engineer.

If your company manufactures a variety of different product lines, in medium to high volume, and/or sets up and produces to a job order type of system, assembly process documentation that is complete and in the greatest level of detail is especially important. It is a proven fact that good assembly process planning and documentation significantly reduces operator learning time (especially important where the production run is relatively short). It also teaches correct methods to operators and thereby reduces costs of assembly labor. The assembly process documentation package is essential to the operation of an ongoing production control and timekeeping system. The assembly process routing provides the steps or sequences the material, parts, assemblies, and work in process must follow to build the product; it provides the time standards for each operation, and the assembly parts list for each operation provides the information needed by production control to pull and kit material for production.

In the small plants where production runs may be small to nonexistent, assembly process planning with minimum documentation is required and can be justified for the reasons mentioned earlier. Even where no formal production control system exists and the production supervisor draws his material from the stockroom in one batch issue for the entire job, pictorial visual aids, workstation layouts, a tool list, and an assembly process routing should be provided.

Determining Operation Sequences

In assembly process planning, operation sequences usually parallel the indented parts list or engineering tree chart, since that represents how the product goes together, or is assembled. This initial assembly process sequence plan would have an assembly operation for each major and minor subassembly and for the final assembly. It should be strongly emphasized that this is an initial breakdown, and normally would be followed by a more thorough analysis of the steps required to assemble the various subassemblies and the final assembly. This detailed analysis is normally done in the preproduction planning phase (see Chapter 3), and would be in the form of an operation process chart as shown in Fig. 8.5.

Earlier, it was stated that assembly processes included soldering, wiring, press fitting, brazing, shrink fitting, welding, adhesive bonding, riveting, and mechanical fastening. Within each of these assembly processes there is a series of sequences required to accomplish the process without regard to the product configuration, material, quantity to be produced, or rate of production. Examples are the steps in creating a circuit card assembly, in assembling a wire harness, in assembling the frame of a truck, or in the installation of fitting on a sailboat. The detailed instructions for the sequence would spell out the differences peculiar to the product at hand. Often in assembly work, standard sequences or operations are possible for any product where these processes of assembly are used. The result is a considerable saving in manufacturing engineering time and the elapsed time required to prepare and release an assembly process plan to the shop. Such standard processes enable preprinted process planning documentation, which may require only a part number and quantity to be entered before it is ready for release. A good example of this would be in the manufacture of cables for electrical or electronics equipment, where there are preprinted diagrams of the various electrical connector pin configurations, requiring only that the manufacturing engineer sketch in the wires terminated to the pins for the specific cable application.

Routings, Work Instructions, Visual Aids

Assembly process routings, such as the one shown in Fig. 8.6, list the operations in the sequence in which they must occur to assemble the item or product called out in the heading. In addition, they list the standard times for each operation, the performing department, and the latest revision level of the process instruction sheets currently in effect. The issue or revision level of the process information establishes configuration control of the product on the shop floor, since normally the assembly department does not work to engineering drawings. The importance of this cannot be emphasized too strongly. It is the responsibility of the manufacturing engineer to always have the latest revisions to the engineering drawings incorporated in the assembly process documentation, especially when the job is active on the production floor. In many companies the inspection department will use the

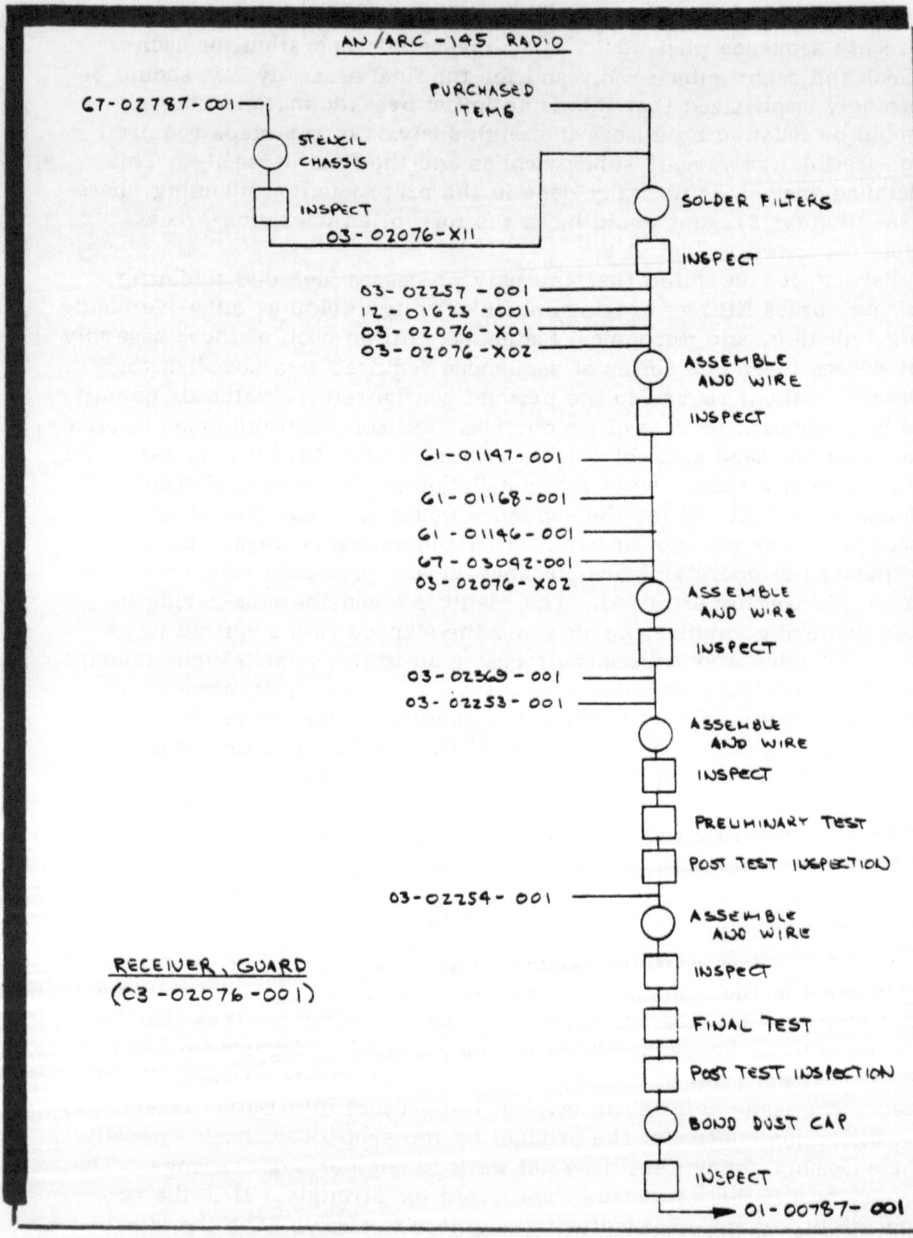

FIG. 8.5 Operation process chart for radio guard receiver. (Reprinted with the permission of *Industrial Engineering* Magazine, Vol. 2, No. 10, October, 1970. Copyright © Institute of Industrial Engineers.)

ASSY. PROCESS SUMMARY

Project	Proc. Engr.	Page	of
ARC 145	R. AREND	1	1

Part No.	Reason for Change:	Sum. Iss.	Dwg. Rev.	P/L Rev.	Date
03-02112-X01		8	E	E	12-4-84

Part Name	Wire Run List	Next Higher Assy.
WIRING HARNESS	05-05468-000	03-02112-001

Oper.	Description	Std/Hrs	Dept	Process Sheet Issue 1	2	3	4	5	6	7	8	9	10
005	C/L WIRE								ZEROX MASTER				
010	TIN		731	3	4	4	3						
015	HOT STAMP												
020	TWIST PAIR		731	1									
025	INSPECT (SAMPLE		741										
030	C/L SLEEVING		731	3									
035	SERVICE SHIELDS, PRELIM.		731	3	2	2							
040	TAG SHLD WIRES		731	3	2	2							
045	INSPECT		741										
050	ASSEMBLE BUSS J7		731	1									
055	INSPECT		741										
060	ASSEMBLE BUSS J18		731	1									
065	INSPECT.		741										
070	ASSEMBLE BUSS J39		731	1									
075	INSPECT		741										
080	PREWIRE J40		731	2	2								
085	INSPECT		741										
090	FORM CABLE		731	3	2	3	4	4	4	2	2	4	3
				3	3	3	3	2	3	3	4	2	2
				2	1	2	2	1	1				
095	INSPECT		741										

FIG. 8.6 Assembly process summary for wire harness. (Reprinted with the permission of *Industrial Engineering* Magazine, Vol. 2, No. 10, October, 1970. Copyright © Institute of Industrial Engineers.)

process documentation to perform in-process inspections of the product; this is especially true where detailed process work instructions are used and the process documentation is used for shop configuration control.

As indicated earlier, the assembly process summary or routing is used by production control to move material or work in process to the next operation or sequence. A copy of the assembly process summary travels with each batch of parts and material, and in effect becomes a routing sheet or shop traveler. For this to happen, the sequences must be stamped off, either by inspection or by the operator as they are completed on a batch of assemblies. If line production is involved, such a routing or traveler is unnecessary, as the progression on the assembly line is the routing followed by the assemblies. The process routing is an especially valuable tool in the job shop, where the shop is laid out by function or process and does not follow the product flow.

Assembly work instructions are the heart of the manufacturing engineering documentation package, and explain how the product is to be assembled in production. The assembly work instructions should be available at the operator workstation, preferably on an easy-to-see holder mounted on the workbench. The assembly process routing would list all of the operations for assembly in the sequence in which they must be performed; the assembly work instructions or assembly methods sheets for each listed operation would explain in detail exactly how to perform the operation. Figure 8.7 is an example of an assembly methods sheet for the wiring of a connector which becomes part of a wiring harness for a marine shortwave radio receiver.

An assembly operation parts list should be included with the assembly work instructions or assembly methods sheets for each operation. This tells production control and the operator what parts and materials are required to perform the assembly operation for one unit. Figure 8.8 shows one version of such a parts list. In addition, there must be a list of standard and special design assembly tools needed to perform the operation. In this case, it is included on the operation parts list. It can also appear as a separate callout on the work instruction or methods sheets.

An extremely important part of the assembly process documentation is visual aids. Visual aids can be anything from an actual mock-up of the product, to a black-and-white or color photograph, to a three-dimensional isometric or exploted-view drawing, to a simple sketch, or a tracing lifted directly from the engineering drawing. Figure 8.9 shows a light table being used by a manufacturing engineer to trace parts of the engineering drawing for making up a visual aid. If regular office copy machines are used, visual aids can be constructed by cut and paste methods. Figure 8.10 shows an inexpensive example prepared in this way. It should be kept in mind that all process documentation

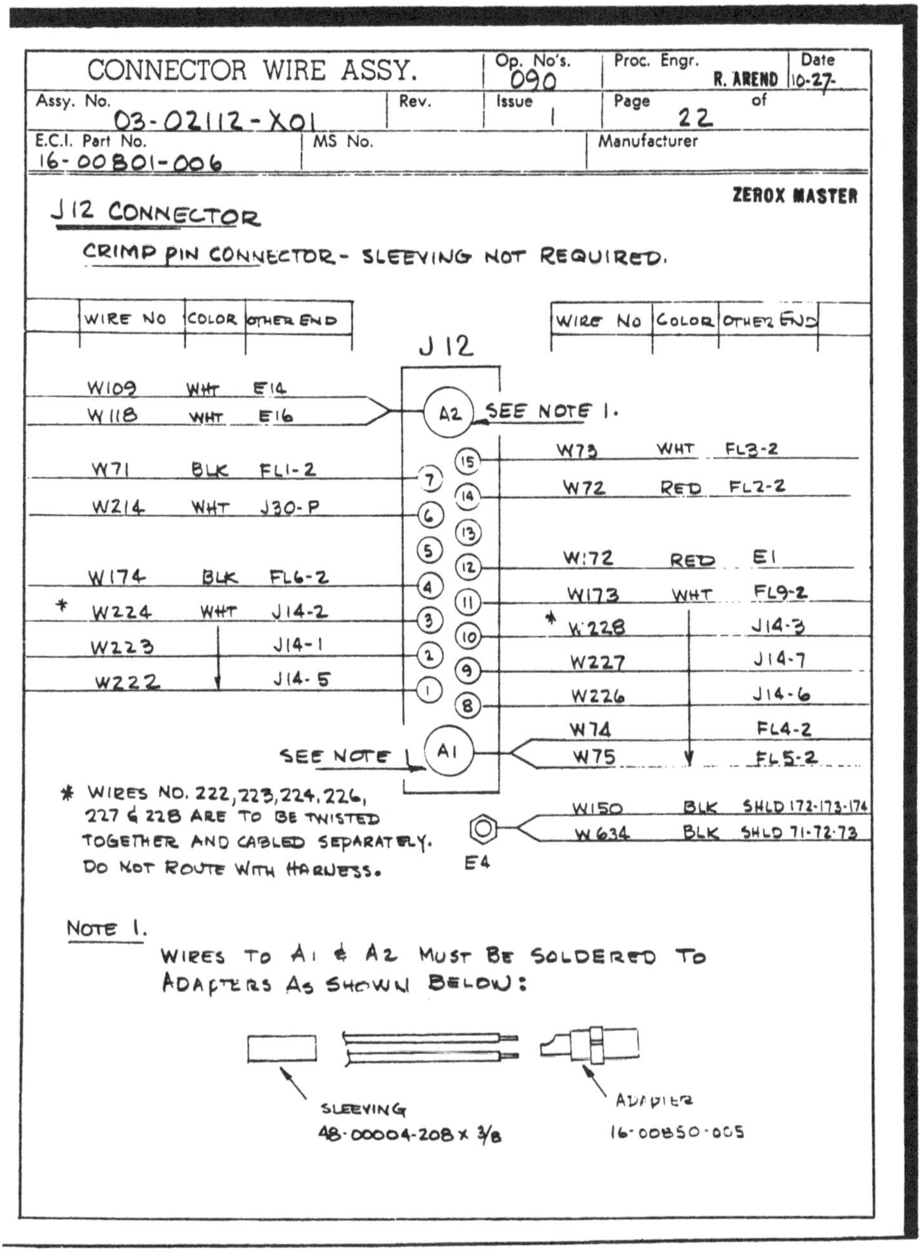

FIG. 8.7 Assembly methods sheet for wiring a connector. (Reprinted with the permission of *Industrial Engineering* Magazine, Vol. 2, No. 10, October, 1970. Copyright © Institute of Industrial Engineers.)

ASSY. OPER. PARTS LIST		Op. No's. 005	Operation Description ASSEMBLE AND WIRE		
Assy. No. 03-02076-001	Rev.	Next Assy./Oper. 015	Proc. Engr. J.WOOD		Date 7-28
Assy. Name RECEIVER, GUARD			Issue 1	Page 1	of 5

Seq.	E.C.I. Part Number	Description	Remarks	Qty.
	TOOLS:	ASSEMBLY VISE		
		W SOLDER IRON – TIP		
		INSERTION TOOL	IT-1-16-00852-004	
1	OP002	CHASSIS ASSY	POSITION IN VISE	1
2	16-00852-004	JACK (J1, J2, J3)	INSERT JACKS INTO	3
			CHASSIS USING TOOL	
			IT-1-16-00852-004	
			REF. SKETCH "A"	
3	03-02257-001	AMPL, RF (A3)	APPLY LOCTITE TO	1
4	23-01442-202	SCREW	SCREW THREADS,	2
	22-00065-005	LOCTITE, BRN	POSITION AND ASSEMBLE	A/R
			A3 TO CHASSIS	
			REF. SKETCH "A"	
5	31-00008-030	BUSSWIRE (1")	CONNECT LEADS	1
6	10-00275-044	RESISTOR (R1)	TO TERMINALS	1
7	11-00392-001	CAPACITOR (C1)	REF. SKETCH "B"	1
8	09-00415-011	COIL (L1)		1

FIG. 8.8 Assembly operation parts list. (Reprinted with the permission of *Industrial Engineering* Magazine, Vol. 2, No. 10, October, 1970. Copyright © Institute of Industrial Engineers.)

FIG. 8.9 Manufacturing engineer uses light table to make tracing for visual aid.

costs money, and consideration should be given to length of production run, anticipated changes during production, and what is really needed to instruct assembly operators. One other very important consideration is that a visual aid supplemented by minimal notes and instructions is far superior to lengthy written work instructions. In many companies, as explained earlier, visual aids that highlight key assembly details are all that are used.

Workstation and Line Layouts

Workstation layouts are very important from the standpoint of assembly operator methods. They tell the assembly supervisor how to set up and configure the individual workstations for optimum productivity and flow of work. Workstation layouts are usually in the form of a plan view of the workstation, and show where to place tooling and fixtures, parts bins, work instructions, tote pans for staging completed work and for placing incoming work and any other information pertinent to the operation and setup of the work station.

Line layouts are used where progressive assembly lines are to be used to build the product, and the plant layout drawings do not show which workstation goes where. Again, these would be used by supervision in setting up the line to conform to the assembly process flow, and to ensure optimum methods and work flow.

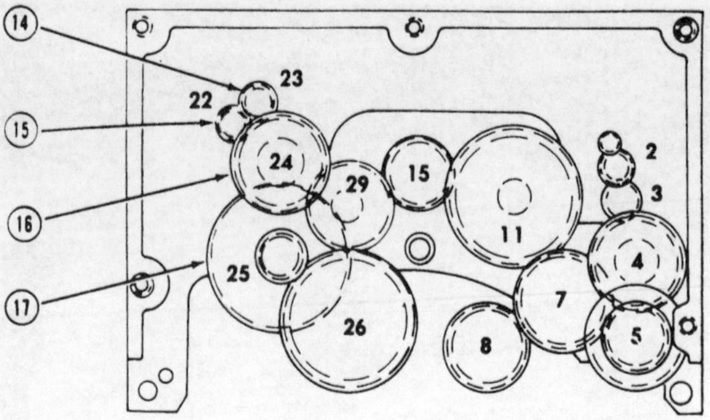

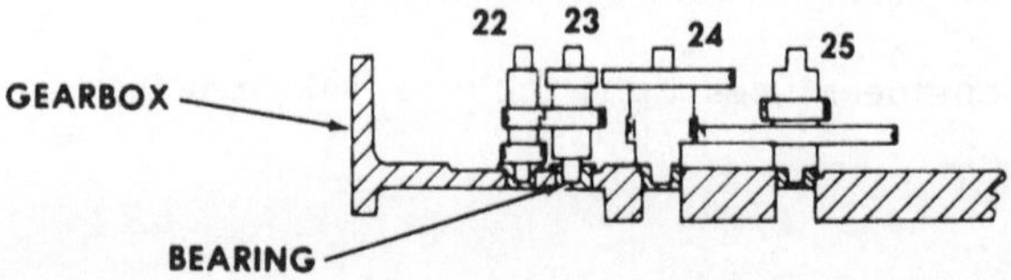

SEQ	PART NUMBER	DESCRIPTION	QTY	REQUIRED
14	39-00286-001	GEAR 23	1	POSITION TO HOUSING PER SKETCH
15	39-00287-001	GEAR 22	1	POSITION TO HOUSING PER SKETCH
16	39-00276-001	GEAR 24	1	POSITION TO HOUSING PER SKETCH
17	39-00275-001	GEAR 25	1	POSITION TO HOUSING PER SKETCH
18	—	—	—	SET HOUSING ASIDE AND POSITION COVER TO WORK AREA

FIG. 8.10 Inexpensive visual aid prepared by cut and paste methods.

8.2 MANUFACTURING METHODS ANALYSIS

In developing the manufacturing process, whether for fabrication, machining, forming, finishing, or assembly, the manufacturing engineer must specify the most economical methods for the job and for the work at hand. In order to do this he must understand and be able to apply the fundamental techniques of methods analysis, motion economy, and work simplification. The presumption of good methods, and the ability of the manufacturing engineer in methods analysis, is so basic that not to be proficient in this art is tantamount to being incompetent as a manufacturing engineer.

Even the best and most thorough process planning will sometimes overlook details, or specify methods that can be improved upon later, after the product is in production. Part of the job of the manufacturing engineer is to be alert for these opportunities to improve the process and the flow.

The purpose of this part of this chapter is to provide the basic information needed by the manufacturing engineer to gain a degree of proficiency in manufacturing methods analysis.

8.2.1 Work Simplification

Work simplification can be defined as the organized application of common sense to find easier and simpler ways of doing work. Work simplification provides a systematic, commonsense approach to making work easier and at the same time lowering costs. The basic premise of work simplification is that there is always a better way to do any task. The work simplification pattern includes five basic steps:

1. Selection of a job to be improved
2. Recording of the job details
3. Analysis of the job details
4. Development of improvements
5. Installation of the improvements

Selecting the job to be improved requires careful consideration and study. Efforts expended for improvement should be made first where the returns will be the greatest. Priority should be given to bottlenecks, choke points, trouble spots, jobs that require excessive or inordinate amounts of time, or generally where unsatisfactory conditions exist. The following list provides assistance in making this selection:

1. Greatest cost: Work that involves the greatest expenditure of funds, man-hours, and use of equipment
2. Greatest work load: The largest volume of work being performed by the activity

3. Number of persons assigned: Work that requires large numbers of people to perform similar tasks
4. Walking: Jobs that require a great deal of walking around
5. Bottlenecks: Work not flowing smoothly
6. Schedules not met: Failure to meet deadlines resulting in work backlogs or overtime
7. Excessive fatigue: Work that requires great physical effort or is being done with frequent rest periods
9. Unsafe and unpleasant work practices: Work that results in numerous accidents, or is undesirable because of extreme conditions such as dust, noise, fumes, vapor, or extremes in temperature

Every job is made up of three parts:

1. Make ready. This is the time and effort put into the setup, or getting ready to work.
2. Do. This is the actual work accomplished.
3. Put away. This is the time and effort put into cleaning up after the "do" part of the job.

For example, if we have a carpenter making a wooden box, we might expect the job to break down as follows:

1. Make ready. Open bin, pick up nails, close bin, pick up hammer, set aside nails at the bench.
2. Do. Hammer nails.
3. Put away. Set aside hammer, pick up box, set aside box to storage.

Anything that reduces the time required for the make-ready and put-away parts of the job reduces the nonproductive time associated with the job.

Recording of the job details can best be accomplished through the use of process chart techniques, specifically the flow process chart and the flow diagram. The flow process chart is a device for recording each step of a job in a compact manner, as a means of better understanding it and improving it. The chart represents graphically the separate steps or events that occur during the performance of work, or during a series of actions. The process chart may be used to record the flow within a unit, a section, a department, or between departments. The flow process chart has no bounds.

No matter how complicated or intricate the series of operations may be, a flow process chart can be constructed if you take one step at a time. The flow process chart, however, like other methods of graphic representation, may need to be modified to meet the requirements of a particular situation. For example, it may show in sequence the total activity of a production operator, or it may show in sequence the steps that the worker, part, or material goes through. The chart should be either the operator type or the material type, and the two types should not be combined.

Careful study and analysis of such a chart, giving a graphic picture of every step in the process, is almost certain to suggest improvements. It is not uncommon to find that some operations or a part of an operation can be eliminated, that one operation can be combined with another, that better routes for the parts can be found, more economical machines used, delays between operations eliminated, and other improvements made, all of which go to produce a better product at a lower cost.

To make a flow process chart requires careful adherence to the rules outlined below:

1. State the activity being studied. Make certain you are really naming the activity you have chosen to study.
2. Choose the subject to follow. Decide on the person or the material and follow him or it through the entire process. When you pick the subject stick with it.
3. Pick starting and ending points. This is to make certain you will cover all the steps you wish to cover, but no more or less.
4. Write a brief description of each detail, step by step, no matter how short or temporary.
5. Apply the symbols. The description determines each symbol. Draw a connection line between the proper symbols.
6. Black in the "do" operation. Shade in the symbols for the operations that you decide are the do operations. This will help you later in your analysis, when you begin challenging.
7. Enter distances. Whenever there is a transportation, enter the distance traveled.
8. Enter time, if required. Often this is not necessary. However, if it will help, note the time required or elapsed.
9. Summarize. Add up all of the facts and put them in the summary block. The summary should indicate the total number of operations, transportations, inspections, delays, storages, and distance that is traveled.

Figure 8.11 shows a flow process chart that is completely filled in, with circled numbers used to illustrate each of the rules listed above.

The chart symbols and their meanings are as follows:

> Operation: An operation occurs when an object is intentionally changed in any of its physical or chemical characteristics, is assembled or disassembled from another object, or is arranged or prepared for another operation, transportation, inspection, or storage. An operation also occurs when information is given or received, or when planning or calculating takes place.

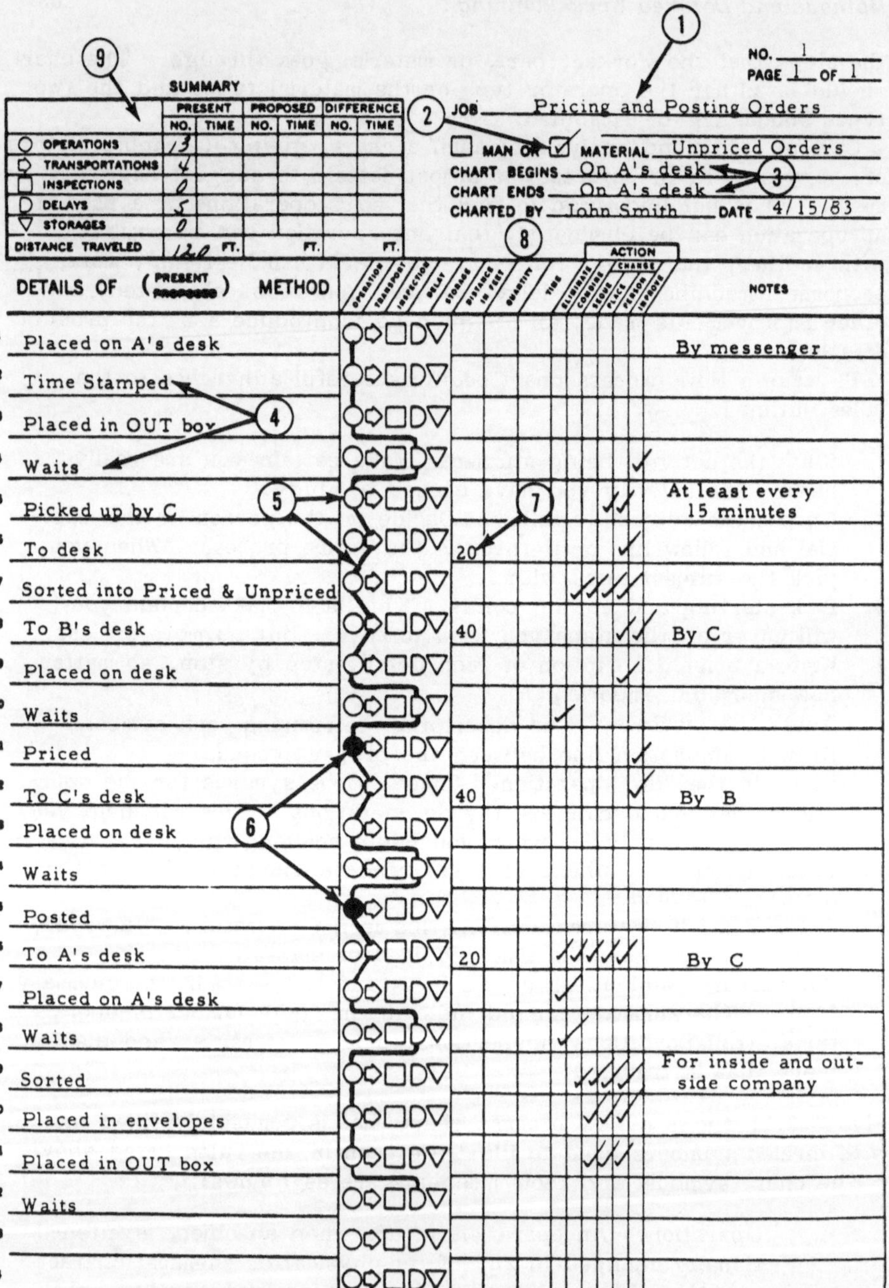

FIG. 8.11 Flow process chart example.

Transportation: A transportation occurs when a person moves from one workplace to another or when an object is moved, except when such movements are part of the operation or are caused by the operator at the workstation during an operation or an inspection.

Inspection: An inspection occurs when an object is examined for identification or is verified for quality or quantity, or any of its characteristics.

Delay: A delay occurs to an object when conditions, except those that intentionally change the physical or chemical characteristics of the object, do not permit or require immediate performance of the next planned action.

Storage: A storage accurs when an object is kept and protected against unauthorized removal.

It is sometimes helpful to supplement a flow process chart with a flow diagram. A flow diagram is simply a layout of the area involved in the job being studied, over which you indicate by a line in the path of the object or person followed in the flow process chart. It is often desirable to indicate the action taking place by using the same symbols used on the chart. They may, if desired, by keyed to each other by item numbers.

The third step of the work simplification pattern involves questioning every part, aspect, or detail of the job. Here we examine each operation and ask some very pointed questions. Do not be satisfied until you have asked all possible questions and received the related answers. The first thing to do is question the entire job being studied. Why is it done? Is the job really necessary? If the answer is that it is, then question each "do" operation. If you can eliminate a do operation, you also eliminate the make-ready and put-away that go with it. The do operations are those that add value to the product or process being studied.

There is always a better way, and our task is to find it:

Why? The overriding question, it establishes the reason for the job. The answer defines and justifies the purpose of the job.

What? What is done? What are the steps? What does each step do? What makes the step necessary?

Where? Where should this step be done? Can it be done more easily, with less time, energy, or transportation, by changing the location of the employees or equipment?

When? When should this step be done? Is it done in the right sequence? Can the job be simplified by moving this step ahead or back?

Who? Who should do the job? Is the right person handling it? Would it be logical to give the job to someone else?

How? How is the job being done? Can it be made easier? Can it be done better with different equipment or a different layout?

This question-and-answer approach will suggest improved methods:

Why, What } leads to elimination

Where, When, Who } lead to combine, change sequence

How } leads to simplicity

Careful consideration of the possibilities presented in looking for ways to eliminate, combine, change sequence, and simplify in the questioning approach brings us finally to the better method and provides us with the answer to "How should the job be done?" The simplest way is the best way. Figure 8.12 shows the proposed improvement flow process chart for the pricing and posting of orders shown in Fig. 8.11.

8.2.2 Principles of Motion Economy

It is the purpose of the following discussion to explore the rules or principles of motion economy which have been and are now being used successfully in manufacturing methods studies. These principles form a basis or code, or body of rules which, if applied correctly, will make it possible to greatly increase the output of manual factory labor with a minimum of fatigue. These principles will be examined under the subdivisions of operator tasks, the workplace, and tools and equipment.

Operator Tasks

The principles of motion economy as related to the tasks of the operator are as follows:

1. The two hands should begin as well as complete their motions at the same time.
2. The two hands should not be idle at the same time except during rest periods.
3. Motions of the arms should be made in opposite and symmetrical directions and should be made simultaneously.

These three principles are closely related and should be considered together. It seems natural for most people to work productively with one hand while holding the object being worked on with the other hand. This is extremely undesirable and should be avoided. The

NO. 2
PAGE 1 OF 1

SUMMARY

	PRESENT NO.	PRESENT TIME	PROPOSED NO.	PROPOSED TIME	DIFFERENCE NO.	DIFFERENCE TIME
○ OPERATIONS	13		10		3	
⇨ TRANSPORTATIONS	4		2		2	
□ INSPECTIONS	0		0		0	
D DELAYS	5		3		2	
▽ STORAGES	0		0		0	
DISTANCE TRAVELED	120 FT.		42 FT.		78 FT.	

JOB Pricing and Posting Orders

☐ MAN OR ☑ MATERIAL Unpriced Orders

CHART BEGINS On A's desk

CHART ENDS On C's desk

CHARTED BY John Smith DATE 4/15/83

DETAILS OF (~~PRESENT~~ PROPOSED) METHOD	OPERATION / TRANSPORT / INSPECTION / DELAY / STORAGE	DISTANCE IN FEET	QUANTITY	TIME	ACTION: ELIMINATE / COMBINE / CHANGE: SEQUE / PLACE / PERSON / IMPROVE	NOTES
Placed on A's desk	○⇨□D▽					
Time Stamped and Sorted	○⇨□D▽					
Placed in OUT baskets	○⇨□D▽					2 wire baskets
Waits	○⇨□D▽					Shorter wait
Picked up by B	○⇨□D▽					
To desk	○⇨□D▽	14				
Placed on desk	○⇨□D▽					
Priced	●⇨□D▽					New rotary price file
To C's desk	○⇨□D▽	28				By B
Placed on desk	○⇨□D▽					
Waits	○⇨□D▽					
Posted and Sorted	●⇨□D▽					
Placed in Envelopes	○⇨□D▽					
Placed in OUT basket	○⇨□D▽					
Waits	○⇨□D▽					Picked up by messenger

FIG. 8.12 Flow process chart showing improved method plan.

two hands should work together, each beginning a motion and completing a motion at the same time. Motions of the two hands should be simultaneous and symmetrical.

In many kinds of work more can be accomplished by using both hands than by using one hand. For most manufacturing assembly operations, it is advantageous to arrange similar work on the left- and right-hand sides of the workplace, thus enabling the left and right hands to move together, each performing the same motions. The symmetrical movements of the arms tend to balance each other, reducing the shock and jar on the body and enabling the operator to perform his task with less mental and physical effort. There is apparently less body strain when the hands move symmetrically than when they make nonsymmetrical motions, because of balance.

The fourth principle of motion economy states that hand and body motions should be confined to the lowest classification with which it is possible to perform the work satisfactorily.

The five general classes of hand motions show that material and tools should be located as close as possible to the point of usage. The motions of the hands should be as short as the work permits. In the listing of classifications shown below, the one requiring the least amount of time and effort is shown first:

1. Finger motions
2. Motions involving fingers and wrist
3. Motions involving fingers, wrist, and the forearm
4. Motions involving fingers, wrist, forearm, and the upper arm
5. Motions involving fingers, wrist, forearm, upper arm, and shoulder (cause posture change)

It should be pointed out that finger motions have been found to be less accurate, slower, and more fatiguing than motions of the forearm. Evidence seems to indicate that the forearm is the most desirable member for performing light work. In highly repetitive work, motions about the wrist and elbow are superior to those of the fingers or shoulders.

The fifth principle of motion economy states that momentum should be employed to assist the worker wherever possible, and it should be reduced to a minimum if it must be overcome by muscular effort. The momentum of an object is defined as its mass times its velocity. In the factory environment, the total weight moved by the operator may consist of the weight of the material moved, the weight of the tools moved, and the weight of the part of the body moved. It should be possible to employ momentum to advantage when a forcible blow or stroke is required. The motions of the worker should be so arranged that the blow is delivered when it reaches its greatest momentum.

The sixth principle of motion economy states that smooth, continuous, curved motions of the hands are preferable to straight-line motions involving sudden and sharp changes in direction. Abrupt changes in direction are not only time-consuming but also fatiguing to the operator.

The seventh principle of motion economy states that ballistic motions are faster, easier, and more accurate than restricted or controlled movements. Ballistic movements are fast, easy motions caused by a single contraction of a positive muscle group with no antagonistic muscle group contracting to oppose it. A ballistic stroke may be terminated by the contraction of opposing muscles, by an obstacle, or by dissipation of the momentum of the movement, as in swinging a sledgehammer. Ballistic movements are preferable to restricted or controlled movements, and should be used whenever possible.

The eighth principle of motion economy states that work should be arranged to permit an easy and natural rhythm wherever possible. Rhythm is essential to the smooth and automatic performance of any operation.

Rhythm, as a regular sequence of uniform motions, aids the operator in performing work. A uniform, easy, and even rate of work is aided by proper arrangement of the workplace, tools, and materials. Proper motion sequences help the operator to establish a rhythm that helps make the work a series of automatic motions, where the work is performed without mental effort.

The ninth principle of motion economy states that eye fixations should be as few and close together as possible. Where visual perception is required, it is desirable to arrange the tasks so that the eyes can direct the work effectively.

Workplace

The first principle of motion economy related to the workplace states that there should be a definite and fixed place for all tools and materials. The operator should always have tools and materials in the same location, and finished parts and assembled units should be placed in fixed positions or locations. For example, in the assembly of mechanical hardware, the hand should move without mental direction to the bin containing flat washers, then to the bin containing lock washers, then to the bin containing bolts, and finally to the hex nuts. There should be no thinking on the part of the operator required to do any of this.

The second principle of motion economy related to the workplace states that tools, materials, and controls should be located close to the point of use. In the horizontal plane, there is a definite and somewhat limited area which the worker can use with a normal expenditure of effort. This includes a normal working area for the right hand, one for the left hand, one for each working separately, and another for both hands working together. Figure 8.13 shows this and the dimensions of normal and maximum working areas in the horizontal and vertical planes. Both the standing and sitting positions are included. It also shows normal bench work surface heights, which can have a significant adverse effect if not correct. Figure 8.14 shows in greater detail the areas of easiest reach for the left and right hands, for both hands working together, and the area in which small objects can be most easily picked up.

The third principle of motion economy related to the workplace states that gravity feed bins and containers should be used to deliver the material close to the point of use. This can sometimes be accomplished by using parts bins with sloping bottoms which feed parts by gravity to the front of the bin, eliminating the need for the assembly operator to dip down into the bin to grasp parts.

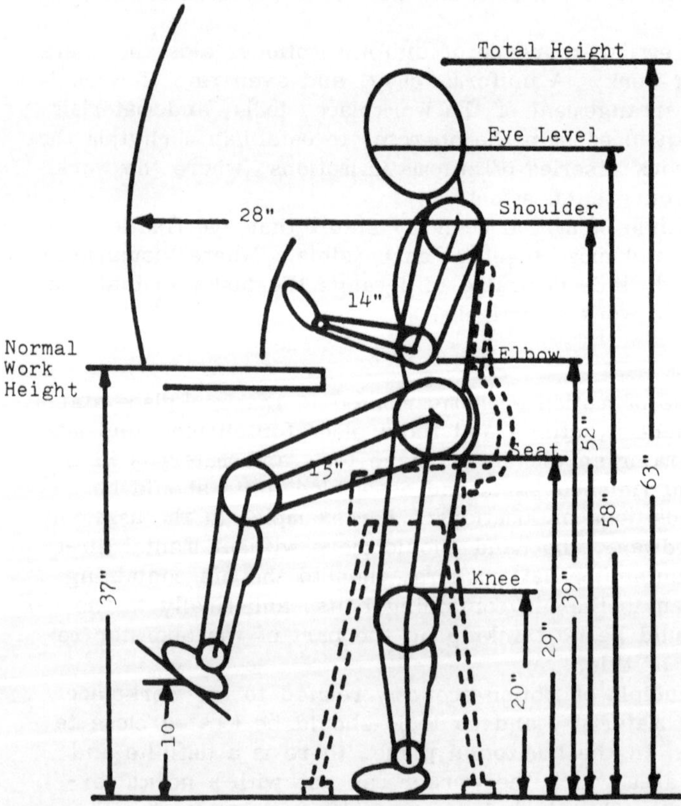

FIG. 8.13 Normal and maximum working area and heights.

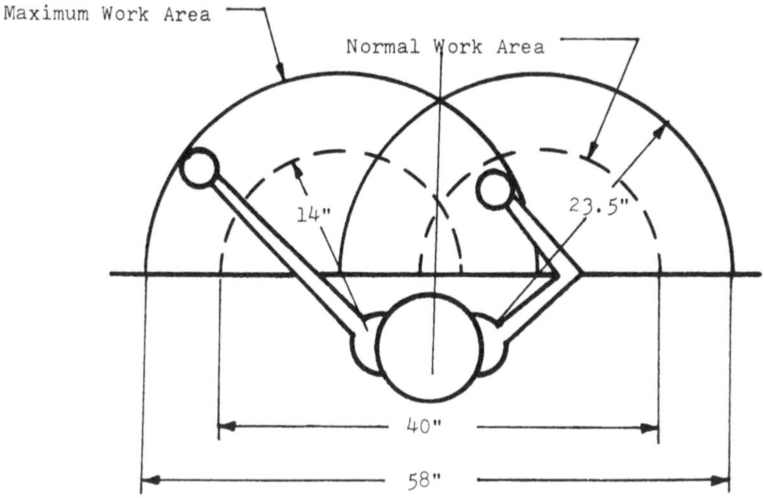

FIG. 8.14 Areas of easiest reach for each hand and both hands working together.

The fourth principle of motion economy related to the workplace states that drop deliveries should be used wherever possible. This requires configuring the workplace, for example, so that finished units may be disposed of by releasing them in the position in which they are completed, delivering them to their next destination by gravity. Besides the saving in time, this frees the two hands so that they may begin the next cycle simultaneously without breaking the rhythm.

Other principles of motion economy related to the workplace include:

5. Materials and tools should be located to permit the best sequence of motions.
6. Provisions should be made for adequate lighting.
7. The height of the workplace and chair should be arranged so that alternate sitting and standing at work are easily possible.
8. A chair of the type and height to permit good posture should be provided for the operator.

Tools and Equipment

Principles of motion economy related to the design of tools and equipment include the following:

1. The hands should be relieved of all work that can be done more effectively by a jig, fixture, or foot-operated device.
2. Two or more tools should be combined wherever possible.
3. Tools and materials should be prepositioned whenever possible.
4. Where each finger performs some specific movement, the load should be distributed in accordance with the inherent finger capacities (arrangement of typewriter keys).
5. Levers, crossbars, and hand wheels should be located in such positions that the operator can operate them with the least change in body position and with the greatest mechanical advantage.

An excellent example of a production equipment item that utilizes the principles of motion economy is the hand arbor press. Figure 8.15 shows a two-ton bench arbor press. A special feature of the hand arbor press is its ability to exert a tremendous force. This is achieved through a combination of the mechanical advantage of a long lever, and the gear ratio of the pinion drive. The force derived from an arbor press is not an impact force, but a variable, vertical force under the complete control of the operator. Thus the surface of the part or the tool is not subject to shock or damage.

The table of an arbor press is machined square to the vertical line of ram. This feature ensures the proper working angle when tooling or when parts are put into the press. This and other features, make an arbor press an invaluable piece of general purpose shop equipment.

8.2.3 Workmanship Standards

Workmanship standards tell the operator what is good, acceptable work or practice, and what is not. These standards can be line drawings supplemented with narrative text, color photographs of acceptable and unacceptable work supplemented by narrative text, or actual models of prototype units, known as standards. Figure 8.16 shows an example of a line drawing workmanship standard for taping the core of an iron core transformer in the magnetic components industry. It should be noted that instructions are superimposed where needed, indicating the important points to watch for when tape-wrapping the core. Also, it should be noted that this workmanship standard tells only the requirements for making a good and acceptable tape wrap, and that it is necessary to refer to a product specification to determine tape material and number of layers required.

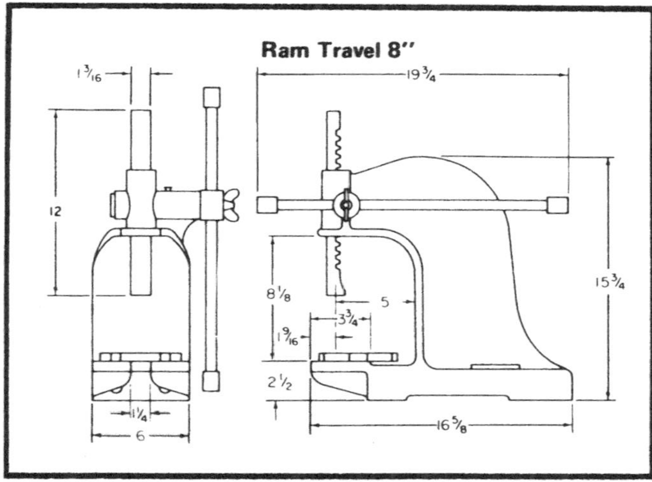

PRESSURE (Tons)	#2	Lever Type 2
	#2-12PW	With 12" Pilot Wheel(2) 1½
LEVERAGE		Lever Type .... 35:1 With 12" Pilot Wheel 23:1
PLATE . . 5$\frac{1}{16}$" Dia. x ½" Thick . . Slots ⅝, ½, ¾, 1 and 1⅛"		
SHIPPING WEIGHT (Lbs.)		Lever Type .... 60 With 12" Pilot Wheel 91
FLOOR STAND .... 32" Height .... No. 2		
FLOOR STAND SHIPPING WEIGHT (Lbs.) 85		

FIG. 8.15 Two ton hand arbor press basic specifications (Courtesy of Greenerd Press & Machine Company, Inc., Nashua, N.H.)

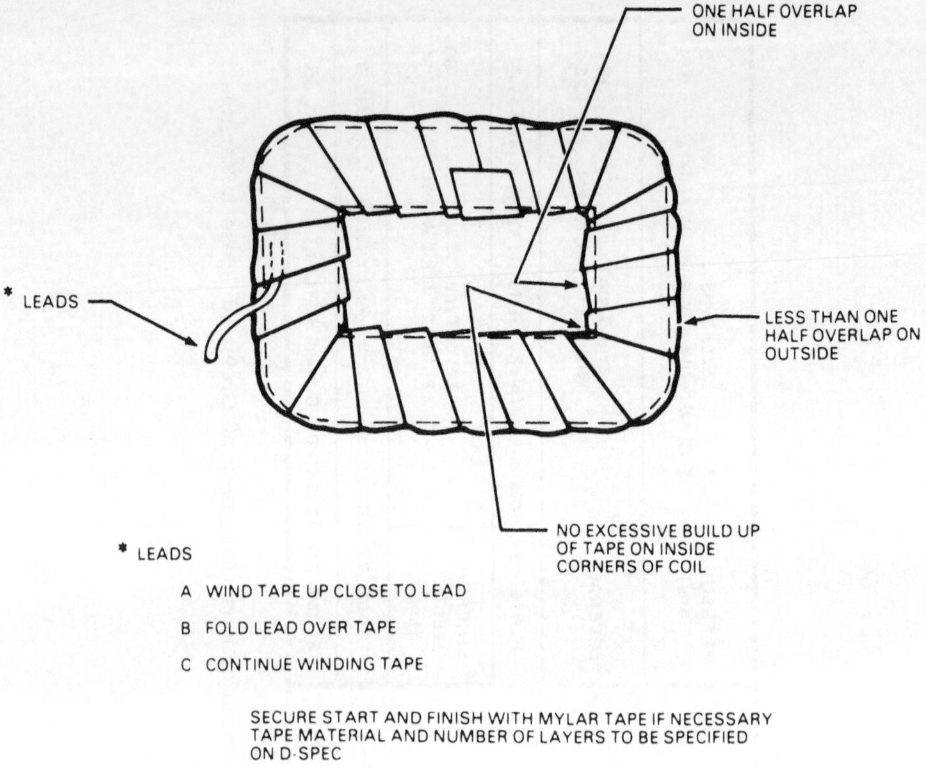

FIG. 8.16 Workmanship standard for taping iron core transformers.

8.3 SPECIAL MANUFACTURING INSTRUCTIONS

Frequently it becomes necessary for the manufacturing engineer to look beyond the operations in his own plant, and to make certain that material and parts received from outside suppliers come into the plant ready to become part of the product being produced. In the majority of cases, if the vendor meets the requirements of the purchase order and the engineering drawings (if applicable), the parts or material will be ready for processing when they arrive at the receiving dock. In some instances, however, it may be required that the vendor only partially complete a part or subassembly, or that he produce a certain part on the low side of the drawing tolerance if the parts are to fit together in assembly.

To ensure that this information is correctly transmitted to the vendor, the manufacturing engineer may prepare special manufacturing instructions which are called out on the purchase order along with engineering drawings and specifications. The special manufacturing instructions along with the drawings and specifications would then become the acceptance criteria for the parts or material when it goes through receiving inspection.

Special manufacturing instructions would be prepared by the manufacturing engineer at the time the make or buy determination is made and before the detailed internal work instruction documentation is prepared. These instructions can be in the form of simple handwritten information given to the buyer to incorporate directly into the purchase order, or a formal document which becomes an attachment to the purchase order. The important point to remember is that the manufacturing engineer needs to review the items to be purchased, and make certain that if these items are made exactly according to drawings and specifications, they can enter directly into production without any problems or unplanned processing steps.

QUESTIONS

1. What is the first step in determining the manufacturing process for a fabricated part and why is this step so important?
2. In planning the manufacturing process for a fabricated part, what role does the material specification play?
3. What are the critical areas for a machined part in determining its sequence of manufacture?
4. In piecepart machining, what would be considered as qualifying and requalifying operations? Give some examples.
5. What is the significance of part geometry in planning fabrication processes?
6. The degree of control that can be maintained during the process cycle for a machined part, and the logical process order are largely determined by what three basic considerations?
7. What are the most important restrictions imposed by the process itself in fabricated part process planning?
8. In fabricated piecepart process planning, what is the single most important consideration?
9. Why is it important to perform as many operations as possible on a single setup of the workpiece when planning the process for a machined part?
10. An initial assembly process plan should, as a minimum, contain what operations or sequences?
11. Why should an assembly operation parts list be included with the operation work instructions for each assembly operation or sequence in the product assembly process plan?
12. Why are visual aids so important in preparing the assembly process plan?

13. What are the three parts of every job whether in manufacturing, in assembly, in inspection, in test, or anywhere?
14. Why is it so important for the manufacturing engineer to know and understand the principles of motion economy?
15. What is the first and most important principle of motion economy related to the workplace?
16. What is the first and most important principle of motion economy related to tools and equipment?
17. What is the primary advantage offered by a hand-operated arbor press, and how is it achieved?

9

Mechanization and Automatic Assembly

9.1 MECHANIZATION

Mechanization is a first step toward automatic assembly and ultimately automation. Mechanization is moving-belt assembly lines, power assist tools, powered material handling, and the movement of parts, subassemblies, and other work in process by such devices as wire-guided vehicle systems, powered and free overhead conveyor systems, and gravity feed devices. Figure 9.1 shows a moving-belt assembly line equipped with powered speed tables or transition devices and static curved roller conveyor sections. This line is also equipped with tool balancers for heavy powered tools which ensure that the tool is always in the proper position when needed; it is pulled down directly to the assembly operation, then released by the operator to be retracted by the spring-loaded tool balancer to the ready position for the next assembly.

Figure 9.2 shows an example of mechanization in a wire-guided vehicle equipped with a section of powered roller conveyer to enable it to perform load and unload operations automatically. It also provides power for the nonpowered load and unload stations between which it travels. Figure 9.3 shows a standard power and free conveyor system equipped with a special carrier designed to carry artillery shells. The numbering system and the gap between the two rows of shells are to accommodate special X-ray equipment. Many food-processing plants, such as those that manufacture potato chips and other types of chips, use mechanization and mechanized production extensively in their manufacturing process. Mechanization, however, is not automation. Extensive requirements exist for manual assembly operators to, in effect, perform the work as before, but with the help and assistance of mechanized equipment. There is no central computer

FIG. 9.1 Moving belt assembly line. (Courtesy of Speedways Conveyors, Inc., Buffalo, N.Y.)

control of the process or the factory, but certain equipment may be computer controlled. For many companies and manufacturing processes, mechanization of the manufacturing process may be all that is required or that can be economically justified.

9.1.1 Mechanization Versus Automation

To put mechanization versus full automation into proper focus, we must look at the trends today in the industrial world in general, and the United States in particular. Batch production in lot sizes of 50 or less represents the majority of parts manufactured in U.S. metalworking industries, according to informed sources. It is estimated that the discrete parts industry represents 35% of all U.S. firms and accounts for about 75% of all U.S. trade in manufactured goods. The same sources estimate that large firms which produce in excess of 10,000 parts that are identical per year can make economical use of highly automated transfer lines, but as many as 75% of the manufacturers of discrete parts in the United States fall below that level, and need more adaptable systems.

In assembly of products such as automobiles, washing machines, small appliances, television sets, and the like, where the volume is high, the production run is long, and the design is stable, a fully automated line can be easily justified. Again, however, this is not

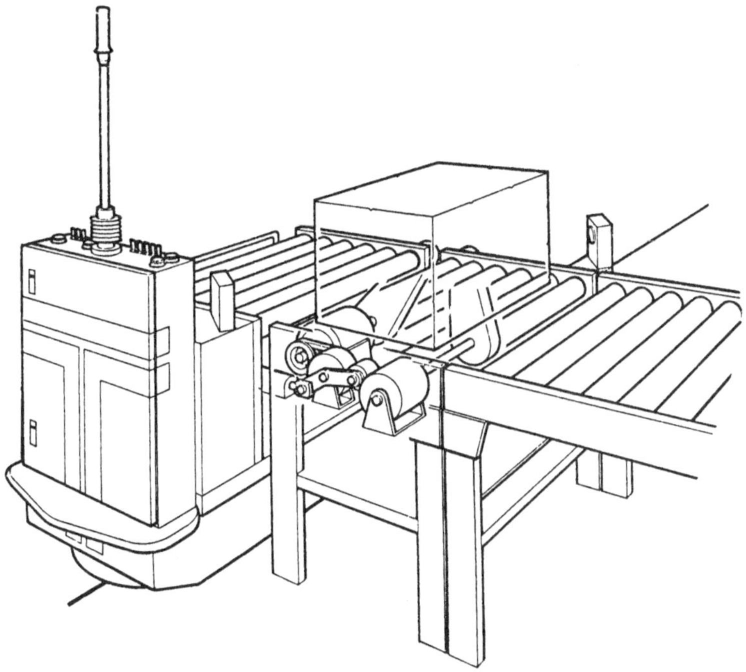

FIG. 9.2 Wire guided vehicle with powered roller conveyor. (Courtesy of Jervis B. Webb Company, Elmhurst, Ill.)

the case in the majority of manufacturing and assembly firms today. An approach other than full automation is needed.

The manufacturing engineer has the problem of finding a suitable middle ground between conventional parts fabrication, or manual assembly methods, and full automation. There are methods of making small-batch manufacturing operations automated that are feasible and can be justified economically.

When considering product assembly, the manufacturing engineer must consider the many factors that affect the choice of an assembly system. For a new product line, these factors would be:

1. The cost of assembly
2. Peak production rate required
3. Availability of labor
4. Market life of the product
5. Stability of product design

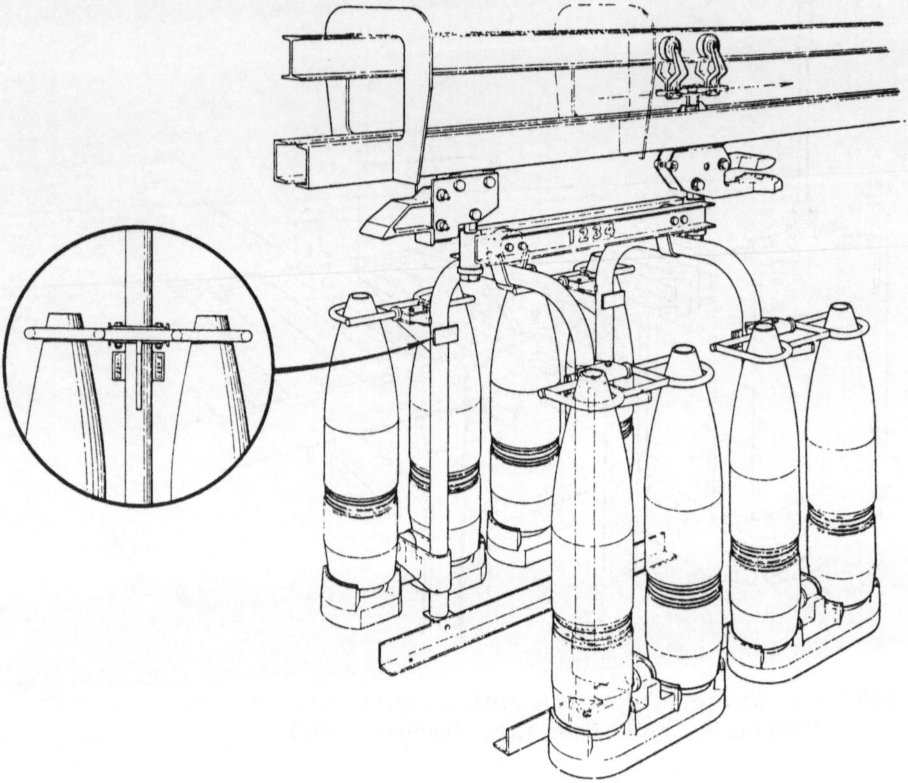

FIG. 9.3 Power and free conveyor system with special carrier for product transport. (Courtesy of the Jervis B. Webb Company, Elmhurst, Ill.)

If the automation of an existing operator assembly line is planned, consideration must be given to the reassignment of those operators who would be displaced. If labor is plentiful, the degree of automation depends upon the reduction in assembly cost and the increase in production rate brought about by the automation of the assembly line. It should be kept in mind that the capital investment in automatic machinery, like that in special-purpose tooling, must be amortized over the market life of the product unless the equipment can be adapted to assemble a new product. If the market life of the product is short, dedicated automatic machinery is clearly not justifiable, but standard, general-purpose automatic assembly equipment which is adaptable to a variety of products may warrant further investigation.

Figure 9.4 shows such a standard, general-purpose assembly system, known as a transporter line. Transporter lines can be used for a variety of different product assemblies. They significantly reduce inventory, save on indirect labor and supervisory cost, and increase direct labor productivity. A transporter line is normally laid out to begin with kitting for the various operations; an order picking carousel is utilized to pick the parts needed and have them placed in tote boxes along with the operator work instructions. The tote boxes, which make up a kit of parts for a given number of assemblies, are placed in a flow storage rack in a first in, first out mode. As a reserve kit of work is required at a particular workstation, a red light goes on at a central dispatch station. The dispatcher checks the operation and picks a kitted tote box from the front of the flow storage rack. Each stand-alone assembly or assembly operation is stored in a separate pigeonhole for simple identification in picking.

Once the required box has been taken out of the flow rack, the dispatcher depresses the console button adjacent to the station number requiring the spare kit of parts. The red light goes out, the white light is energized, and simultaneously a two-way power cylinder swings a diverter arm across the top belt conveyor and bed of the transporter. The tote box is released and travels at 150 ft/min until it glides around the diverter arm to the reserve station. The tote box depresses a microswitch flipper assembly at the workstation, which closes the diverter arm, and the white light at the console is now extinguished.

Once a tote box of work has been completed, it is returned by the bottom level of the transporter to the rear of the flow storage rack. Here the assembly in the tote box is inspected, and then is placed in the appropriate pigeonhole for the next operation.

9.1.2 The Stepped Approach to Mechanization and Automatic Assembly

Once assembly tooling has been developed for a relatively low rate of production, the standard approach in most companies is to build multiples of this tooling as the production rate increases. A much better approach would be to use power assist tools in place of manual hand tools, to begin using motion economy devices to hold the work while the operator productively uses both hands instead of one to perform work, and to use gravity to deliver parts and set aside completed assemblies. As the production rate continues to build, the next step is to break the operations down to a series of short-cycle steps, employ a moving belt or some similar means to establish a balanced flow of work between stations, and provide a forced flow of continuing assembly operations.

As production rate requirements continue to increase, consideration should be given to using automatic assembly stations to perform at

FIG. 9.4 Transporter line at Western Electric, showing flow storage rack, as partially completed assemblies proceed to next work station. (Courtesy of Speedways Conveyors, Inc., Buffalo, N.Y.)

least part of the work. This approach is shown in Fig. 9.5, where illustration 1 shows a three-machine riveting machine consisting of a fixed center pneumatic double riveter, and a single-head pneumatic riveter at the right. Three rivets are clinched simultaneously when the operator presses a palm button. With the three-headed arrangement, the operator produces 170 assemblies per hour. This represents a 35% increase in assembly rate.

In illustration 2 of Fig 9.5 the operator at top center places a loaded work holder in a special pneumatic slide fixture in the "out" position and then presses two palm buttons. The slide fixture automatically moves the work holder into five successive riveting positions, automatically inserts semitubular rivets over roll set pins and clinches the rivets simultaneously. The slide fixture then returns the work holder to the "out" position, where the base assembly is automatically ejected. Exchanging the empty fixture for a loaded fixture, the riveter operator repeats the process.

In layout illustration 3 of Fig. 9.5, (a) shows the simplest arrangement—one operator at one single-head riveting machine. Stock boxes are located in front of the operator on either side of the machine head. The riveted assembly is placed by the operator in a tote box or dropped in a chute to slide into the box. Reducing the amount of reaching necessary to pick up the parts minimizes materials handling time. In (b), a multiple-machine layout, the grouping of two or more separate machines is shown. Stock for each machine can be assemblies riveted at an adjacent machine. In (c), stocks of parts are still supplied in boxes to each riveter, but the riveted assemblies go on a conveyor for distribution to storage or subsequent assembly stations. In (d), riveting and manual assembly operations may be combined along a conveyor. In (e), a typical machine layout employs a rotary indexing table, which indexes automatically at a predetermined rate and controls the cycling of the single-head riveter. The three operators at loading stations around the table simultaneously place parts on eight identical fixtures at the table's indexing rate, far higher than the rate of one operator loading the machine directly.

In illustration 4 of Fig. 9.5, the rotary table is shown as 12 identical fixtures indexed rapidly to the loading position directly in front of the operator. The operator's eyes and hands avoid all unnecessary motion to minimize parts handling time. Parts in the bins to the left and right of the small rotary table drop down into the spill trays in front of the operator, only a few inches from the riveting fixtures.

The examples outlined above show how it is possible to move by a stepped approach from manual assembly to automatic assembly. The methods discussed all minimize direct labor cost by:

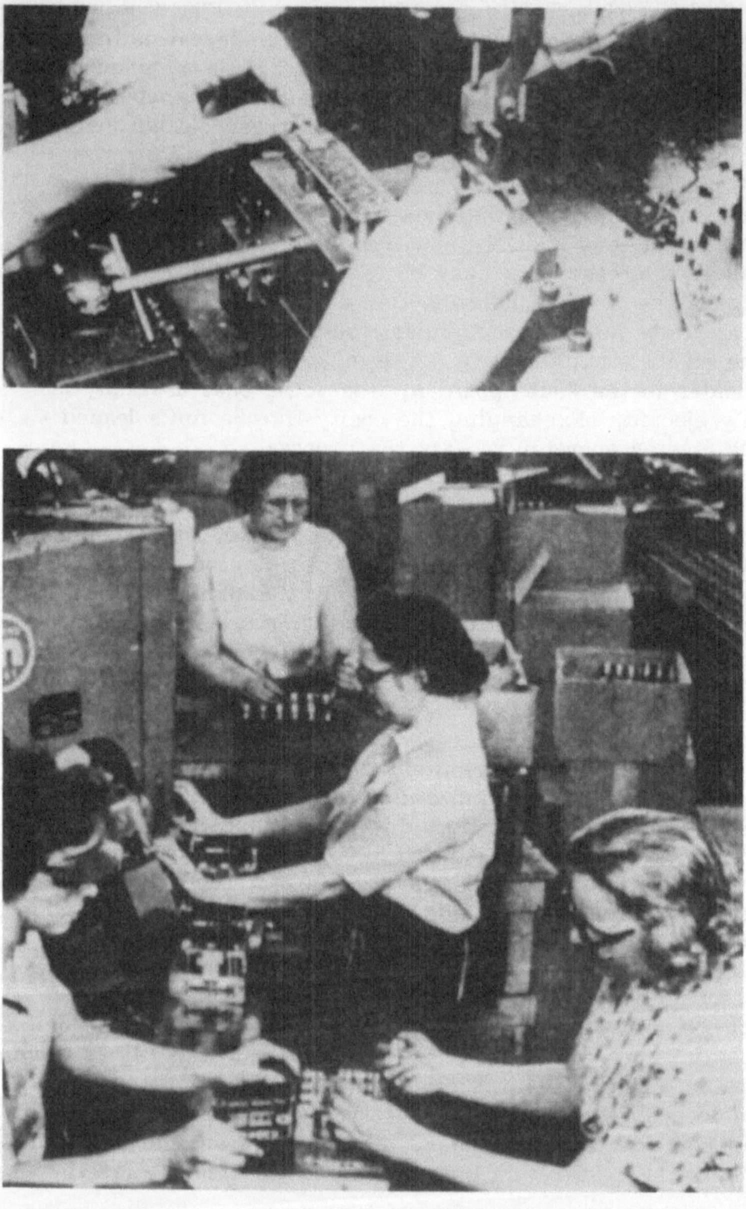

(1)

(2)

FIG. 9.5 Stepped approach to assembly automation. (Courtesy of Milford Rivet and Machine Company, Milford, Conn.)

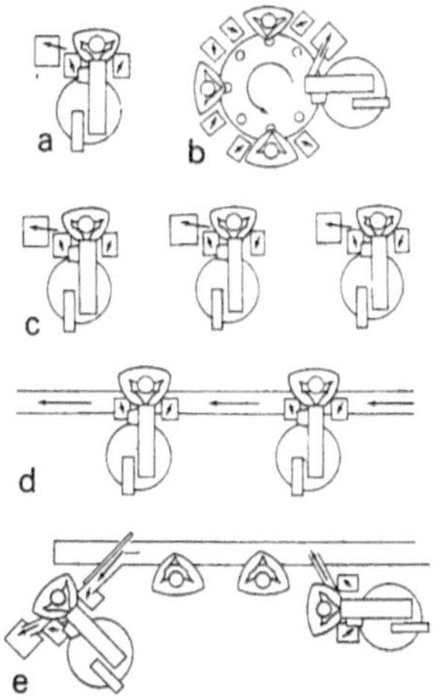

(3)

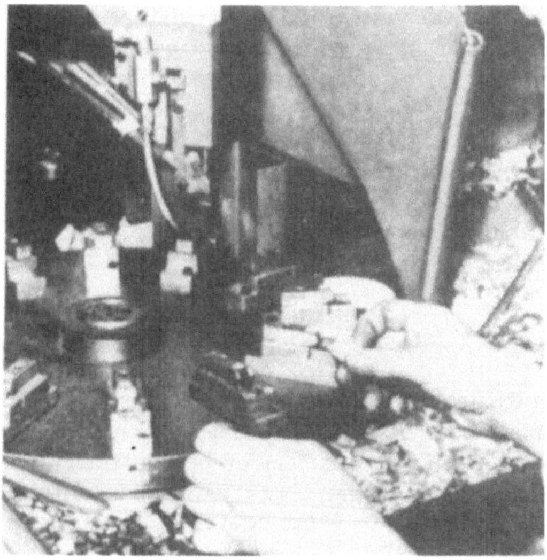

(4)

1. Reducing materials handling time through the use of motion-saving locating fixtures and an optimum machine layout. (This is an approach that costs little or nothing.)
2. Increasing operator productivity through multiple-head riveting. (May require substantial capital investment.)
3. Overlapping machine cycle and handling time through the use of rotary tables and other automatic work transfer devices. (May require substantial capital investment.)

Proper fixturing and tooling plus machine layout deserve a great deal of attention as a first step in the move to full automatic assembly.

Automated parts handling systems are an important part of any plan for mechanization or automatic assembly. Automatic parts handling systems reduce manpower, equipment requirements and help cut scrap and rework. Figure 9.6, shows a part of the five highly automated parts handling systems recently installed at Borg-Warner Automotive in Muncie, Indiana. These systems have substantially reduced the manpower and equipment required for just-in-time manufacturing and assembly of transmission transfer cases for Ford.

In addition, a significant reduction in scrap can be attributed to the programmable-controlled conveying systems, which assure a steady, gentle flow of transfer case components through a series of intricate machining, finishing, and heat-treat operations. The conveyor systems synchronize the production of five of the nine major steel components required for assembly of transfer cases for Ford 4-wheel drive and other vehicles.

9.2 AUTOMATIC ASSEMBLY

Automatic assembly offers a multitude of advantages, not least of which is the replacement of the human operator by a mechanical means of assembly. More precisely, it is replacement of the human operator with automatic workheads, where the tasks being performed are very simple, but the human operator is retained for the more complex tasks that would be uneconomical to automate. Other advantages of automatic assembly include:

1. Reduction in the cost of assembly
2. Increased productivity
3. A more consistent product
4. Safety; removal of the human operator from hazardous operations

Automatic assembly machines consist of a transfer system for moving the assemblies from workstation to workstation, automatic workheads to perform the assembly operations, vacant workstations for human operators to perform the more complex assembly operations,

FIG. 9.6 Automatic loader picks up automatic transmission reduction hubs from conveying line and places in an induction heat-treat machine. (Courtesy of Simplimatic Engineering Co., Lynchburg, Virginia.)

and inspection stations to check that the various operations have been completed successfully.

The automatic workheads are fed either manually with magazine or individually fed parts. Magazine-stored component parts or parts from an automatic feeder through a feed track may also be used to feed the workheads.

The workheads consist of fastening devices or parts-placing mechanisms. Some examples would be:

1. Nut and screw running heads
2. Welding heads
3. Riveting heads
4. Pick and place mechanisms
5. Push and guide mechanisms

9.2.1 Automatic Assembly Equipment

The manufacturing engineer must have some knowledge of the components that make up an automatic assembly machine. Let us first consider the basics of transfer systems. Assembly machines are usually designed according to the system adopted for transferring the work carriers. For example, an in-line assembly machine is one where the work carriers are transferred in-line along a straight slideway, and a rotary machine is one where the work carriers move in a circular path. In both types of machine, the transfer or work carriers may be continuous or intermittent. With continuous transfer, the work carriers are moving at constant speed while the workheads index back and forth. The assembly operations are carried out during the period when the workheads are moving forward, keeping pace with the work carriers. The workheads then return quickly to their initial positions, ready to repeat the operations during the next cycle.

Intermittent transfer is the most common system employed for both rotary and in-line machines. In such a system, the work carriers are transferred intermittently and the workheads remain stationary. The transfer of all work carriers occurs simultaneously and the carriers then remain stationary to allow time for the assembly operations. These kinds of machines are called indexing machines. Vibratory bowl feeders are one of the most important types of hopper feeding devices for automatic assembly. In such a feeder, the path along which the parts travel is formed as a helix and is located around the inside wall of a shallow, cylindrical, bowl-shaped hopper, secured to a heavy base by sets of inclined leaf springs. Figure 9.7, shows such a feeder. The bowl is vibrated by an electromagnet mounted on the case, and a support system ensures the movement of the bowl in such a way that it provides torsional vibration about the vertical axis coupled with linear vertical vibration. All parts of the inclined track vibrate along a short, almost straight path, inclined to the horizontal at a greater angle than that of the track. The vibratory motion causes parts in the bowl to climb up the track to the outlet at the top of the bowl. Boothroyd et al. (1982) summarize design guidelines for vibratory bowl feeders as follows:

1. For a given track acceleration, the mean conveying velocity in a vibratory bowl feeder is inversely proportional to the operating frequency.
2. For a given frequency, an increase in vibration amplitude, or track acceleration, will increase the conveying velocity. However, with large accelerations the behavior of the parts becomes erratic and unpredictable.
3. For maximum conveying velocity, the track angle should be as small as possible. However, feeding around the flat bowl bottom

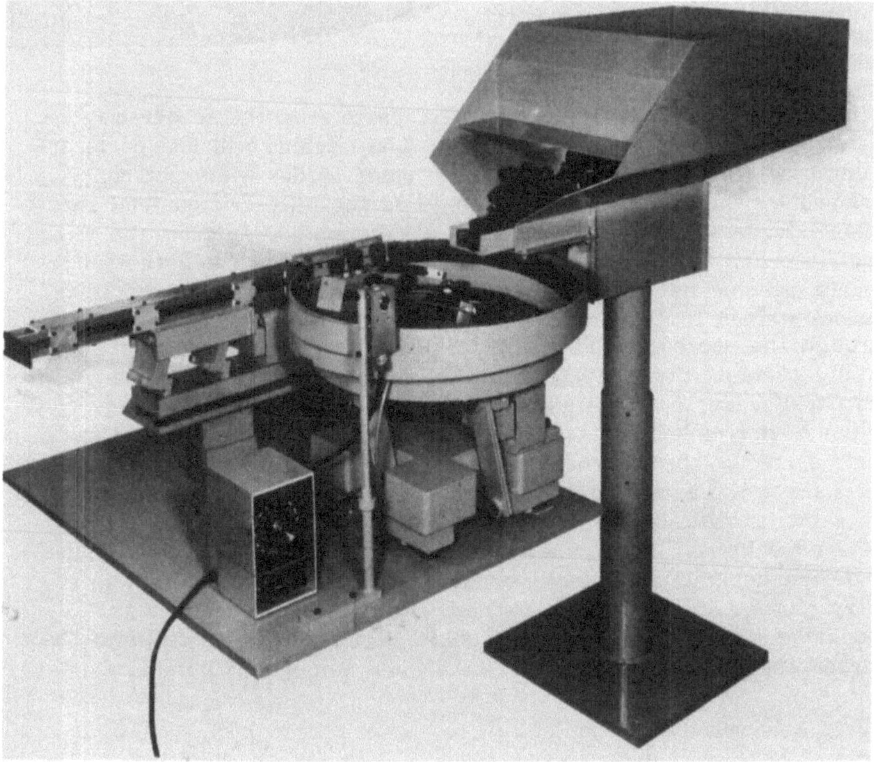

FIG. 9.7 Vibratory bowl feeder with horizontal power track. (Courtesy of Parts Feeders, Inc., East Hartford, Conn.)

will always be faster than on the inclined track, and therefore the parts will generally be pushed up the track by the circulation of those on the bottom.

4. An optimum vibration angle exists for any given conditions.
5. Higher coefficients of friction between the parts and the track will generally give higher conveying velocities.
6. For any bowl feeder, a support spring stiffness may be chosen that will give a reasonably low load sensitivity. However, with many commercial feeders, this will necessitate an increase in the power from the electromagnet in order to maintain a sufficiently high feed rate.

Other nonvibratory parts feeders include by classification:

Reciprocating feeders
Rotary feeders
Belt feeders

In automatic assembly, the output of parts from the feeder may be restricted by the machine being fed. The machine will use parts at a uniform rate. The feed rate of the parts feeder when not constrained by the machine is referred to as the unrestricted feed rate.

Parts feeders are usually placed some distance away from the workhead in order to provide free access to the workhead and the assembly machine itself. This requires that the parts be transferred and maintained in orientation between the feeder and the workhead through the use of a feed track. Since most parts feeders do not supply parts at the discrete intervals required by the workhead, the parts feeder must be adjusted to overfeed slightly, and a metering device called an escapement is required to ensure that the parts arrive at the workhead at the correct intervals. After the parts leave the escapement, a partsplacing mechanism places the parts in the assembly.

Feed tracks may be either gravity or powered tracks. The choice of design depends in large measure upon the required direction of entry of the part into the workhead. In the feed track design, it should be remembered that the track may not always be full and that the feeding of parts should still take place under this condition.

9.2.2 Economic Consideration

One of the serious faults to be found with automatic assembly machines is loss in production due to machine downtime resulting from less than perfect parts. In manual assembly line stations, the operators will discard defective parts quickly, and little or no loss of production occurs. A defective part fed to an automatic workhead can, on an indexing machine, cause a stoppage of the entire machine until the defect is found and cleared. The downtime from this can result in a serious loss of production and a consequent increase in assembly cost if several automatic workheads are involved. The quality level of the parts to be used in automatic assembly is a prime consideration when an automatic assembly machine is planned and designed.

The decision to build or purchase automatic assembly machines should be based upon a thorough study and analysis by the manufacturing engineer. The objective of this study is to predict the performance and economics of the proposed automatic assembly machines. In automatic assembly these predictions are likely to be subject to greater error than most other feasibility studies. This is primarily due to the fact that the machine will be one of a kind, and its performance will be very dependent on the quality level of the parts to be assembled. Another reason is that similar machines will not generally be available for study.

Certain vital information is needed before any study can be made. Maximum and minimum production rates during the probable life of the machine must be known. The range of variations in these figures can be crucial, because a single automatic assembly machine is very inflexible. Operators required on the machine must all be present when the machine is running, and if the machine is stopped due to a falloff in demand for the product, they must be employed in some other duties. In general automatic assembly machines are suitable only when the volume of production is known to be steady. They can be applied profitably only when the production volume is high. It has been suggested that where the volume is 500,000 assemblies a year or higher, the application of automatic assembly has an excellent chance of success. Also, the labor costs of the existing assembly process must be high if automatic assembly is to be a feasible alternative. Much more information will be required and many other factors will combine to determine the final answer. For example, the quality level of the parts must be analyzed as in Table 9.1, which shows the hypothetical quality levels of parts that make up a power plug assembly. With this information, the manufacturing engineer can divide the parts that are defective into two groups: those parts that cannot be assembled, and those parts that can be assembled, but would be rejected by a human operator. The number of parts falling within the first category will allow estimates to be made of the automatic assembly downtime and those falling in the second category will allow estimates to be made of the number of unacceptable or defective assemblies that will be produced by the machine.

Another consideration would be parts feeding and assembly. An estimate should be made of the degree of difficulty with which the individual parts can be automatically fed and assembled. It should be noted that for each operation, four possibilities exist:

1. Automatic feeding and assembly
2. Manual feeding and manual assembly
3. Automatic feeding and manual assembly
4. Manual feeding and automatic assembly

After estimating the effective production rate for each type of machine, it should be possible to compare their economics. Such an analysis is shown in Table 9.2, for the hypothetical power plug assembly. A cursory analysis touching only on the factors covered above may be sufficient, without further analysis or calculation to decide whether further consideration should be given to the use of automatic assembly machines. It should be very clear that automatic assembly is not for everyone, and even where the economics are favorable, a cautious approach is advisable.

TABLE 9.1 Quality Level of Parts (Hypothetical)

Parts	Fault	Number of faults in assembling 10,000 plugs	Percentage faults
Base subassembly	Chipped	10	0.10
	Earth pin will not assemble	170	1.70
	Live pin will not assemble	20	0.20
	Neutral pin will not assemble	30	0.30
Earth pin subassembly	No screw	41	0.41
Neutral pin subassembly	No screw	59	0.59
Live pin	Fuse will not assemble	123	1.23
	Fuse assembles unsatisfactorily	21	0.21
Fuse clip subassembly	Fuse will not assemble	115	1.15
	Fuse assembles unsatisfactorily	17	0.17
Fuse	Damaged	18	0.18
Cover	Chipped	10	0.10
	Cover screw hole blocked	200	2.00
Cover screw	No thread or slot	20	0.20

From Boothroyd et al., *Automatic Assembly*, Marcel Dekker, New York, 1982.

TABLE 9.2 Economic Analysis of Power Plug Assembly

	In-line index	Rotary index	Free transfer
1. Cost of transfer device and work carriers (k$)	88	88	216
2. Cost of workheads, etc. (k$)	75	75	75
3. Total cost of machine [(1) + (2)] (k$)	163	163	291
4. Effective production rate (assemblies per minute)	14.7	13.8	16.9
5. Number of operators for three shifts	9	6	9
6. Number of assemblies producted in three 8-hour shifts	21,150	19,830	24,360
7. Number of operators required to produce (6) manually	21.2	19.8	24.4
8. Cost of equipment for manual operation [(7) × 10/3] (k$)	70.7	66.0	81.3
9. Effective cost of machine [(3) − (8)] (k$)	92.3	97.0	209.7
10. Number of operators saved [(7) − (5)]	12.2	13.8	15.4
11. Capital outlay per operator saved [(9) ÷ (10)] (k$)	7.57	7.03	13.62

From Boothroyd et al., *Automatic Assembly*, Marcel Dekker, New York, 1982.

9.3 FLEXIBLE MANUFACTURING SYSTEMS

Flexible assembly systems can provide cost reductions for short production runs similar to those that would be expected from mass production. Flexible assembly systems allow work to be selectively routed to selected assembly stations, bypassing stations where no work is required to assemble the product at hand. These flexible systems are used to greatest advantage where more than one line of product is being assembled. Properly planned and applied, these systems can:

1. Reduce work in process
2. Improve product quality
3. Optimize the use of floor space
4. Minimize capital expenditure requirements
5. Provide the needed flexibility to change or relocate the assembly system at minimum cost.

With flexible assembly systems, changes can take minutes instead of hours or days. Layouts that include flexible parts routing to workstations set up along the length of a powered belt transporter are an excellent example. Flow racks and carousels are often used for storage of parts kits needed for assembly. Automatic load and unload devices would interface the transporter with the flow racks or carousels. Standard power and free systems of conveyors, as well as the relatively new automatic guided vehicle systems, can be used as building blocks for a flexible assembly system. Bar code scanners can be used for automatic routing systems in identifying parts carriers, a computer or other controller activates a switch in a conveyor spur, and the parts carrier moves to the proper location.

9.3.1 The Concept

The concept of moving assembly operations work along flexible-path routing is not new. Automotive, appliance, and consumer electronics manufacturers have used power and free conveyor systems for many years. These are flexible systems because work can be routed to specific workstations. Parts or subassemblies in carriers are diverted to spurs off the main conveyor line, where certain assembly operations or testing can be performed.

Flexible assembly systems can be made up of computer-controlled monorails, inverted power and free conveyors, shuttle cars, or car-on-track conveyors, where sections of the conveyor track can be made to pivot automatically to selectively route the work as it moves down the conveyor. The needs of each assembly shop dictate which system is best suited for it. There are three ways to move work through assembly: balanced assembly line flow, intermittent or semi-flexible flow, and fully flexible movement.

In balanced assembly line flow, work in process moves from one workstation to the next, until a fully assembled product emerges from the end of the production line. All work goes through every assembly workstation in the system. This can be in a straight line on a moving conveyor system, or the work can be indexed from one station to the next as on a rotary table. In the semiflexible systems, work in process moves to spurs that are set up off a central delivery path. Transporter/carousel systems used by electronics Manufacturers and power and free systems used in assembly are examples. Buffers are required to allow the cycle time at each station to be based upon the average cycle of all stations in the system. Every time a new product is introduced to the system, a new average cycle time must be calculated.

There are a number of ways in which totally flexible movement for an assembly system may be provided; one way is by using automatic guided vehicles or free-path delivery equipment to deliver work in process to assembly stations. Guided vehicles are not restricted to a central delivery path, since guide paths can be established in any desired geometry. Figure 9.8, shows the concept for such a system. Buffers are often used in systems employing guided vehicles to keep assemblers and equipment fully productive. By using guided vehicles, stations can be bypassed and vehicles can be sent directly to rework and repair stations, or back to stations out of sequence in the system. However the movement is handled in a flexible assembly system, the key to efficient operation lies in the controls used for routing and tracking work in process.

There are two ways to use automatic guided vehicle systems for fully flexible assembly. The first method uses the vehicles as mobile work platforms. A workpiece is loaded on the vehicle, which then moves from one assembly station to the next, while assemblers and machines perform operations on the workpiece. The vehicles can deliver the work to test or repair stations, and then finally carry it to the output end of the system. In this method, the work stays on the vehicle throughout the assembly operations. In the second method, or conventional method, the vehicles are transport devices only. which load and unload work at each assembly workstation. This method should be used when the cycle time is long at each station, so as not to tie up vehicles.

9.3.2 Planning the System

In planning a flexible assembly system, it must first be decided whether a totally flexible system or one that is semiflexible is desired. It must be decided whether workstations will be manned or unmanned, fully manual to fully automatic, and whether computer control and automated material handling will be required.

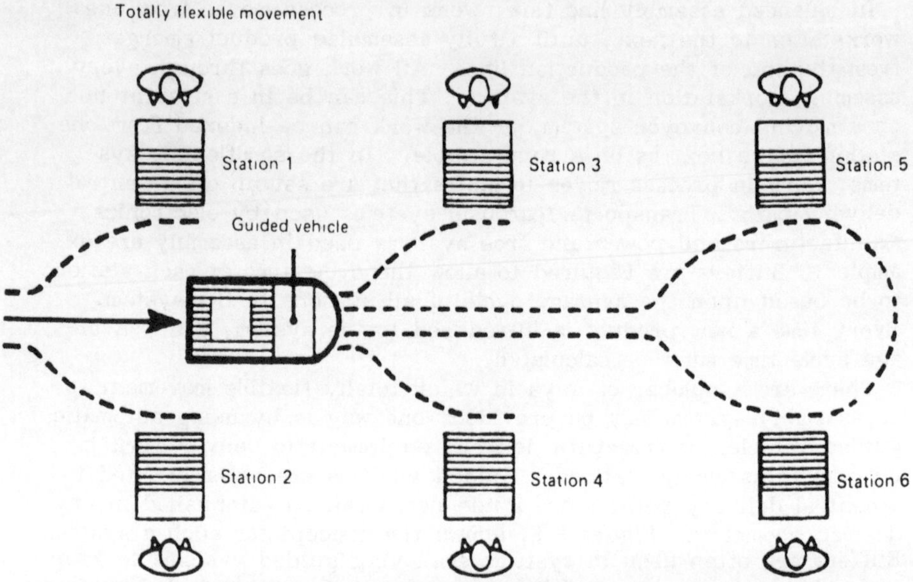

FIG. 9.8 Total flexible assembly system utilizing wire-guided vehicles. There are no restrictions to workflow. An example is a guided vehicle used for auto assembly. (Reprinted with the permission of Industrial Engineering Magazine, Vol. 16, No. 6, June 1984. © Institute of Industrial Engineers.)

A computer control system, for example, can be programmed to perform many key functions in the system. These include:

1. Sequencing and tracking different work to selected workstations based upon variable routing instructions
2. Providing added control to machines and robots at workstations
3. Collecting real-time data regarding operation of the workstation equipment and production
4. Communicating with a host computer system
5. Performing other process monitoring functions as required

The automated material transport system is, in effect, the system integrator. Work in process must be moved quickly and reliably from one workstation to the next, and stored workpieces must be easily accessible and readily available when needed. Choosing a material transport system for a given task or application is not easy. The first step would be to examine every aspect of the application. This will be a major factor in the selection of the right transport system. Areas to consider include the following:

1. Distances to be covered
2. Amount of floor space available and required
3. Intended throughput
4. Physical characteristics of parts and subassemblies
5. Anticipated degree of automation
6. Types of interfaces encountered:
   a. Manual
   b. Semiautomatic
   c. Fully automatic
   d. Other transport systems

After the material transportation needs of the assembly system have been determined, the existing material-handling systems should be analyzed in term of their operating characteristics, ability to interface with other types of equipment, and compatibility with the proposed system.

Another key aspect in considering the installation of a flexible assembly system is the ability to expand as production expands. Additional elements can be added to the system in a series of planned steps. Empirical analysis has shown that in a favorable financial climate, a firm can save approximately 12% in capital recovery costs if a flexible assembly system is implemented on an incremental basis as compared to total implementation.

Manufacturing engineers must be the driving force behind all the initiatives a business may take to modernize its manufacturing operations. If this is not the case, it has been the author's experience that it just will not happen. Flexible assembly systems offer the small manufacturer with short production runs and multiple product lines the opportunity to obtain many of the advantages and economies of mass production with a minimal capital investment that can be programmed over several years or longer, as the planned system is brought on-line in planned increments or building blocks. The key is the advance or preproduction planning that must be done by manufacturing engineers with the support of management.

9.3.3 Economics

A properly designed flexible assembly system offers a number of economic advantages:

1. Can provide for as many as 52 inventory turns per year
2. Allows a decrease in work-in-process inventory by as much as 60%
3. Will allow integration of both MRP and factory shop floor control systems
4. Increase direct labor productivity
5. Saves floor space due to work-in-process reduction
6. Enables balance of work flow

7. Allows control of production with accurate, daily, perpetual inventory

The greatest economic advantage of such systems is the significant reduction in setup times. It can take minutes to set up for a new or different product configuration. The work flow can be changed simply by routing work differently through the system. Costly and time consuming plant rearrangements are no longer necessary.

Other economic advantages of flexible assembly systems include the following:

1. *Optimum system productivity:* Each operator and each machine operates at maximum output potential, which, in turn, increases the overall system yield.
2. *Flexible routing:* Routing of the assemblies through the system can be accomplished through computer control and can be modified on the basis of what is actually occurring within the process.
3. *Line balancing is unnecessary:* Unlike the progressive assembly line, there is no production line balancing problem.
4. *Modular expandability:* As production requirements increase, the manufacturer using a flexible assembly system can easily expand capacity by adding new production work centers and by modifying the routing of the material-handling system.
5. *Multiple vendor capability:* Most flexible assembly systems can utilize equipment from different vendors in the various operations of the assembly process. This allows manufacturers who wish to initiate a flexible assembly system program to utilize existing machines and equipment, as will as to select the new equipment which most nearly meets their requirements.

Flexible assembly systems can provide the small, job order firm and the large manufacturer the means to compete successfully with high-volume producers of the same or similar product lines.

9.4 ROBOTS AND AUTOMATION

The Robot Institute of America, in November 1979, defined a robot as a reprogrammable multifunctional manipulator designed to move material, parts, tools, or specialized devices through variable programmed motions for the performance of a variety of tasks. This definition was also accepted by the international coordinators at the eleventh ISIR in Tokyo in October 1981. In order for a robot to be useful in manufacturing, it must have the following attributes:

1. A hand to grip a workpiece

2. An arm to move the hand in three planes
3. A wrist with two or three articulations
4. Sufficient power to move the limb and workpiece around
5. Manual controls so that an operator can control limb motions
6. A memory to store a sequence of instructions
7. A means of executing a sequence of instructions stored in memory
8. The ability to function at speeds equal or greater than those of a human operator
9. Reliability

Machine load and unload is the most popular robot application in industry today, and from recent surveys it is clear that this trend will continue for some time. Such applications require only simple pick-and-place robots, rather than the more complex controlled path robots. One of the primary tasks facing the manufacturing engineer will be to analyze the right level and type of robot to apply to specific tasks. Robots can perform many different tasks in the manufacturing world. The capability of a robot is determined by its controller. Depending on the type of control, industrial robots can be classified as:

1. Non-servo-controlled
2. Servo-controlled
3. Point-to-point
4. Continuous, or controlled path

A non-servo-controlled robot is limited to pick and place only, or two positions. A servo-controlled robot is capable of stopping at any point in a gripper path rather than at the end, like a non-servo robot. Most robots on the market today are point-to-point controlled. This type of control system can be applied where only a given number of points are approached, as in the loading of a part, material handling, or as the holder for spot-welding electrodes. Such robots are difficult to program, because the information must be posted point by point. In a continuous path robot, the path of the arm is represented by a large number of points in close proximity. Table 9.3 lists the advantages and disadvantages of the different industrial robots based on their control type.

There are three basic robot programming techniques. These are:

1. Manual lead-through programming
2. teach pendant programming
3. textual language programming

The manual lead-through programming technique is based upon a lead-through device in th form of a special attachment to the manipulator wrist of a special teaching arm. The robot control program

TABLE 9.3 Advantages and Disadvantages of Robot Control Systems

Robot type of control	Advantages	Disadvantages
Non-servo	High reliability	Needs safeguard
	Programmable	Limited movement flexibility
	High repeatability	Can stop only at the end points
	Relatively high speed	
Servo	Move and stop at any point of arm path	More expensive than non-servo
	Accelerate and decelerate	Less reliable than non-servo systems
	High accuracy	
	Smooth path	
	Execution of more than one program	
Point-to-point	Flexible	Tedious to program
	High load capability	
	Large working range	
Continuous	Smooth path	Position accuracy less than point-to-point systems
	Varied arm speed	
	Small size and weight	

is manually fed into the memory by using keyboards, switches, and similar devices. This technique is now rarely used, except to correct programs created with the other programming techniques.

Teach pendant programming is the most popular method for programming robots today. In this technique, the robot are is moved along the prospective working path, and all the necessary information is recorded. In the case of point-to-point controlled robots, the arm is moved from position to position by means of a special control element. Continuous path controlled robots are programmed by manual guidance of the robot arm. In both situations, any nonrecorded information can be entered by keyboard.

The textual language programming technique offers the advantage of being able to prepare a robot control program in advance, without tying up production equipment. This technique has additional advantages over the other two methods. One of these is better program definition, which comes about from the use of a high-level computer language. This technique offers various software tools, which simplifies programming effort and reduces errors. The same subroutines can be used to define frequently repeated tasks.

9.4 Robotic Systems

The unmanned robotic cell in a flexible manufacturing system is built around robots that handle parts and material, perform assembly tasks, operate tool, and perform other manufacturing operations. Figure 9.9, shows a manufacturing cell hardware schematic of a cell built around a Cincinnati Milacron T^3 material-handling robot. The robot controller, the material-handling system and its controller, the fine resolution system and its controller, and the automatic riveter and its controller receive signals through a multiplexer from a central DEC PDP-11/34A computer. Such systems are rapidly becoming commonplace in larger companies, and the aerospace industry in particular. A representative layout of such a cell with the load and unload functions automated by a robot is shown in Fig. 9.10, the sequence of operations is:

1. Unload raw stock from unloading receptacle and place on cutoff saw.
2. Cut off to length.
3. Unload workpiece from cutoff saw and move to three-axis milling machine.
4. Machine workpiece.
5. Unload workpiece from milling machine and move to precision drill press.
6. Drill and countersink holes.
7. Unload workpiece from precision drill press and move to mobile inspection cart.

This type of robot is essentially a controlled hand. The ability to pick up an object with accuracy, move it to a specific location, and place it in the correct position for the next operation is the robot's primary attribute.

In designing a robotic system for manufacturing, the flexible system concept described earlier should be the base upon which the largely unmanned robotic system should be built. Figure 9.11, shows graphically how the robotic module could be expanded to an entire system or line as quantities increase. Beginning with the module, the system would grow to a cell, then a grouping of cells, then to an entire flexible manufacturing system, and beyond that to the flexible manufacturing line-the ultimate in flexible manufacturing systems. To be effective, the flexible manufacturing system must take full advantage of available manufacturing and computer technology. It must allow a high rate of production and information flow among the manufacturing subsystems, which requires the availability of an efficient material-handling system and computer system. The critical features of the material-handling system include the following:

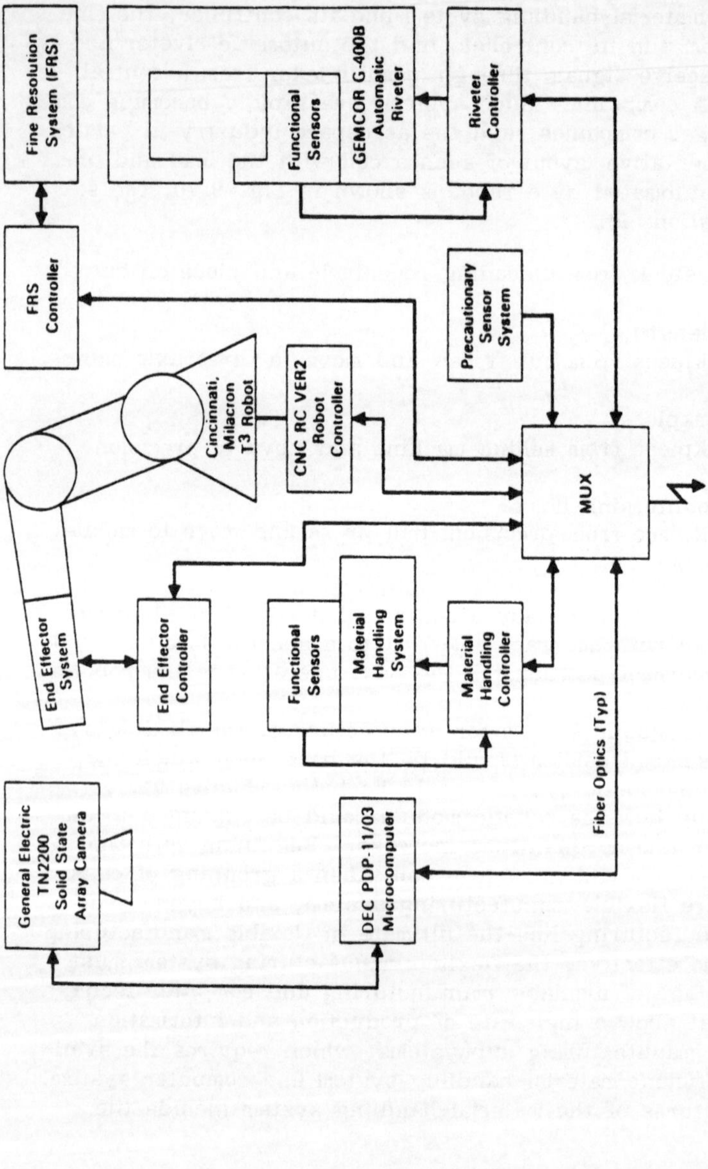

FIG. 9.9 Manufacturing cell hardware schematic. (Courtesy of the McDonnell Douglas Corporation, St. Louis, Mo.)

FIG. 9.10 Manufacturing cell with material handling robot. (Courtesy of Cincinnati Milacron Corporation, Cincinnati, Oh.)

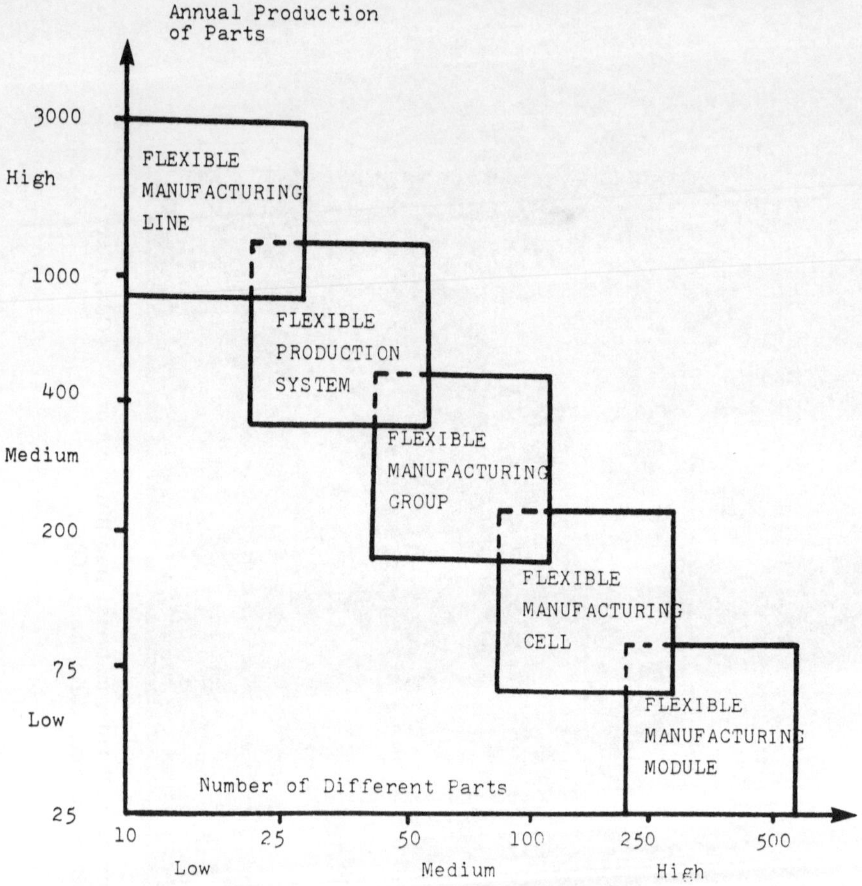

FIG. 9.11 Flexible manufacturing system building blocks versus production of parts. (Courtesy of Professor Andrew Kusiak, Technical University of Nova Scotia, Halifax, Nova Scotia, Canada.)

1. It should have the ability to operate at different speeds.
2. It must be able to support fast, minimum cycle time requirements.
3. Acceleration and deceleration must be controlled.
4. It must have the capability for high-precision repeatability.
5. It should be computer-controlled.
6. It should have the capability for load accumulation.
7. It must provide flexibility.

The economics of the robotic system is a proven reality. Robots typically cost anywhere form $10,000 to $200,000 or more, yet because of their exceptional productivity, payback periods are 2 years or less. A robot can normally do the work of as many as six human operators. Robots cost about $4.00 an hour to operate, in contrast to hourly labor costs for workers in the United States. As a result, tenfold reductions in costs are often possible with robots.

Robot capabilities continue to be refined and extended, and experts agree that robots used in combination with N/C machine tools are the key elements for future automated factories. Robots of the future are seen as evolving in two directions. In massive factories that manufacture complex products such as automobiles and aircraft, the individual robot may dissolve as a separate entity as sensors, computers, manipulators, and machine tools are combined. In smaller plants, the robot will remain a separate unit and with even larger memory and simpler programming to perform more diverse functions. A single robot, for example, may feed a machine tool, move pallets and raw material, and inspect parts.

9.4.2 Robot Applications

Robots can be divided into the following groups according to the tasks they perform:

1. Handling robots
2. Tool-operating robots
3. Assembly robots

Handling robots basically perform the function of transferring an object such as a piece part or tool from one place to another. Such a robot would have the capability of high-speed motion and would have long arm reach in order to perform relatively easy handling operations. Accuracy and wrist ability are of relatively minor importance in handling robot applications.

The most common application of handling robots is loading and then unloading machine tools with parts. Rotational parts are loaded on pallets. There is a practical weight limit for robotic handling or prismatic parts. Both parts and tools can be loaded and unloaded by robots utilizing automatic loaders and tool changers. Attempts have been made to apply handling robots to changing parts holders, also to remove chips from machines. The loading and unloading of pallets with parts is performed manually because of the great variety of shapes of parts and their diversified orientation. A major application of industrial robots involves the operation of tools such as drills, welding guns, and painting guns. Tool-operating robots offer high movability, easy positioning, and usually high freedom, where the average handling robot offers only four or five.

A new application of robotics is being developed by the National Institute of Standards and Technology in Gaithersburg, Maryland. A deburring cell has been developed which demonstrates the deburring of parts based on their description and a graphically developed process plan. This cell, a cleaning and deburring workstation, is capable of deburring roughly prismatic aluminum and brass parts.

The cell consists of a workstation controller, two robots, various quick-change deburring tools, a rotary vise for part fixturing, and a part transfer station. The workstation controller is a set of software modules that support process planning and task scheduling. A six-axis hydraulic robot is used for part handling and buffing. A six-axis electric robot is fitted with a quick-change wrist, that can select between a rotary end brush, a countersink tool, a hole brush, and a gripper. The vise can be rotated to any of three perpendicular orientations, which reduces the amount of robot motion required for part refixturing and deburring. The part transfer station consists of two roller tables which can transfer trays of parts between the cell and material storage. Figure 9.12, shows the floor layout for this workstation.

Some of the more popular applications of tool-operating robots include the following:

1. Painting or coating application such as automotive underbody protection
2. Spot-welding and arc welding
3. Machining, drilling, tapping, and trimming

The Microbot Alpha II robot system is a current example of a tool-operating robot. In its dispensing system configuration, the robot can perform printed circuit board solder masking, adhesive and potting material application, as will as other liquid-dispensing functions. The robot is capable of a repeatability of ±0.015 in. at payloads up to 1.5 lb. Figure 9.13, shows the Alpha II robot.

Assembly robots perform such operations as parts transfer and parts fitting. The average time to perform these operations is much less than in the handling associated with machining and fabrication. Assembly robots must have high accuracy and excellent acceleration and deceleration. In order to achieve this, assembly robots are very often of short reach and carry small loads. Assembly applications offer the greatest promise for future robotic applications. Such factors as parts variation, variations in parts orientation, defective parts, and different products assembled on the same flexible assembly line are challenges that need to be addressed in an assembly robot research project.

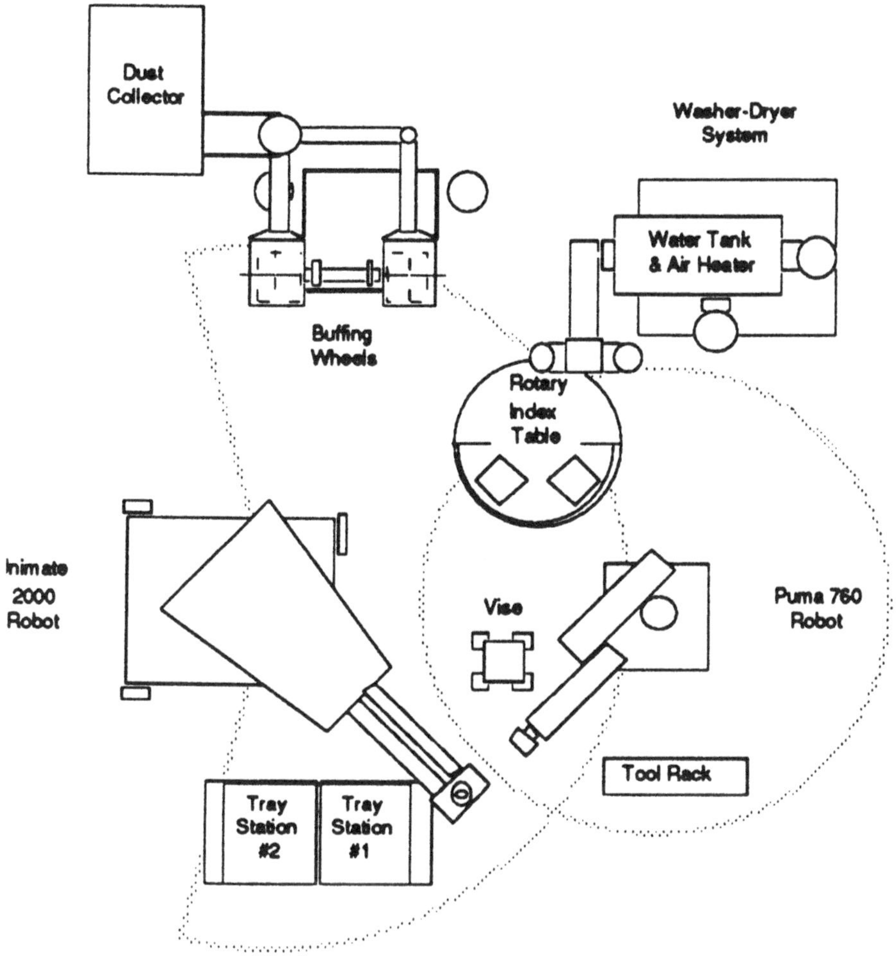

FIG. 9.12 Cleaning and deburring workstation utilizing robots. (Courtesy National Institute of Standards and Technology, U.S. Department of Commerce, Gaithersburg, MD.)

Robotic assembly has two basic parameters: (1) the weight of the parts being lifted and (2) the repeatability needed to consistently perform the assembly. As defined by the Robot Industries Association, repeatability is the closeness of agreement of repeated po-

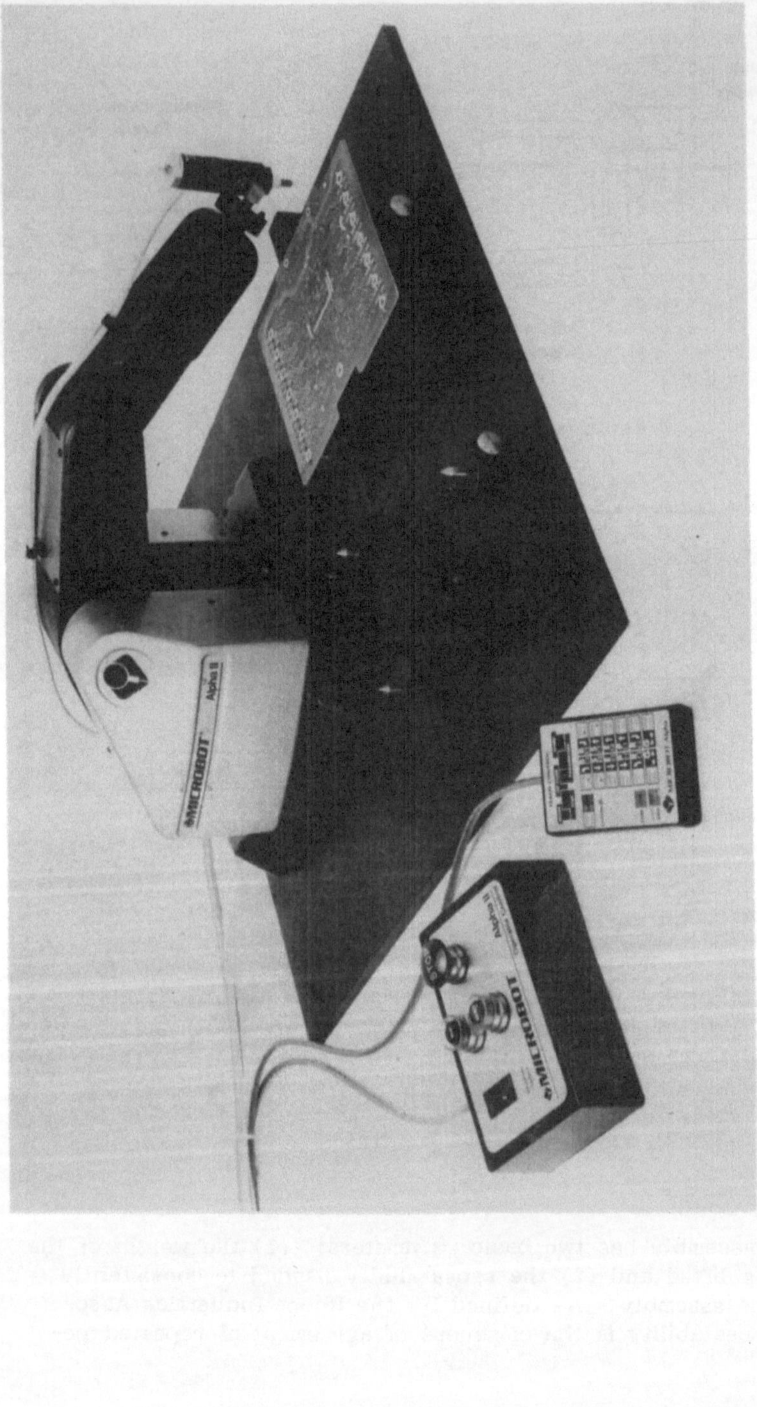

FIG. 9.13 Alpha II handling robot. (Courtesy of Microbot, Inc., Mountain View, California.)

sition movements, under the same condition, to the same location. Economic considerations add two more parameters: (3) speed of the manipulator and (4) size of the work envelope. These four prime parameters, together with cost, constitute the major trade-offs that the buyer of an assembly robot system must face.

When considering robotic systems for light assembly, the three basic components of the robot—the manipulator or arm, the robot controller, and the end of the arm tooling—should be examined. The manipulator must have high accuracy, repeatability, and velocity. End of the arm tooling should be capable of handling several different parts or a family of parts. Tactile sensing on the finger tooling will assist, in addition to allowing the robot system to control the amount of force applied to the gripper. The robot controller often has the power of a programmable controller built in, and is often capable of accepting positional information from sensors to control the manipulator position in real time. The ability to communicate with the factory communication networks will allow the automated cells to be interconnected with a flexible system for automated assembly.

An example of a state-of-the-art light assembly robot that is available today is the Anorad Corporation small assembly robot, which employs basic X-Y-Z (16 × 12 × 5 in.) linear motion, and features a motorized four-position turret equipped with an elector-optical position centering system, gripper, screwdriver, and nutdriver. The individual tools are servo-driven under computer control with feedback signals for torque and force sensing. Figure 9.14, shows this robot set up to perform component insertion onto printed circuit boards.

Vision systems are often associated with state-of-the-art robotic systems, although they are not truly part of such a system. Even though vision systems have been available for a number of years, widespread use linked with robotics has not occurred. One reason is that the early vision systems needed too much time to process and respond with the desired information. Often, this delay was considered unacceptable. Another, more significant reason was that the system needed lighting that was structured to work effectively. This meant that special illumination also had to be purchased and maintained.

Recent developments in vision systems have greatly reduced the processing times to highly acceptable levels, and have made it possible for the systems to work effectively under regular lighting conditions. Figure 9.15, shows a vision system used in conjunction with a Puma robot. Developments such as this will lead to affordable vision systems which will be able to control a moving manipulator in real time.

FIG. 9.14 Small assembly robot. (Courtesy of the Anorad Corporation, Hauppauge, New York.)

9.5 SUMMARY

Mechanized assembly, automatic assembly, and flexible assembly systems offer the manufacturing engineer and his company several viable options in selection what is best for the product line or lines, the circumstances of the company, and the length of the production run. These concepts and technologies are not new, but when applied correctly, and used in conjunction with computer control and other advanced technologies and concepts, much can be done to solve the problems traditional in short-run, job order assembly.

The key is understanding: Understanding that mechanization is not automation, that automatic assembly does not imply a computer-controlled process or robotics, although robots may be used in a automatic assembly situation. Understanding that there may be

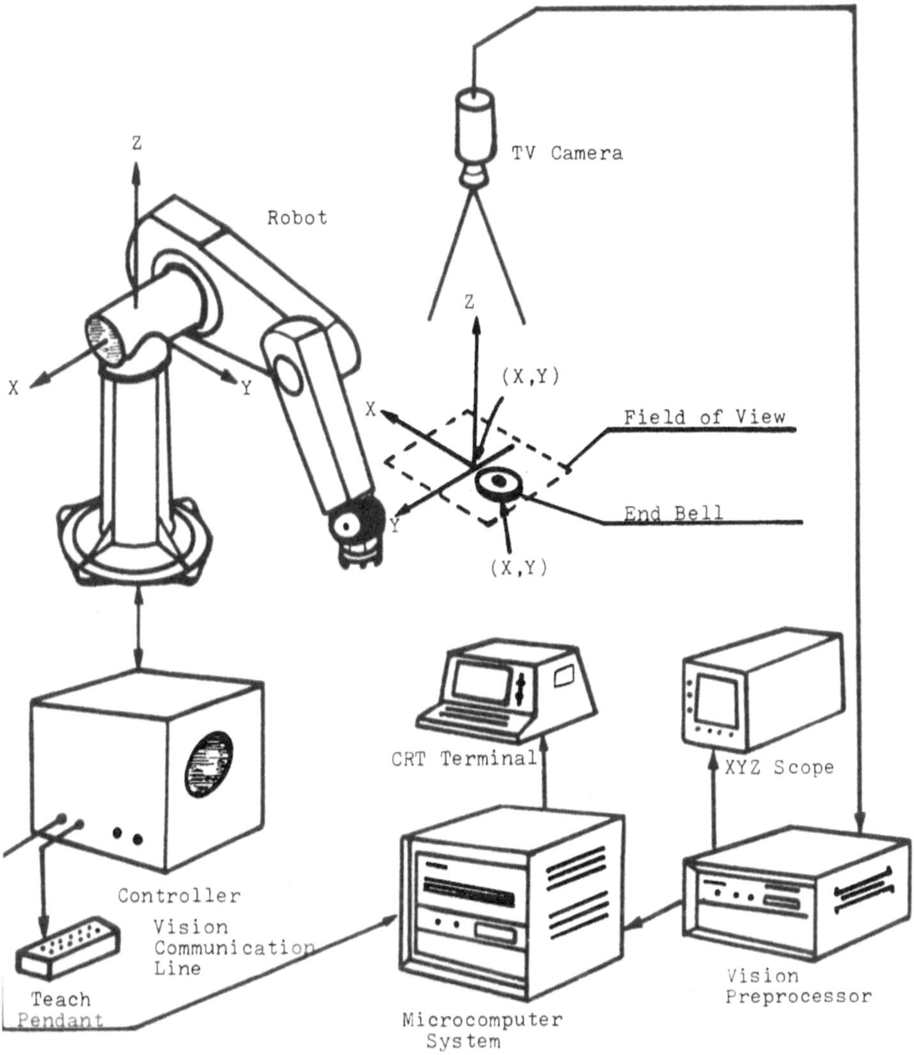

FIG. 9.15 Vision system for robotics.

situations and applications where the moving-belt assembly line, with its balanced, manual workstation, is better than a transporter line with its flexible assembly capabilities.

The various manufacturers of equipment for these systems will be continuously bringing newer and better hardware and software into the market place, and while it is helpful to be knowledgeable about these advances, the manufacturing engineer should have the capability of selecting the best system for the application at hand, preparing the specifications for such a system, and the selecting the best hardware and software that satisfies the specifications and the requirements of the specific application.

QUESTIONS

1. What is the difference between mechanization and automation? Give an example of each.
2. What are some of the key factors that must be considered when selecting an assembly system?
3. If the market life of the product is short, and dedicated automatic assembly equipment is clearly not justified, what other options are available to establish a degree of process automation?
4. What is a transporter line, and what are some of its advantages in product assembly?
5. What is automatic assembly, and what are some of its advantages?
6. What are some examples of automatic workheads and how are they supplied with parts for assembly?
7. What is a vibratory bowl feeder and how does it operate?
8. What are some design guidelines for vibratory bowl feeders?
9. What is the primary advantage in using a flexible assembly system?
10. List the equipment that might be used to construct a flexible assembly system.
11. In a flexible assembly system, what functions could be performed by the computer control system?
12. One of the economic advantages of a flexible assembly system is multiple vendor capability. What does this mean?
13. There are four basic classifications of industrial robots based upon its controller. What are these classifications, and what do they mean?
14. How does the teach pendant robot programming technique work?
15. What is the primary attribute of a material handling robot?
16. What are the primary considerations that need to be examined when designing a robotic system for light assembly?

10
Manufacturing Engineering Systems

10.1 CAD/CAM OVERVIEW

Computer-aided design (CAD) and computer-aided manufacturing (CAM) use digital computers, with their high speed and accuracy, as the integrating force throughout the entire process from engineering design to product manufacture. CAD/CAM evolves from and brings together such technologies as numerical control and computer technology. Tied together by the common usage of digital data, and fed by a continuous stream of electronics developments, numerical control and the digital computer have reached the maturity to permit a level of integration that brings us to the threshold of the long-envisioned automatic factory.

The various computerized functions under CAD/CAM fall into three general areas:

1. Design/drafting, or computer-aided engineering
2. Planning/scheduling, or management information systems
3. Fabrication, or manufacturing automation

A conceptualized or idealized CAD/CAM system is shown in Fig. 10.1. This shows how the user interacts with the computer via a graphics terminal, designing and controlling the manufacturing process from start to finish with information stored in a shared data base. With the advent of interactive graphics, the problem of the user entering data and instructions into the computer with stacks of coded punched cards or reels of tape went away. No longer was the user required to be experienced in computer programming and operation to use the machine. With interactive graphics, the user communicates with the computer in display-screen pictures. No

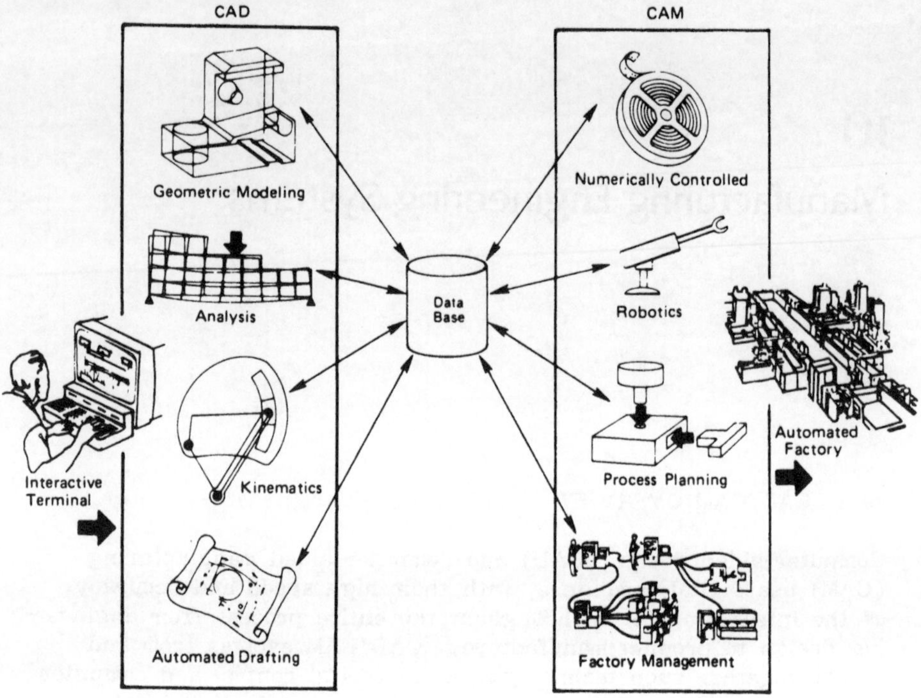

FIG. 10.1 Conceptual CAD/CAM system. (From Knox, CAD/CAM Systems Planning and Implementation, Marcel Dekker, 1983.)

knowledge of computers is required to operate these systems, and the communication is in real time, which means that the computer's response to the user's instructions is almost instantaneous. In CAD/CAM, the requirement is for the solution of three-dimensional mechanical design and manufacturing problems. Interacting with the computer via keyboard and lightpen or other pencil-like devices, the designer specifies points and lines on the display screen to quickly construct a display-screen drawing or model. This is in reality the representation of the diagram stored in the computer data base.

With a stroke of the pen, or by pushing a button, the designer can move, magnify, rotate, flip, copy, or otherwise manipulate the entire design or any part of it. For example, by pushing a button the designer may issue a flip command to produce a mirror image of a display for modeling symmetrical parts, or a translate command may be used to create models for linear parts. A cross section may be used to create models for linear parts. A cross section may be defined and then translated linearly to create a surfaced model. In a similar manner, circular parts can be easily modeled with a rotate command in which a cross section is rotated about a central axis to model parts. These operations take advantage of the computer's capability to repeat detailed operations quickly and flawlessly.

In working with a complex model, the designer can temporarily erase portions of it from the screen to see the area under construction more clearly. The deleted area is recalled later to complete the model. Portions of the model may be enlarged to view and add minute details accurately. The model may be moved and rotated on the screen for viewing at any angle. Mechanisms such as linkages and gears in the assembly may be animated on the screen to ensure proper check for interferences and ease of operation. When the design is complete, the system may automatically add dimensions and notes. The automated drafting part of the system may then be used to produce detailed engineering drawings.

After part geometry is defined with the completed model, the designer can have the computer calculate physical properties such as weight, volume, surface area, moment of inertia, and center of gravity. The finite-element method may be used to determine the stress, deflections, and other structural characteristics. After such an analysis, the display screen may show color-coded stress plots, the deflected shape of a part subjected to a given load, or an animated mode shape that shows how the structure might vibrate and deform in operation.

The result is that with a CAD/CAM system engineers and designers can view complex forms from various angles at the push of a button instead of having to construct costly, time-consuming models and mock-ups. Changes can be made quickly and inexpensively at the keyboard without requiring drawing changes, or modification of a physical model. In addition, the computer displays can produce realistic simulations of product operation before any hardware is actually produced.

After design completion, the resultant geometric data stored in the computer memory may be used to produce numerical control instructions for making the parts on automated machine tools, or to generate the artwork and drill tapes for automated printed circuit card fabrication. Many CAD/CAM systems can now produce NC instructions automatically for a range of different part types. Tool paths may be simulated on the monitor to validate, verify, and modify the program more rapidly.

An engineer at a computer-aided design graphics terminal can design a part, analyze stresses and deflections, study part dynamics, and then produce engineering drawings automatically. From the geometric description provided by computer-aided design, manufacturing engineers may produce numerical control instructions, generate process plans and work instructions, program robots, and provide plant management information with a computer-aided manufacturing system. With an integrated CAD/CAM system, a design is created and the manufacturing process is controlled and executed with a single computer system. These highly sophisticated systems are becoming a way of life at increasing numbers of plants throughout the industrialized world. As CAD/CAM capabilities have increased, costs for these systems have decreased to the point where a company can install a system for a few hundred thousand dollars, which puts them within the budget of many manufacturers.

In summary, CAD/CAM is comprised of distinct functional areas. The experts group CAD functions into four major categories:

1. Geometric modeling
2. Engineering analysis
3. Kinematics
4. Automated drafting

CAM technology also centers around four major areas or categories:

1. Numerical control
2. Process planning
3. Robotics
4. Factory management

For CAD/CAM to realize its full potential, the technology of the CAM functions must be advanced to the level of CAD sophistication, and CAD/CAM functions must be combined into a truly integrated system. Efforts to do this are under way, and the technology is changing so rapidly that it is difficult to remain abreast of the state of the art without a continuing and sustained effort.

10.2 NUMERICAL CONTROL AND COMPUTER NUMERICAL CONTROL

Numerical control is the operation of a machine by a series of coded instructions comprised of numbers, letters of the alphabet, and other symbols. These are translated into pulses of electrical current or other output signals that activate motors and other devices to run the machine. On machine tools, numerical control commands may range from positive positioning of the spindle in relation to the workpiece to auxiliary functions such as selecting a tool station on a turret, or controlling the speed and direction of spindle rotation. These commands, gathered together and logically organized so that they will direct a machine tool in a specific task such as the complete machining of a workpiece, comprise an N/C part program.

Such a program may be stored and utilized at some later time to obtain exactly the same results as the first time it was used.

Numerical control is a concept of machine control, not a method of machining. A part program is a clearly defined process for machining a workpiece. That process is then converted into coded instructions in a form that can be acted upon by a machine control unit, which converts the coded instructions into electric signals that then activate machine tool functions and cutting tool movements. It is important to remember that numerical control will not add any machining capabilities that are not built into the machine.

10.2.1 Numerical Control

Numerically controlled machine tools in production are used to machine parts:

1. With complex machining requirements
2. Requiring exceptional precision
3. In the development stage where many changes are required
4. Normally requiring expensive tooling
5. Requiring fast or slow machine speeds
6. Made from expensive raw materials
7. Required in small batches

The original intent of numerical control of machine tools was to make the machining of small lots of parts economical. The benefits of using numerically controlled machine tools are significant gains in production lead times and reductions in tooling cost and time, resulting in fewer tool designers, less complex tools, and, of course, less tool storage. The quality of manufacturing is increased, enabling closer tolerances, less inspection, fewer inconsistencies, and less scrap.

The sequence of events in developing an N/C program to machine a part on a numerically controlled machine tool begins with the part programmer (manufacturing engineer). He studies and analyzes the drawing or workpiece to determine how to make the part economically, what operations are required, in what sequence, on what machines, with what tools. He must conceptually visualize the steps and machine motions necessary to machine the particular part. He then must go ahead and document them in logical order on his part programming manuscript. This is done whether the program itself is being done manually or with the assistance of a computer. The manuscript data are then converted to a medium, usually punched paper tape, that can be acted upon by the machine control unit, or controller. Machine tools in and of themselves cannot read coded commands. The actual reading is done by a program reader that is part of an overall machine control unit, which we will call the controller. This unit actually converts coded instructions to obtain the resultant machine tool action.

The function of the controller, with its program reader, is to act upon the coded instructions developed by the programmer, and then generate output signals which control the servomechanisms used to actually drive and direct the machine tool. On the machine itself, as a slide motion occurs, a resolver or other feedback device continually determines the precise position of the slide. This position is fed back to the controller and compared with the control output signal. Signals to the drive motor drive the feedback device and the machine slide until a null condition with the control output signal is obtained.

The effective operation of the machine tool with numerical control is solely the responsibility of the programmer. It is he who decides the sequences, tools, and overall motions of the machine in machining the workpiece. The machine operator oversees the operation of the machine, loads and unloads the work, and provides information feedback and little else. Numerical control forces the programmer to logically envision and clearly state all machine actions. Numerically controlled machine tools do not think or exercise judgement, but they do follow instructions. Ill-conceived instructions will be followed with the same amazing dedication and efficiency as will logically organized program commands. An N/C program instruction calling for a move of 7.000 in. when the programmer means 0.700 in. will be faithfully followed since the machine controller has no inherent intelligence about what is right and what is wrong.

Numerical control is used in a broad spectrum of metalworking and other manufacturing operations. N/C positions the work or drives the tools on practically every type of machine tool, including drills, boring machines, milling machines, machining centers, lathes, turning centers, and grinders. In sheet metal shops, N/C is found on turret punch presses, brakes, shears, and flame cutters. It is used on electrical discharge machines (EDM), riveting machines, welders, and inspection machines. In electronic fabrication it is used on printed circuit drilling machines and on wire wrapping machines.

Numerical control tape formats are governed by the Electronic Industries Association standards RS-273 and RS-274, which define variable block tape format for positioning and straight-cut controls and for contouring and positioning controls, respectively. Figure 10.3 shows an RS-273 tape format, which contains a program for drilling four holes. All tapes use the binary coded decimal method of individual digit coding, as established by EIA standard RS-244. Five of the eight channels in the standard eight-channel tape are assigned values 1, 1, 2, 4, and 8, so that any numerical quantity from 0 to 9 can be designated on one horizontal level of the tape.

A 7, for example, would be a combination of the 4, 2, and 1 punched holes. A 3 is a combination of the 2 and 1 punched holes, and so on.

A multidigit quantity, as for an X-axis position, is expressed in consecutive horizontal rows, introduced by the binary code for the symbol X to identify the following data. Any numerical digit, letter, or symbol code has its own combination of holes in a single row on the tape.

The decimal value of any digit is determined by its relative position within a given numerical quantity. A controller may require, for example, that any X-axis designation be stated in five digits, with the first two being the numbers to the left of the decimal point and the next three those to the right, as shown in Fig. 10.2. Some controllers will allow all leading zeros to be omitted. Others permit all trailing zeros to be omitted, while others allow adding the decimal point to the tape, which then permits both the leading and trailing zeros to be omitted.

10.2.2 Computer Numerical Control

Computer numerical control (CNC) is a system in which the machine is connected to a dedicated minicomputer that stores the N/C instructions in its memory. Computer numerical control offers a number of advantages over standard numerical control. These include the following:

1. Instructions may be stored and handled more efficiently with a data cartridge, or floppy disk memory.
2. The minicomputer system may be programmed to perform functions beyond simply controlling the machine (machine-control data can be recorded and modified right at the computer itself).
3. Real-time and off-line diagnostic capabilities may be built into the CNC system.
4. Machining data and operator instructions may be displayed on a screen as part of the system.

The most sophisticated systems use direct numerical control (DNC). Often these systems have been referred to as distributed numerical control, because control of the machine tools is distributed among different computers. In such a situation, individual CNC units are linked to a central mainframe computer that supplies the part programs to the individual CNCs through communication lines, which also provide for feedback of production and machine tool status from the shop floor, allowing the coordination of machine tool operation from a central location.

DNC systems provide rapid, real-time feedback on problems as they occur, enabling immediate corrective action to be initiated.

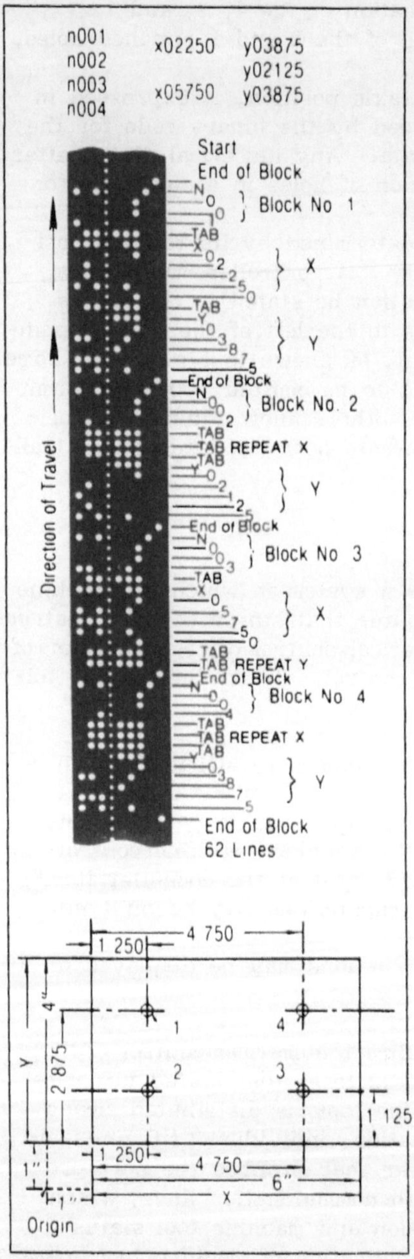

FIG. 10.2 EIA RS-273 tape format, which contains a program for drilling four holes. (Reprinted courtesy of Modern Machine Shop, Cincinnati, Ohio.)

This single feature of DNC systems offers significant advantages. The ability to store massive amounts of data on programs and to retrieve the data quickly is another significant advantage of DNC systems. Several different configurations of machine tools and computers are possible in a DNC system. Some systems may use a hierarchical arrangement of different levels of minicomputers performing various functions between the machine tools and the computer mainframe. Other DNC systems have eliminated the minicomputer altogether, and have a direct interface between the central computer and the machine tools. When a central computer is linked directly to the machine tools, two identical DNC computers are sometimes used, permitting the system to continue operating if one computer should fail.

Figure 10.3 provides a conceptual overview of how numerical control, computer numerical control, direct numerical control, machining centers, and transfer lines interface and operate separately as a system.

The CNC unit has several characteristics not found in the hardwired N/C controller, such as a CRT multifunction screen that displays the full operational or parametric data as a part is being machined. It also aids in job setup, program verification, program editing and update, and maintenance and troubleshooting. The CNC usually has an internal memory for program storage, with the capability to read in and store up to 2000 ft of part program tape, and with a floppy disk, multiple program part storage capacity may be as high as 16,000 ft of taped program. Many CNCs have a manual data input capability, allowing the part programming to be done by the machine operator. This capability has been most popular in grinder, lathe, and sheet metal applications that require relatively short part programs, where non-producing machine time would not present a problem. Many CNCs have a full alphanumeric keyboard to facilitate manual data entry.

The wave of developments in computer graphics is having a major impact on CNC. Controllers, already introduced or proposed, will display on the CRT two- or three-dimensional graphic views of the tool cutter path or of embedded macros and subroutines, for example, and will in some cases do so in color. Such a capability will greatly enhance the menu concept of programming and give some additional significance to shop-level part programming.

The minicomputer has given rise to new programming aids in the CNC which greatly simplify part programming and reduce the length of the part program tape. This is brought about in two ways:

1. With subroutines permanently stored in CNC memory
2. With shop-floor programming language stored in CNC memory

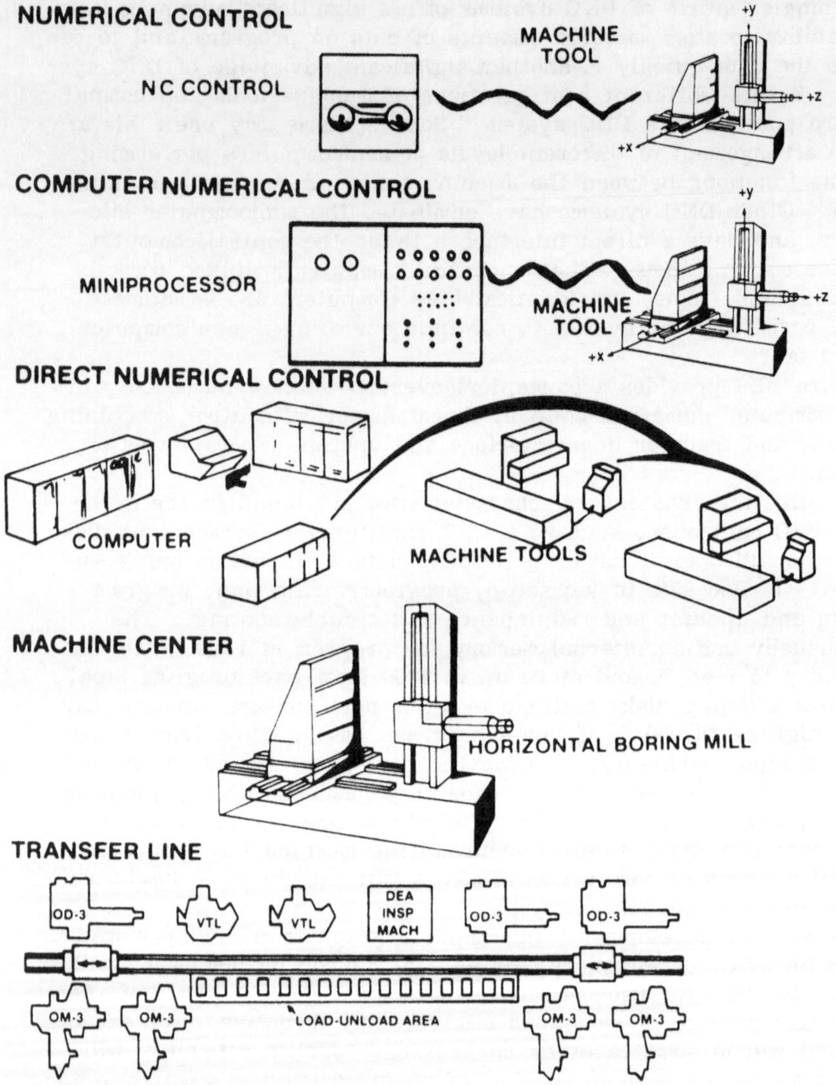

FIG. 10.3 Several types of N/C machine tools exist today. Normal N/C uses paper tape and controller. A computer may be used to control machine tools (CNC). A large computer may be used to control several minicomputers (DNC). Machining centers have up to five-Axis N/C machines. Transfer lines are using robots as well as N/C tools. (From Knox, CAD/CAM Systems Planning and Implementation, Marcel Dekker, New York, 1983.)

Often CNC executive programs are written so that they will execute common subroutines such as bolt hole circles and pocketing sequences from a single descriptive statement. Such capabilities are a definite aid when manual programming is used.

Some CNCs have enough computational capability built in that it is unnecessary to enter specific coordinate positions and other machine commands. It is only necessary to enter source data, which consist of workpiece descriptions or generalized statements about how the cutting tool should move. The computational capability within the controller itself will calculate the specific coordinate commands and then execute them. Such a capability is called shop-floor programming.

Another advance in control that features shop-floor programming language will be a multitask capability that allows the CNC to operate a machine from one part program, while an operator/part programmer creates another program. In some cases, a split-screen display will show data for the part being machined, while also assisting in development of the new program.

CNCs are being developed that do far more than execute standard instructions for making a part. The capability to sense cutting conditions and adapt to them without intervention of a human operator will soon be forthcoming. An example of this would be compensation for temperature variations sensed by sets of thermistors in the machine tool or the workpiece.

10.3 CELLULAR MANUFACTURING SYSTEMS

A new type of manufacturing system has emerged in recent years, made possible by the advent of the computer and of the technology of numerical control. It offers both cost and quality-control advantages over traditional manufacturing systems. This system is known as cellular manufacturing.

Cellular manufacturing systems are also referred to as work cell systems. The cellular system groups processes, people, and machines to manufacture a specific group of parts. The emergence of the cellular manufacturing system, and its highly automated form, the flexible manufacturing system, has resulted in a new type of production system which is capable of producing high-quality products at low cost.

Group technology is a systems-based rationale for solving the reorganization problems involved in setting up cellular manufacturing systems. It provides a computer-oriented data base and tools the manufacturing engineer can use to design the work cell. A group technology analysis develops the families of parts which can be manufactured by a flexible, cellular grouping of machines. The machines in the cells can be retooled so that one can rapidly change from one lot of parts to another, eliminating set up time or reducing it significantly. The elimination or near-elimination of setup time significantly alters the economics of lot or batch production to permit the economical production of very small lots.

Cellular manufacturing has existed for many years, but it has not been properly defined or well understood, and has not been recognized as a particular type of manufacturing system. A manufacturing cell can be defined as a cluster or collection of machines designed and arranged to produce a specific group of component parts.

Little exists in the way of rules or theory for designing cellular manufacturing systems, but the first rule should be that the design be as flexible as possible, so that it can be expanded to include other components or modified to handle additional members of the family. The objective is to link the cells into a large, manned or unmanned, integrated manufacturing system. Cells can be grouped into two general categories:

1. Manned cells: Contain machine tools which are conventional, numerically controlled, or computer numerically controlled, and production operators who have been trained and are skilled in the operation of more than one piece of equipment within the cell. Manned cells are efficient because the number of operators can be adjusted, and minimized to meet the desired output. The U-shaped cell offers the greatest advantage because the range of jobs which the multifunctional operator can cover can easily be increased or decreased as needed. As the production requirements of the cell are reduced, the number of workers is reduced accordingly, reversing Parkinson's law, and preventing operators from expanding the jobs to fill the increased time available.

The cells are tied to each other by a material-handling system, or are directly linked or combined into larger integrated lines.

In manned cellular systems, the worker is released from the machines, so that the utility of the worker is no longer tied to the utility of the machine (this may mean fewer workers in the cell than there are machines). The objective is to improve the utilization of the workers by making them multifunctional, capable of operating all the machines in the cell.

2. Unmanned cells: Contain CNC machine tools or other kinds of automated equipment, and few if any workers within the cell. Unmanned cells are classified by type of arrangement:

Fixed automated: Such cells are classically represented by the automatic transfer line, in which the quantities or lots are large and the production runs long. These systems are generally arranged in straight lines, in circles, or in U shape. They usually contain a conveyor which both locates the part and transports it from the machine station. The line is balanced such that the part spends the same amount of time at each station. The volume of parts is very large and the variety very small. These are not flexible cells.

Flexible automated: These cells involve the flexible manufacturing system concept, and the robotic cell. The flexible manufacturing system is generally laid out in a straight line, or a rectangular design with a computer-controlled conveyor to transport the parts to any machine, in any order. These would be CNC or DNC machines, and therefore can change tools and programs to handle different parts. Parts can be introduced into the cell or system in any order; consequently, this system works on a family of component parts with medium to large lot sizes. These systems are large and expensive and typically contain up to 12 machine tools. System cost parallels that of the automatic transfer line.

The robotic cell generally has few machines, and is arranged so that a material-handling robot can load and unload the machines and change the tools in them, if required. Robotic cells are, for the most part, circular in design to take advantage of the spherical or circular spatial range of the robot. Some of the newer machining centers have a material handling robot as an attachment to the machine, to load and unload parts, eliminating the operator.

In setup design, the objective is to eliminate setups between the different component parts in the parts family, or to be able to use essentially the same fixtures for every part in the parts family, with quick action modification, or one-touch setup, between parts. In manned cells this objective can readily be accomplished, as demonstrated by numerous Japanese and American manufacturers, with the result that it is now possible to economically produce in very small lots.

Group technology is a manufacturing management philosophy which identifies and makes use of the sameness of items and processes used in manufacturing industries. It groups units or components into families of parts which have similar design or manufacturing sequences. Figure 10.4 shows such a grouping of parts. Machines are then arranged into groups or cells to process the parts family. Portions of the functional system are converted, in steps, to the cellular system or to a flexible manufacturing system. Finding families of parts is one of the first steps in converting the functional system to a cellular or flexible system.

Most companies that have converted to cellular manufacturing use a parts coding or classification system to sort parts into classes or families on the basis of their similarities. Coding assigns symbols, letters, or numbers to specific part elements based upon differences in shape, function, material, size process, and so on. Most coding systems are computer-compatible, so that computer sorting of the

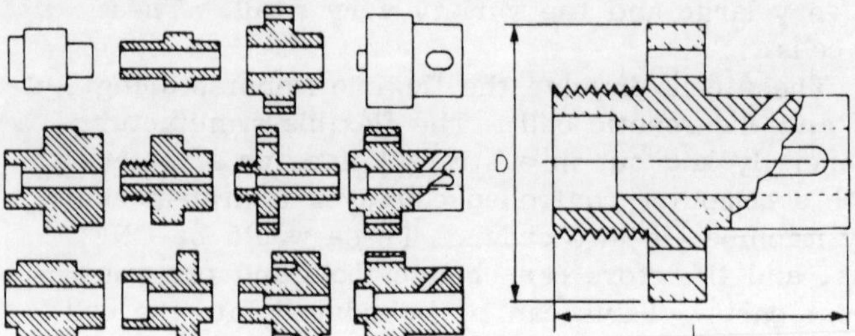

FIG. 10.4 Utilizing group technology, parts grouping utilizing parameters of diameter, length, material, and weight. Acceptance parameters: depth, 4.8 in., length, 12.6 in., material: ferrous casting; weight: 200 lb maximum. (From Knox, CAD/CAM Systems Planning and Implementation, Marcel Dekker, New York, 1983.)

codes then generates the classes or families of parts. The composition of the parts families, and therefore of the cells designed to manufacture them, is dependent upon what characteristics it is decided to sort on.

Depending upon their material flow, different families will require different layout arrangements. In short, the manner in which the parts family is determined influences the design of the cell. In some parts families, every part will go to every machine in exactly the same sequence, no machine will be skipped, and no reverse flow will be allowed. This would be a cellular system in its purest form, except for the machining center in which all parts are made on one machine.

Other parts families may require that not all parts go to all machines, or that the forward sequence of order through the machines be changed, or that back flow in the group be provided for. Flexible manufacturing systems are designed to accommodate these situations. The following rules may prove helpful in the design of manufacturing cells and flexible manufacturing systems:

1. There should be sufficient volume of work in the parts family to justify establishment of a cell. Some families may be too small to form a group of machines, but would be excellent candidates for processing on a single machining center (the smallest form of cell).
2. The makeup of a parts family should permit a satisfactory machine utilization situation.
3. The processes in these systems should be technologically compatible.

4. The required system capacity must be determined from the quantities of parts needed and the production schedule, which determines when they are needed. Problems of balance and machine utilization must be solved.
5. The physical reorganization of the existing manufacutring systems will require redesign of the production system. This will require resolution of certain problems: product design will be affected in terms of new parts and standardization of old parts, and planning and scheduling for cellular manufacturing systems will be different.

In advanced forms of unmanned cellular layouts, the microcomputers of the CNC machine tools are networked together with a robot for material handling within the cell. All the machines in the cell are programmable, and therefore this kind of automation is very flexible, as well as more economical than traditional job shop processing methods.

Progression from the traditional job shop to the shop with manned cells, to clusters of CNC machines, to an entire system of linked cells must be accomplished in logical, economical, justifiable steps, each building from the previous state. Simply adding robots and computers to an existing job shop is no substitute for a well-planned system conversion to cellular manufacturing. Converting to cellular manufacturing is a systems level change. It requires the best in planning and the full cooperation of everyone involved. All must understand the problems, costs, and limitations of making such a conversion and the long-term effort that will be required. The magnitude of such a change also means that it offers the potential of tremendous savings and other meaningful advantages.

10.4 COMPUTER INTEGRATED MANUFACTURING SYSTEMS

The metalworking sector of the industrialized world is moving forward rapidly toward what we will call the automatic factory. Advances in numerical control and computer technology have made the lightly manned automatic factory technologically possible. As other manufacturing sectors follow the metalworking sector, we begin to enter the world of computer integrated manufacturing systems.

The computer integrated manufacturing system is built around the computer hardware and system software that are required to put it in operation. Data communications and networking capabilities are then needed to link the system to other computers and integrate it into the company manufacturing information network. Software solutions for specific applications rest atop the other components. The final component is one of the most critical: the internal and exter-

TABLE 10.1 Hierarchy of Computers in Manufacturing

Manufacturing application	Computer system hierarchy	CAD/CAM connection	Type of computer system	Response time
Marketing and distribution Financial planning Manufacturing planning	Resource planning		Batch processing	Weeks
Purchasing Inventory control Shop floor control Quality management Maintenance management	Manufacturing control		Transaction processing	Days Hours
Process monitoring and control Machine monitoring and control Facilities monitoring and control Test and inspection	Monitoring & supervisory control	Computer aided engineering Product design Test design Process design		Minutes Fractions of a second
Numerical controllers Program controllers Robot controllers Loop controller Microprocessors	Dedicated equipment control			Real-time/interrupt-driven processing

nal implementation support required to make the CIM system successful and make it fit smoothly into the overall company information flow to produce the expected results.

The foundation of CIM is the four-level hierarchy of computers shown in Table 10.1. A manufacturer cannot simply buy any type of computer and put it to work anywhere. Different types of computers, each configured to perform a specific function and set of applications, must be matched to the work they will be required to do. Once this is recognized and made an important part of the company's CIM plans, CIM can be implemented at whatever pace is comfortable. That is the key to a successful CIM installation. A company should think big, but start small. The first CIM system can go anywhere. Once the initial system is successful, others can be added a step at a time.

10.5 ASSEMBLY WORK INSTRUCTIONS AND VISUAL AIDS

Systems are now available that utilize computer aided drafting packages to create specialized manufacturing instructions from existing drawings and text for display on shop floor terminals for production operators to use in product assembly. This kind of capability is especially valuable in specialized, high technology manufacturing where the operator never really learns what he or she has to do, and must work from written instructions. These operations, particularly in assembly, are characterized by low volume, infrequent operation, continuous changes, and volumes of detailed work instructions or process plans. These assembly work instructions can run from half a million to two-and-one half million pages of specially prepared paper instructions. One such system, called Manufacturing Instruction Display (MID), has three ways of transferring engineering drawing information into the system: CAD-programmed drawings can be transferred directly into the system, paper drawings can be scanned and digitized, and system drawing tools allow the manufacturing engineer to create new illustrations or visual aids from scratch. Once the drawing information has been loaded into the system, the manufacturing engineer can make the final work instructions, including visual aids, in a few minutes from an interactive work station.

The assembly operator can call up the instructions, one screen at a time using function keys, a footswitch or a speech recognition system. A typical factory floor setup would provide each assemply operator with a display terminal mounted at a point that provides maximum visibility and convenience. Authorized workers enter their names and passwords. A screen menu then gives the option of typ-

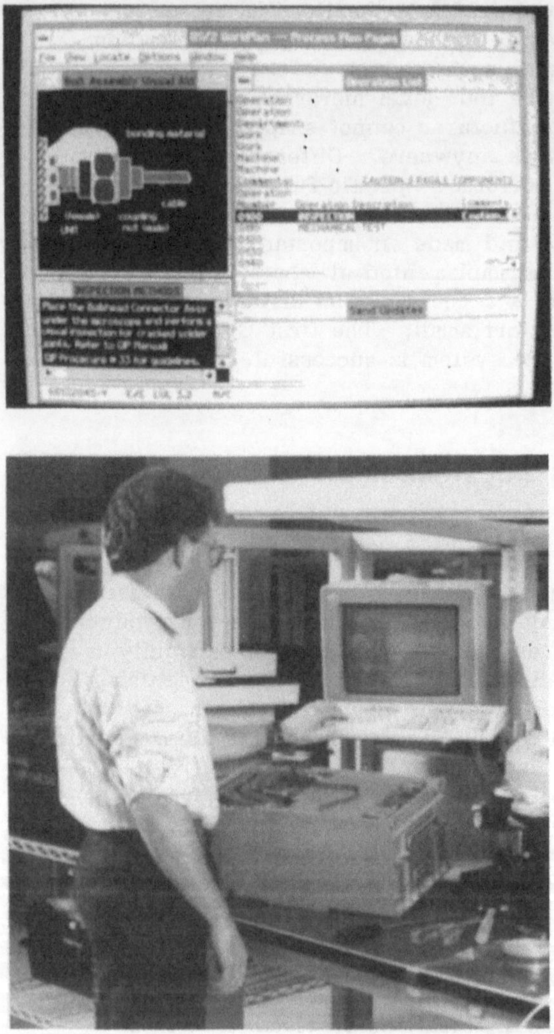

FIG. 10.5 Computer assembly work instructions displayed at work station terminal. (Courtesy of IBM Corporation, Santa Monica, California.)

ing in a job number, or viewing a menu of jobs defined in the control file. When the worker selects a job, the first instruction screen for that task appears on the display. The worker also has the capability of stepping through the screen sequence, or calling up and reviewing the previous screen. Figure 10.5 shows a representative assembly screen display of work instructions and visual aid, and a work station with a display terminal for operator use.

10.6 STATISTICAL PROCESS CONTROLS

The techniques of statistical process control have been known for many years, but only recently have these useful mathematical tools begun to be understood and used in American manufacturing. It is important that the manufacturing engineer understand and be able to apply these tools in his work. Terms such as process capability, control charts for measurements, control charts for attributes, and frequency distributions should be as familiar to the manufacturing engineer as they are to the quality engineer. Statistical process controls are a critical part of manufacturing engineering systems.

10.6.1 Process Capability

The purpose of process capability analysis is to determine the natural variation of a manufacturing process when the effects of all extraneous factors not contributing to the process have been minimized. Process capability may be defined as the minimum spread of a specific measurement variation which will include 99.7% of the measurements from the given process. Since a capability analysis is made to determine natural variation, only 6 sigma prime is used instead of the more familiar plus or minus 3 sigma prime (see Appendix C, Statistical Process Control Statistics).

When making the study it is important to minimize the effects of factors extraneous to the study. These include:

1. Unnatural material variations
2. Process adjustments
3. Process deterioration

When these conditions have been noted and accounted for, the process capability analysis may be made by:

1. Gathering at least 50 observations; preferably 100 or more, if possible.
2. Computing the standard deviation of this data.

One of the easiest methods for determining process capability uses the frequency distribution. It is preferred primarily because it is the easiest to calculate, requiring no square root. An example of how this works is shown in the following actual establishment of a production tolerance for a drilling operation. The drawing specification for a small, flat cover plate allowed a tolerance of plus or minus .001 inch between the center of two drilled holes. Considerable production difficulty was being experienced in meeting this tolerance and a large number of cover plates were rejected at final inspection. The manufacturing engineer claimed that the required tolerance could not be met with existing drilling equipment. The designer who had placed the tolerance on the drawing was equally certain that it could be met.

To resolve the question on a factual basis, a frequency distribution analysis was made on a typical production lot. Both the manufacturing engineer and designer assured themselves that the drill press was properly set up, that its drills were sharp, and that the drill jigs were in good condition. The resulting frequency distribution of the tolerance variations showed a variation of plus or minus .0025 inch in the center distances of the cover plates from the sample lot under examination (see Fig. 10.6). Since all had agreed that the sample was fully representative, it was decided that the ±.001 inch tolerance could not be met with existing equipment. Three alternatives were available:

1. Continue to produce cover plates to the current drawing tolerance with the existing equipment and treat the resulting rejects as a part of the job cost.
2. Widen the tolerance.
3. Buy new drilling equipment.

Further investigation of the machine on which the cover plate was assembled by the designer showed that a between-centers tolerance of ±.003 inch would be fully as satisfactory as one of ±.001 inch. The drawing was changed accordingly and no further production difficulties were experienced on this part.

10.6.2 Process Control Charting

Statistical process control involves determining the truly critical dimensions or parameters (usually several, not many) and establishing control charts for either measurements or attributes to monitor the data for out-of-control, or possible-out-of-control trends, and take corrective action before parts or assemblies are rejected. The mathematics for statistics and control charting is covered in Appendix C.

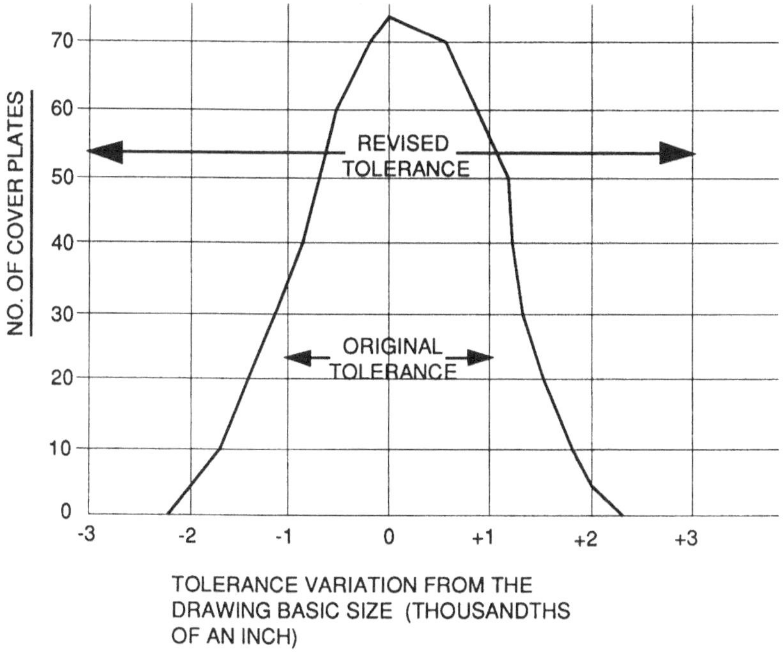

FIG. 10.6 Center distance tolerance study.

Control charts for measurements can be established by computing the control limits in relation to the average value for the process, or the so-called process average method:

1. Select the quality characteristic to be controlled (length, width, hardness, etc.).
2. Gather data by selecting an adequate number of samples of the part in question and measuring the chosen characteristic (sample of 25, sample size of 5 parts).
3. Compute values of the average and range for each of the 25 samples.
4. Compute the grand average of the averages of the 25 samples.
5. Compute the average of the ranges of the 25 samples.
6. Compute the control limits based upon these sample averages and ranges.
7. Analyze the average and range values for each sample with relation to these control limits. Determine if any factors requiring corrective action are present before finally approving calculated control limits.

8. Determine if control limits are economically satisfactory for the process.
9. Monitor the control charts during production as the basis for control of the characteristic in question to insure that the process average and spread have not changed in any significant degree.

In order to detect shifts in a process average in manufacturing, the best approach is to apply a few simple rules that depend only on extreme runs. Consider that grounds exist for suspicion that the process average has shifted:

-Whenever in 7 successive points on the control chart, all are on the same side of the central line.
-Whenever in 11 successive points on the control chart, at least 10 are on the same side of the central line.
-Whenever in 14 successive points on the control chart, at least 12 are on the same side of the central line.
-Whenever in 17 successive points on the control chart, at least 14 are on the same side of the central line.
-Whenever in 20 successive points on the control charg, at least 16 are on the same side of the central line.

Sequences of points on one side of a center line are often useful in relation to possible action on the control chart, even though they may not be used for hunting for trouble in the process. That is, they might suggest the need to review the position of the central line on the chart to determine whether it ought to be changed. These rules are applicable to both control charts for measurements and control charts for attributes.

Control charts for attributes, when applied to 100% inspection can have any or all of the following purposes (Fraction Defective or p chart):

1. To discover the average proportion of defective articles or parts submitted for inspection over a period of time.
2. To discover out-of-control high spots that call for action to identify and correct workmanship problems (bad soldering).
3. To discover those out-of-control low spots that indicate either relaxed inspection standards or erratic causes of quality improvement which might be converted into causes of consistent quality improvement.
4. To highlight any changes in the average quality level.
5. To suggest places for the use of $\overline{X}$ and R charts for the control of measurements to diagnose quality problems.

There are numerous excellent software packages available on statistical process control that can be run on shop floor personal computers. This eliminates the time-consuming manual calculation and plotting of data. It provides rapid display of trends, enabling corrective action, if required, to be taken in an immediate and timely manner. Data is entered by the operator through a data entry keyboard or by direct readings off the machines.

10.6.3 Summary Advantages of Statistical Process Control

Statistical process control eliminates expensive and often redundant visual inspection at each step in the process. It removes the human element and the problems associated with dependence on shop inspection to find rejects after they occur. Since the data is recorded by operators, there is a significant reduction in number of inspectors needed. With statistical process control the process is always under control, and scrap, rework, and rejected parts or assemblies drop to near zero.

QUESTIONS

1. What are the four major functional areas of CAD and CAM?
2. What is required for CAD/CAM to realize its full potential?
3. What is numerical control and how does it operate?
4. In what manufacturing situations would it be appropriate to use numerically controlled machine tools?
5. What is computer numerical control, and how does it differ from standard numerical control?
6. What is direct numerical control and how does it differ from computer numerical control?
7. How does a cellular manufacturing system work and what are its primary advantages?
8. What is group technology and where is it used?
9. Where would a computerized work instruction system be of great advantage?
10. How would you define process capability?
11. What is statistical process control and what are some of its advantages?

11
Manufacturing Research and Development

11.0 INTRODUCTION

Another important function of the manufacturing engineer is manufacturing research and development. Often, this is overlooked in the day-to-day functioning of the manufacturing engineering organization, and management often looks the other way when it comes to spending money to support this important function. Since there is no immediate return on investment from an accounting point of view, management shows little interest. If the United States is to again become the world leader in manufacturing, this short-sighted approach must change. Many of the more progressive companies now recognize this and have setup and funded an ongoing manufacturing research and development function. This means survival in the long term, and a competitive edge in world markets. The Federal government has also recognized the importance of research and development in manufacturing technology, and the National Institute of Standards and Technology (formerly the National Bureau of Standards) has taken the lead in working with other government agencies to fund work in advanced manufacturing technology, and in cooperation with private industry, is beginning to make significant progress. The resulting technology is transferred to private industry.

11.1 MANUFACTURING RESEARCH AND DEVELOPMENT FUNCTIONS

Manufacturing research and development functions best as part of the manufacturing engineering organization. It should be staffed primarily by manufacturing engineers, industrial engineers and skilled technical people. There should be a clearly defined set of objectives to be attained, and a time-phased plan established with measurable milestones in order to measure progress. A medium-sized company might be expected to have the following functions or activities for manufacturing research and development reporting to the manager of manufacturing engineering:

1. An experimental machine shop
2. A methods laboratory
3. A robotics laboratory
4. A materials and processing laboratory
5. An advanced manufacturing technology group

11.1.1 Experimental Machine Shop

The experimental machine shop has the primary mission of supporting the methods and robotics laboratories, of fabricating experimental tooling, and of developing new machining and cutting tool technologies. It might also be active in new metal forming and fabrication technologies. It would be staffed with a small group of precision machinists and toolmakers, under the direction and guidance of a few highly skilled tool and manufacturing engineers. The equipment would include both conventional and computer numerical controlled precision machine tools. The Automated Manufacturing Research Facility (AMRF) at the National Institute of Standards and Technology is a prime example of the ultimate in an experimental machine shop. Figure 11.1 shows the floor layout of this facility.

The National Institute of Standards and Technology decided in late 1980 to establish a research facility to investigate critical issues in factory automation standards. The first major goal of the AMRF involved the construction of a flexible manufacturing system testbed for the small-batch manufacturing environment. It consists of six manufacturing workstations that occupy a 5000 square foot area in the institute machine shop. The workstations perform horizontal milling, vertical milling, turning, cleaning and deburring, inspection, and material handling using robot carts. The facility is being used by government, industry, and academic researchers to develop, test, and evaluate potential manufacturing interface standards. To ensure that the system interface issue is addressed, the testbed is is designed to contain component modules from a variety of vendors.

A major goal of the AMRF has been the transfer of resulting technology to industry. Technology findings are transferred to the private sector through the Institute's Industrial Research Associate Program, participation of staff in factory standards organizations, conference papers, technical reports, staff visits to industrial sites, public demonstration of AMRF technology, and various other programs established with sponsors and participants.

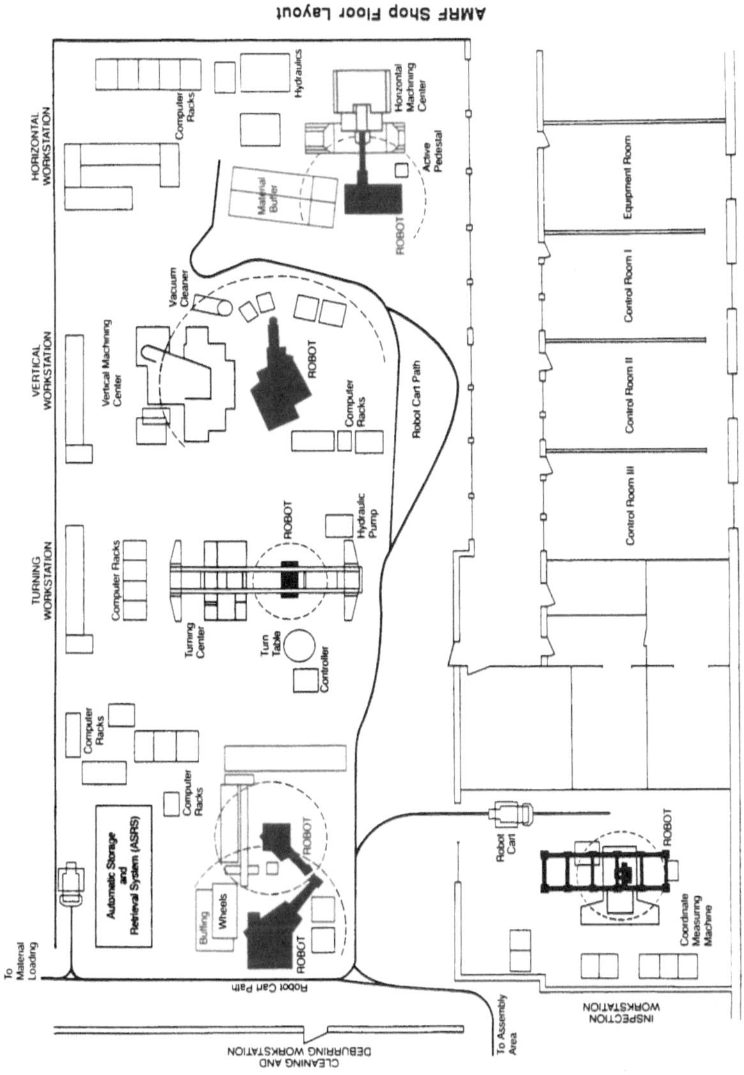

FIG. 11.1 AMRF Shop Floor Layout.

11.1.2 Materials and Processes Laboratory

The primary functions of a materials and processes laboratory would be the testing of new materials and chemical processes for use in manufacturing and in the design of new products, and the preparation of manufacturing process specifications for these new materials and processes. This includes the use of new adhesives for assembly, paints and finishes for manufactured products, chemical plating and finishing including passivation of stainless steels, anodic films for aluminum, chromate films for preparation of surfaces to be painted, plastics and the various composite materials, including pilot set-ups and runs of the associated formulations and processes. Included also are manufacturing process specifications for such standard manufacturing processes as soldering, painting, potting and encapsulation, heat treating of metals, and more.

This laboratory would be staffed with experienced chemical and manufacturing engineers, supported by a small staff of trained and experienced technicians and technical writers. Required facilities would include a chemistry lab equipped with thermal testing, sinks, drains, deionized water, vibration chambers, and more. It might include small pilot lines for plating and chemical processing, for printed circuit fabrication and similar processes. Depending upon products manufactured, there also could be a composites laboratory equipped to do filament winding, braiding, composites lay-up and bonding, vacuum impregnation, thermo-compression bonding and pultrusion. It would also have the capability to test, evaluate and develop processes for adhesive bonding, potting, encapsulation and the manufacture of plastic parts form thermoset resins and filled epoxies.

The actual makeup of capabilities for manufacturing research and development depends upon the company's product lines and plans for the future. The important thing is that there should be a materials and processes capability in the manufacturing research and development organization.

11.1.3 Methods Laboratory

A methods laboratory is absolutely essential in most manufacturing companies, even small companies. This lab provides an enclosed area where new methods, new tools, new workplace layouts, and even pilot production for new products, and new manufacturing systems can be tested and debugged before they are introduced on the production floor. Such an enclosed area for full-time use by the manufacturing and industrial engineers saves many hours and dollars that would otherwise be expended in trying to do a tryout and a shake-down on the production line before the new methods, tools, processes, and systems are really ready.

All of the manufacturing engineering organization, assisted by a few highly trained and skilled production operators would use the methods laboratory to tryout and debug work instructions, line layouts, tools, and so on. The cost for such a facility is minimal, since it would be primarily equipped with shop work benches, stools and basic shop tools. It must be a large enough, fully enclosed and in a secure area under the exclusive control of the manufacturing engineering organization.

11.1.4 Robotics Laboratory

In the modern world of factory automation, the manufacturing research and development function should have an area where robotic and automated manufacturing systems, software, and hardware can be setup, tried out, analyzed and evaluated for company manufacturing operations. This laboratory functions similarly to the methods lab and has essentially the same requirement for a dedicated, enclosed, and secure area away from normal factory production operations. In many companies this work is done in the methods laboratory along with the routine methods lab operations.

This laboratory would be available for use by all of the industrial and manufacturing engineers with an interest, a specific job, or application requiring robotics and factory automation. The National Institute of Standards and Technology funded as part of the Navy's Rapid Acquisition of Manufactured Parts (RAMP) program has set-up an automatic and computer-directed printed wiring assembly workcell which is an excellent example of a combination methods and robotics laboratory project and setup. Figure 11.2 shows a plan view of the workcell. A printed wiring assembly can be defined in this situation as a printed circuit board populated with discrete electronic components. The printed wiring board assemblies are no larger than 18" × 18" and can have through-hole and/or surfacemounted components. The prototype system revolves around an automated storage and retrieval system that serves individual workstations clustered around it. Operations are performed either by computer-directed work stations or by automatic assembly. When fully implemented, the workcell is designed to do the following:

1. Provide machine-interpretable inputs to systems.
2. Have an average lot size of five.
3. Have repeat orders 50% of the time.
4. Have a system throughput of 15,000 parts per year.
5. Implement modular, open-ended architecture with commercial off-the-shelf products.
6. Interface with site-selected support functions.
7. Have a desired output of 15,000 printed wiring assemblies per year.

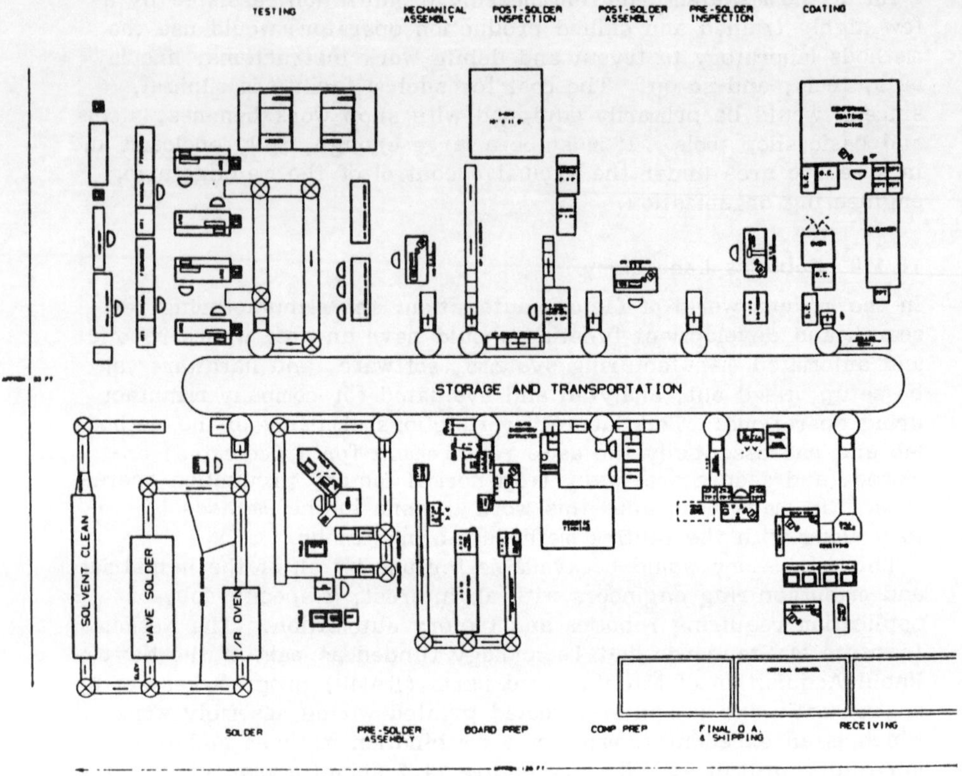

FIG. 11.2 Printed wiring assembly workcell (Courtesy of National Institute of Standards and Technology, Gaithersburg, Maryland.)

11.1.5 Advanced Manufacturing Technologies

Advanced manufacturing technologies are the directing force behind the manufacturing research and development effort in the company. This small group of experienced engineering professionals includes manufacturing and industrial engineers with expertise in specialized areas of manufacturing. Their primary function is to research reports and trade literature, to visit and tour other facilities with new systems and technology, attend seminars and participate in conferences as required to be abreast of state-of-the-art technology in their particular area of expertise, and develop and plan applications to company manufacturing operations. This would also include stocking and keeping a library of current information, reports, catalogs and other documentation of advanced manufacturing technologies. Some projects where manufacturing research is needed are described in the following pages of this chapter.

11.2 MACHINE AND EQUIPMENT UTILIZATION

Before the past several years, little serious effort in the United States was devoted to the reduction of setup times and the more effective utilization of machine tools and production equipment. This needs to be an important area of research and of effective development to minimize these major causes of inefficiency and high production costs, especially in the job shop environment. This task, along with robotics and the implementation of the automated factory, must be led by manufacturing engineers, who are the ones best equipped by training and experience to lead. Effort that is expended on this task offers the potential for very high cost reductions and savings for the time invested. Figure 11.3 shows the average utilization of conventional machine tools in the United States today. This shows

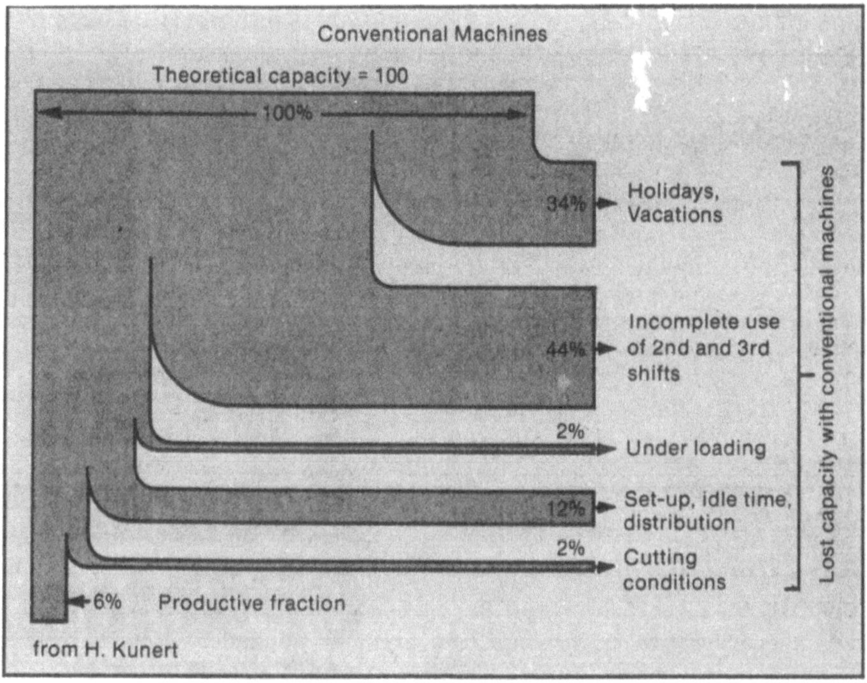

FIG. 11.3 Utilization of conventional machine tools. (Reprinted with the permission of Industrial Engineering Magazine, Vol. 16, No. 6, June, 1984. Copyright © Institute of Industrial Engineers.)

vividly the underutilization of this very important capital equipment and the potential that exists for it to be improved.

11.2.1 Increasing Machine Utilization

The controllable elements that are the cost drivers in machine utilization are:

1. The incomplete utilization of second and third shift capacity. This accounts for 44% of lost or idle capacity.
2. Setup time, and idle time distribution. This accounts for 12% of lost or idle capacity.
3. Cutting tool conditions contribute about 2% of lost and idle capacity.

Holidays and vacations cause 34% of lost time and idle capacity of machine tools. Better planning and scheduling of vacations and of manpower availability will help to some extent in this area, but the real improvements will come from automation and unmanned FMS manufacturing systems. Such systems will also aid in better second and third shift utilization and in the reduction of setup and idle time.

11.2.2 Setup Time Reduction

Figure 11.4 shows an analysis of total machining time and flow time in a conventional manufacturing facility in the United States today. This study by H. Opitz shows that only 5% of total production time is time actually spent on the machine, and of this available machining time only 38% is spent in actually cutting metal; the balance is machine setup and necessary noncutting time. This all means that during 93% of the time a workpiece spends in the machine shop, it is not on a machine, and no value is added. Since work in process is directly proportional to waiting time, if workpieces went directly from one machine to the next with no waiting or time for transport, there could be a 93% reduction in inventory cost, space required, and associated problems. In-process inventory tends to hide manufacturing problems of organization and control that do not really become obvious until the inventory is reduced.

The factors outlined above have a continuing negative impact on productivity. This situation will not improve unless positive action is taken to implement new systems that are less dependent upon labor, further the use of automation, better match the production facilities with product requirements, and increase plant flexibility for fast response and rapid adjustment to dynamic market conditions. Implementation of such systems would do a great deal toward utilizing much of the unused production capacity that exists today. There can be little doubt that significant manufacturing research and development resources should be directed to these problems.

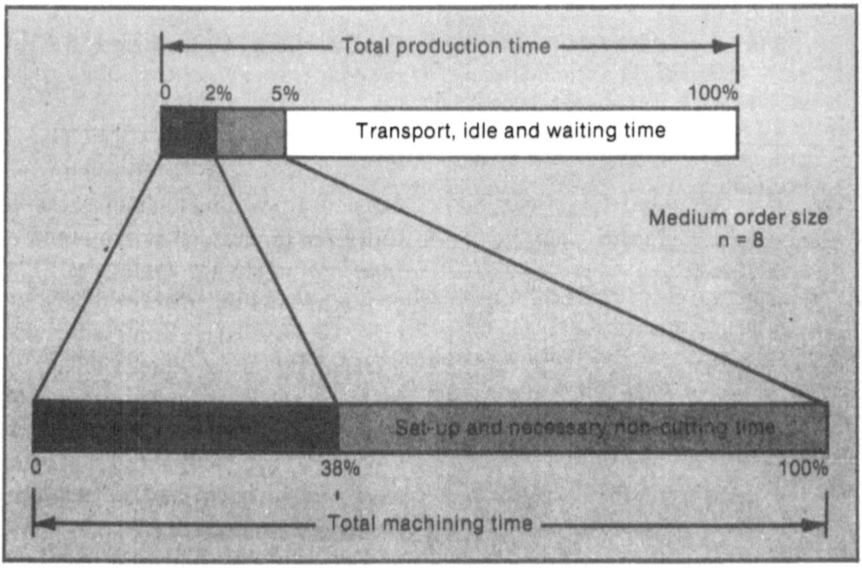

FIG.11.4 Analysis of total machining and shop flow time. (Reprinted with the permission of Industrial Engineering Magazine, Vol. 16, No. 6, June, 1984. Copyright © Institute of Industrial Engineers.)

11.3 ADVANCED MANUFACTURING TECHNOLOGIES

New process technologies are another area where significant resources in manufacturing research and development offer the potential for significant cost reductions. Laser technology, fiber optics, composite structures, electron beam welding, and adhesive bonding are some of the process technologies that offer practical manufacturing uses and applications. Of course, there are many others that also have promise.

11.3.1 Laser Technology

"Laser" means light amplification by stimulated emission of radiation. Laser technology produces a light source that is monochromatic (one color of light), has low divergence, and is coherent in terms of phase. Lasers start with a light source that is nonmonochromatic, incoherent, and somewhat divergent. Through filtering, all other light types (except red) are eliminated. We now have a multiple wavelength of one light type that is still divergent. Next, a uniform distance is fixed between wavelengths. Finally, the monochromatic beam is channeled into one fixed column to the workpiece. The laser has three basic requirements: a laser medium, and excitation source, and an optical resonator. These are necessary for the generation of any laser beam. Several key characteristics of lasers which make them invaluable in manufacturing are:

1. Lasers can produce a very high energy density per square centimeter.
2. Excellent focusing capabilities permit increasing the power density to very high levels.
3. Lasers require no physical contact. Therefore, there would be little or no deterioration or tool wear.
4. Lasers can work in any atmosphere and can be automated and controlled.

Laser manufacturing applications range from intricate hole drilling to accurate high-speed cutting, and spot and seam welding, of sheet metal shapes. One state-of-the-art system that has just come on line is a model 580 laser desoldering machine built by Apollo Lasers. This machine, built especially for the desoldering and removal of clipped leads and conformal coating buildup on assembled circuit cards, uses an 80-watt laser with an 8-mm beam diameter, and premixed $CO_2/He/N_2$ gas at 1 ft per 3 hr as the laser material. The laser wavelength is 10.6 microns, with random polarization.

The laser output beam is guided from the laser head to the X-Y table by means of 90° beam benders. A focusing lens then concontrates the beam to a fine pinpoint for maximum efficiency. The focal point height is adjustable over a 2 in. range. A visible beam pointer is used in this application, as the beam of the CO_2 laser is not visible to the human eye. The underside of the focusing lens is shrouded by a nozzle through which air is pumped continuously. The air performs two functions; it cools the lens so that water cooling is unnecessary, and it prevents smoke or debris from coating the lens by purging the space beneath it. Both of the functions greatly extend lens life. Air jets are provided to blow the clipped wire leads and molten solder off the circuit cards. The air jets operate off filtered shop air and are turned on and off by computer numerical control.

The laser desoldering machine is operated by a single operator, who loads circuit board assemblies onto a tooling plate, adds flux to the solder joints, then places the tooling plate on the X-Y table. Once the enclosure doors are closed, the operator depresses a start switch; then the CNC moves the tooling plate and circuit cards from point to point beneath the laser beam output.

At each solder joint, the laser turns on and melts the solder. When a programmed time has elapsed, air jets above and beneath the circuit card assembly will turn on to blow the clipped leads and solder away. Once all solder joints have been processed, the CNC moves, the X-Y table to the unload position, the tooling plate with processed circuit card assemblies is removed, and the next tooling plate with fluxed circuit card assemblies can be loaded to start the cycle over again. The process time per board is 47 sec, six times faster than these boards can be processed manually with a conventional solder sucker.

11.3.2 Fiber Optics

Fiber optics is that branch of optics dealing with the transmission of light through small filamentary optical elements or fibers, which are transparent. Fiber optics products may be divided into two basic categories: Those which are primarily light carriers and are nonimaging, and those which are imaging. In both cases light is carried by individual fibers independently of neighboring fibers. Only the fiber ends can be used to distinguish the different categories.

Manufacturing applications using monofibroscopes and multifiberscopes to illuminate enclosed areas and cavities and then send an image of the illuminated area have been successful. Such devices enable visual inspection of otherwise inaccessible areas of parts and assemblies. They also make it possible to perform what would otherwise be blind assembly operations.

The possibilities offered by fiber optics in manufacturing have barely been considered, and little effort has been made to exploit this technique.

11.3.3 Other Technologies and Processes

Research into the fabrication of composites and composite structures offers advantages in strength and weight reduction not found in the usual structural materials such as steel and aluminum. In the aviation and aerospace industries significant advances have been made in the process of fabrication for these lightweight and highly heat-resistant materials. Figure 11.5 shows how such a structural member would be built up from layers of glass cloth and graphite epoxy tape to form a longeron for the fuselage of an airframe. Once laid in place, the structure is formed and fused under pressure and heat to its finished state. Composite materials can be machined and assembled using conventional machining and assembly methods. Fabrication costs, however, are not yet competitive with those for the more traditional structural materials, and research and development in this area offers a real potential for application of these materials to the automotive and pleasure boat industries.

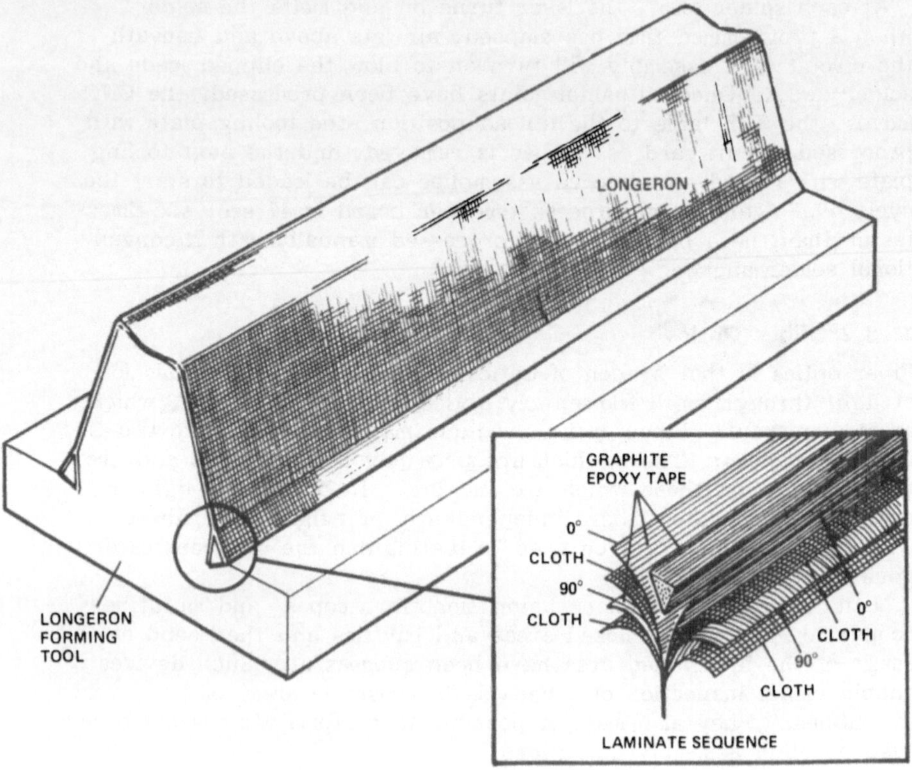

FIG. 11.5 Longeron fabrication using composites. (Courtesy of Martin-Marietta USA, Orlando, Florida.)

Methods of joining and assembly constitute an area where much remains to be done to reduce manufacturing costs. Current state-of-the-art technologies include electron beam and laser welding and much increased usage of adhesives and bonding methods. Advances in the methods of surface finishing and painting have been significant in recent years, but much still remains to be done.

The manufacturing engineer must take the lead in much of this work. A significant part of the resources and manufacturing research and development. More needs to be done by the smaller organizations. This will require selling management on the direct bottom-line impact on profits and production in the world of today that can come from a well-planned and executed manufacturing research and development program. The manufacturing engineer in the smaller company must stay abreast of the latest manufacturing and process methods and technology directly related to his firm's business and product lines. The limited manufacturing research and development resources of the smaller company should be concentrated where they will offer the greatest potential benefits to the products, production processes, and methods of his company, and build upon the latest in state-of-the-art technology available in his industry. The company that does not support such efforts is turning its back on tomorrow in a short-sighted fixation on profits in the marketplace today. Manufacturing technology is advancing at such a rate that those who fail to stay abreast will no longer be in business tomorrow.

QUESTIONS

1. Why is manufacturing research and development so important, and why do some companies fail to commit resources to fund it?
2. What primary functions are normally included under the much broader term of manufacturing research and development, and which of these should even a small company have?
3. Which of the primary manufacturing research and development laboratories would you expect to find involved with a pilot surface mount automated line for circuit card assembly, and prove out of a vapor-phase soldering process?
4. Why is it so important to have an enclosed and secure area for both the methods lab, and the robotics lab? Why not do this work on the production line?
5. How should the individual manufacturing engineer stay abreast of the latest advanced manufacturing processes and technology in his company's product line?
6. Why is the actual metal-cutting time so low in most large and small machine shops?
7. Where is the most time lost in many otherwise highly efficient shops and plants, and what can be done about it?
8. Lasers and fiber optics along with composites are new process technologies in manufacturing. What are some others?
9. Why should the manufacturing research and development function report to manufacturing engineering rather than engineering?

Appendix A

Mathematics and Theory of Learning Curves

Mathematical Characteristics of Improvement Curves

The Relationship Between Average Cost and Increasing Quantity Has Been Observed to Approximate a Power Function

Production cost data from a wide variety of McDonnell Douglas manufacturing programs indicate trends which correlate closely with the equation

$$y = kx^{-n}$$

where y is the cumulative average recurring labor cost, stated in terms of constant economics (usually labor hours), for all units produced through a quantity of x units, and k is the theoretical first unit cost, usually referred to as the "T_1 cost." The exponent (−n), which is a negative number with an absolute value less than one, defines the *rate* at which the average cost decreases as quantity increases.

This exponential relationship, which plots on arithmetic coordinates as a concave curve that "flattens out" as quantity increases (see Fig. A.1.1), plots as a straight line on logarithmic coordinates (see Fig. A.1.2).† These figures (and all others in this section unless otherwise noted) illustrate examples with an arbitrarily chosen first unit cost (k) of 1000 and a curve exponent (−n) of −0.23447.

Reprinted courtesy of D. C. Gibson and the McDonnell Douglas Corporation, St. Louis, Mo.

†Taking the logarithm of both sides of the equation $y = kx^{-n}$ yields the expression $\log y = \log k - n \log x$, which is a straight line on log-log coordinates. Either natural logarithms (ln) or base 10 logarithms ($\log_{10}$) can be used.

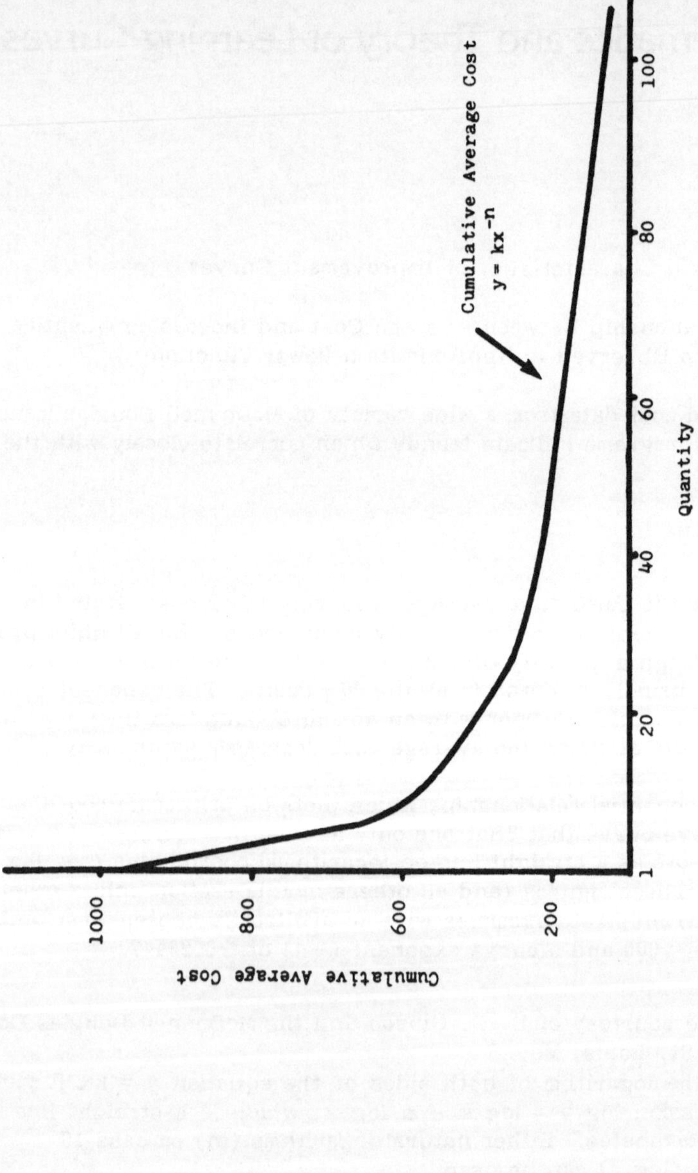

FIG. A.1.1 Relationship between average cost and increasing quantity.

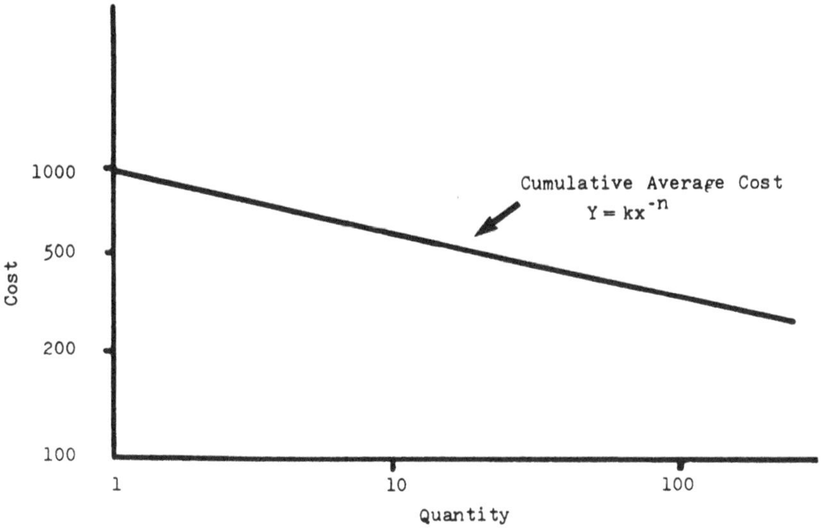

FIG. A.1.2 The exponential cost/quantity relationship.

The Rate of Improvement Is Often Defined in Terms of "Slope Percentage"

It is also characteristic of the above type of equation that each time the production quantity is doubled, the average cost is reduced by a constant percentage. This percentage is often used to describe the "slope" of the improvement curve.

In the case of a "90% curve," the cumulative average cost through two lots of equal size is predicted as 90% of that for the first lot. For this to occur, the second lot must cost 80% as much as the first lot. Doubling the quantity again reduces the average cost to 90% of the average through two lots, or to 81% (90% × 90%) of the initial observation, and so forth.

The slope percentage, p, is defined by the equation

$$p = 2^{-n} \times 100$$

since

$$p \div 100 = \frac{y_{2x}}{y_x} = \frac{k(2x)^{-n}}{k(x)^{-n}} = 2^{-n}$$

Conversely,

$$-n = \frac{\log (p \div 100)}{\log 2}$$

The above equations can be used to derive n if the curve is described in terms of slope percentage, and vice versa.

With a Log-Linear Cumulative Average Cost Trend, the *Total* Cost Curve Is Also Log-Linear

Multiplying the cumulative average cost through any quantity by that quantity yields, by definition, the total cumulative expenditure of costs through that quantity. The equation for the cumulative ("total-to-date") cost is thus

$$yx = xkx^{-n} = kx^{(1-n)}$$

which like the cumulative average cost, plots as a straight line on logarithmic coordinates (see Fig. A.2.1).

The cost of a block of units can be readily determined with the cumulative or total cost equation by determining the total cost through the end of the block and subtracting the total cost for all units prior to the start of that block.*

The Behavior of the Cost Curve at Any Point in Time Is Indicated by the Marginal Cost

The Marginal Cost Curve Follows a Log-Linear Trend Parallel to the Cumulative Average Cost Curve

The first derivative (with respect to quantity) of the cumulative total cost curve is

$$\frac{d}{dx}(yx) = (1 - n)kx^{-n}$$

---

*The cost of a block of Δx units following x_0 units can be readily determined by subtracting the cumulative cost through x_0 from the total cost through $(x_0 + \Delta x)$:

$$y(x_0 + \Delta x)(x_0 + \Delta x) - y_{x_0} x = k[(x_0 + \Delta x)^{(1-n)} - x_0^{(1-n)}]$$

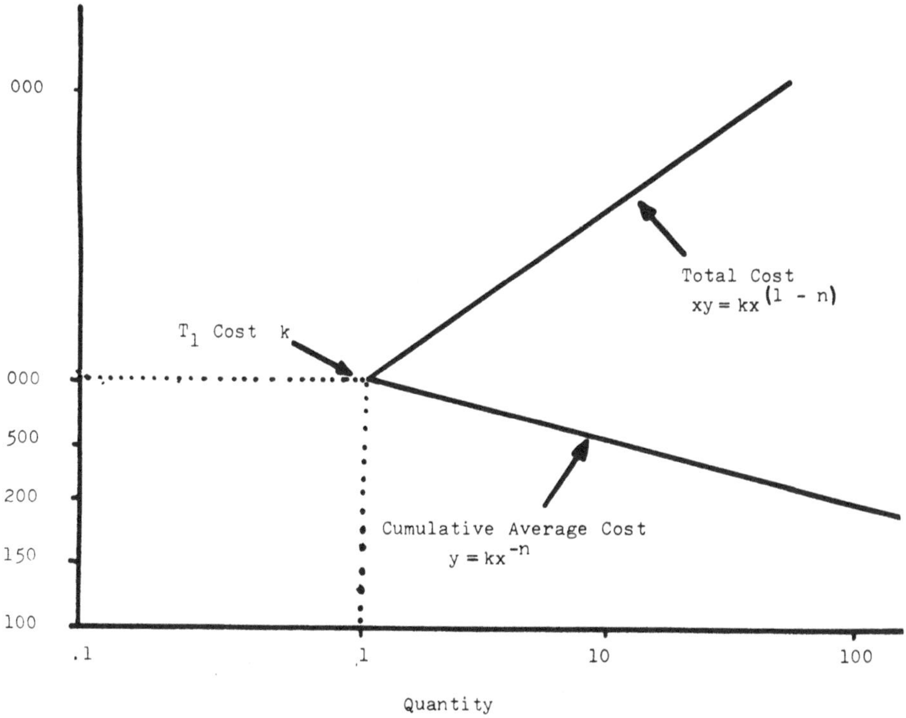

FIG. A.2.1 With log-linear cumulative average cost trend, total is log-linear.

which describes the rate of increase in total cost as quantity increases and can be defined as the *marginal cost curve*. Note that this equation plots on logarithmic coordinates as a straight line which is parallel to the cumulative average cost curve and which has a value equal to all values of x to $(1-n)$ times that of the cumulative average curve, as indicated on Fig. A.3.1.

This *marginal cost* curve represents the instantaneous cost or the cost of "the most minute increment" of additional production.* Strictly speaking, it is not the *unit* cost per se; the relationship between unit costs and the first derivative of the total cost curve is discussed in Section A.4.

---

* That is, the marginal cost at x_0 is

$$\frac{d}{dx}(yx)\Big|_{x=x_0} \lim_{\Delta x \to 0} k[(x_0 + \Delta x)^{(1-n)} - x_0^{(1-n)}]$$

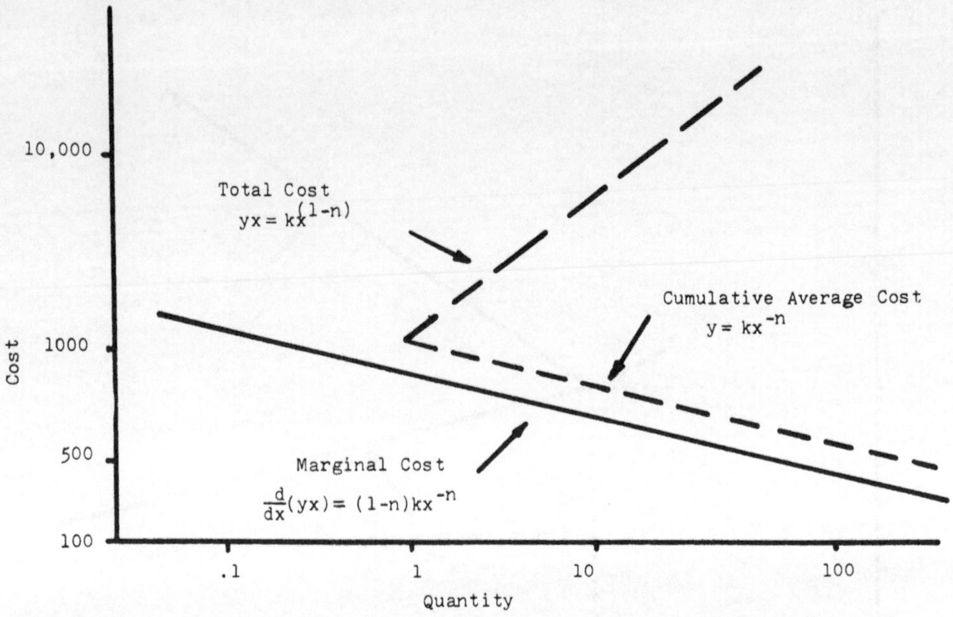

FIG. A.3.1 First derivative of total cost is log-linear and defines marginal cost.

The Marginal Cost Concept Suggests Cost Improvement "Within Units"

The illustration of the marginal cost as a *continuous* function over varying quantity introduces the concept of cost improvement "within units." This phenomenon can be readily illustrated by the example of making a rope ladder: actions involved in braiding rope and in making rungs are repeated successively several times during the making of each ladder, and a progressive reduction in the time required for these actions would be expected during the making of each ladder, contributing to the reduction in time required for each successive ladder. This phenomenon should exist to some extent in most manufacturing operations, for instance, in drilling holes, installing fasteners, terminating wires, and making similar parts such as ribs, stringers, or circuit boards within one end-item unit. The unit cost—cost of one unit—is thus in effect an *average* value of the elements making up the cost of that unit.

At Block Midpoint Quantities, the Marginal Cost Equation Defines Block Average Cost

Within each "block" on the quantity scale, whether that block is comprised of a single unit or several units (or, indeed, of less than a discrete unit), there is a unique quantity measurement X_a such that the cost of the marginal cost curve evaluated at that quantity is the same as the average cost of the units in the block; that is, when the marginal cost curve value at that quantity is multiplied by the block quantity, the resultant product is the total cost of the entire block. This quantity X_a is generally called the *midpoint quantity* and can be calculated by solving the *midpoint formula*

$$x_a = \left[\frac{(x_0 + \Delta x)^{(1-n)} - x_0^{(1-n)}}{(1-n)\Delta x} \right]^{-1/n}$$

where x_0 is the quantity preceding the block and Δx is the block quantity.*

This concept is illustrated in Fig. A.5.1. After the midpoint quantity value is known, the average cost of the units in the block can be determined by solving the *block average cost* formula:

$$\left. \frac{d(yx)}{dx} \right|_{x_a} = (1-n)kx_a^{-n}$$

which yields the value of the marginal cost curve at the midpoint quantity.

In the example shown, the midpoint quantity for a block of 10 units following the first 10 (i.e., units 11–20 inclusive) is calculated to be $x_a = 14.65$ units. Calculating the marginal cost at this quantity yields an average cost of 408 hr for units in the block, or a total cost of 4080 hr for the entire 10-unit block. Note that the same answer can be obtained by subtracting the total cost through 10 units (5828 hr determined by using the cumulative average formula) from the total cost through 20 units (9908 hr).

---

*This formula is derived by solving the equation

$$\underbrace{(1-n)kx_a^{-n}}_{\text{Marginal cost at } x_a} \quad \underbrace{\Delta x}_{\text{Block qty.}} = \underbrace{k(x_0 + \Delta x)^{(1-n)}}_{\text{Cumulative cost through } (x_0+\Delta x) \text{ units}} - \underbrace{kx_0^{(1-n)}}_{\text{Cumulative cost through } x_0 \text{ units}}$$

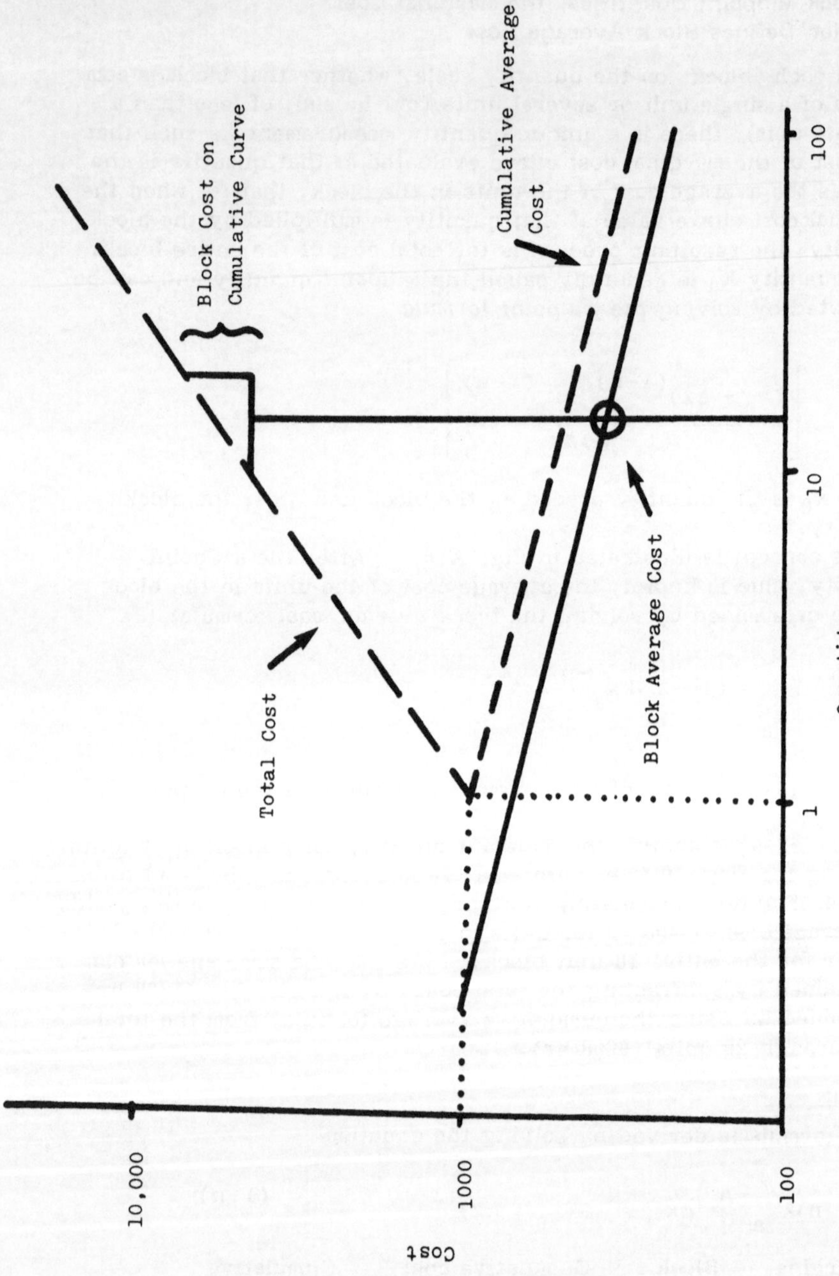

FIG A.5.1 At block midpoints, marginal cost curve defines block average cost.

It may be useful to remember as a general rule of thumb that for the initial block the midpoint occurs at about one-third of the block quantity and for subsequent quantities at slightly less than halfway into the block—usually after about 45% of the block quantity has been completed, depending on the slope of the curve.

At Unit Midpoints, the Marginal Cost Curve Defines Unit Costs

Using this concept of midpoint quantities and block average costs, the "average" cost of the elements making up each individual unit can be readily calculated; the marginal cost curve can thus be used to calculate and illustrate unit costs, as shown using arithmetic coordinates in Fig. A.6.1 and logarithmic coordinates in Fig. A.6.2.

Note that the block average costs when plotted at "units ends" approach the marginal cost curve asymptotically. The marginal cost curve is sometimes referred to as a "parallel unit cost" curve; it is parallel to the cumulative average cost curve, very close to unit values in large quantities, and when evaluated at appropriate midpoint values,

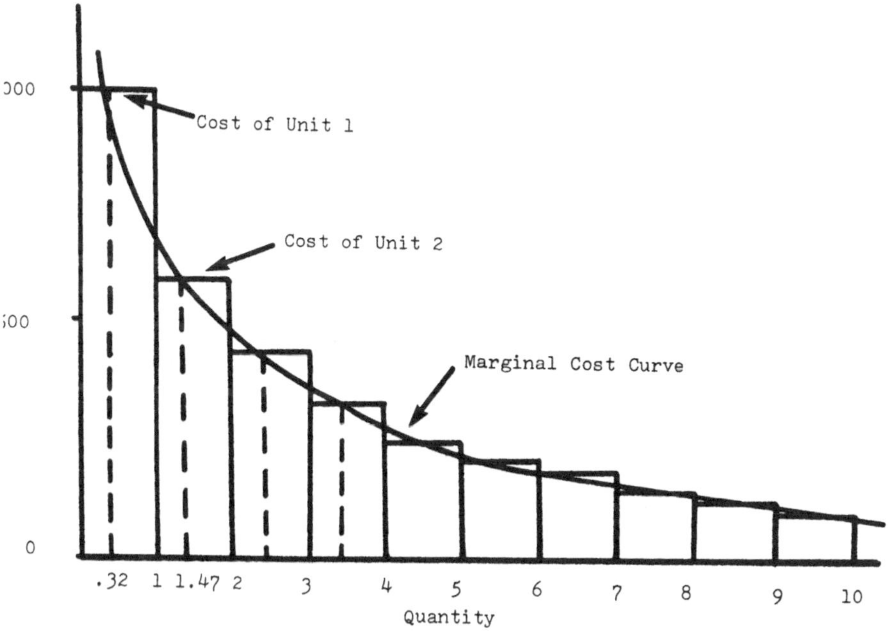

FIG. A.6.1 At unit midpoints, marginal cost curve defines unit costs (arithmetic).

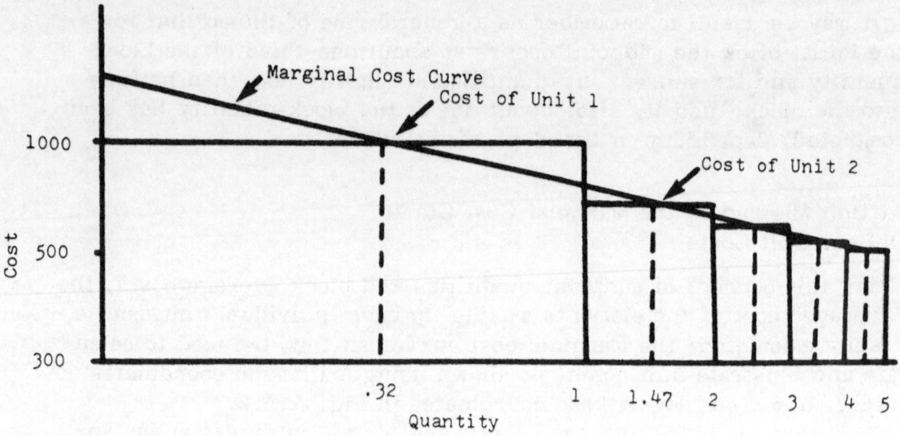

FIG. A.6.2 At unit midpoints, marginal cost curve defines unit costs (logarithmic).

it defines unit costs. For practical purposes, the marginal cost curve can be used directly as a unit curve plot for large quantities.

The Marginal Cost Equation Is a Useful Estimating Tool

1. Projecting the cost of another block (usually future production) from actual or proposed data for a given block and an assumed improvement curve exponent ("n factor")
2. Determining the improvement curve exponent (−n) and T_1 cost (k) from actual data from two or more blocks, using curve-fitting techniques such as least-squares regression analysis to "smooth" the data where appropriate
3. Determining a T_1 value assumption from incomplete data, such as using an assumed curve slope when data for only one block (not necessarily the first) are available.

Examples of these calculations are shown in Chapter 2. For complicated calculations, such as those involving curve-fitting data, programmable hand-held calculators can be used, eliminating the need for actually "working-through" the equations discussed above.

The Sensitivity of Marginal and Unit Curves to Changes Is Useful in Estimating and Performance Measurement

Since the marginal cost curve quantifies cost "at a point in production" rather than "through" that point, the marginal cost curve is more

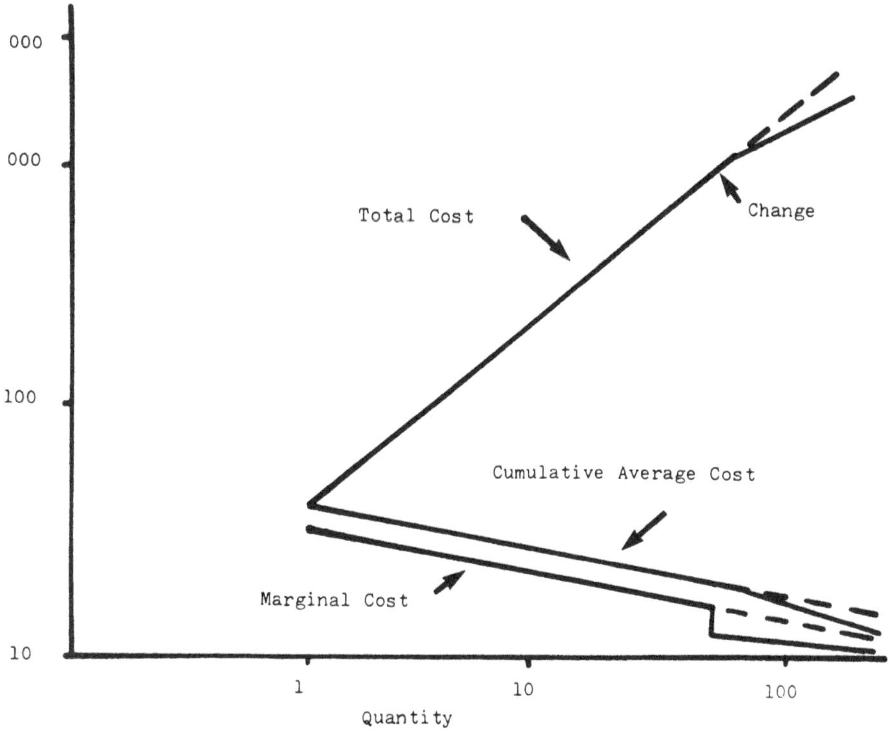

FIG. A.7.1 Marginal cost curve is sensitive to changes.

sensitive to instantaneous variances than either the cumulative average or the total cost curve, as illustrated in Fig. A.7.1. For this reason, the marginal cost curve facilitates early identification of problems and, conversely, early confirmation of the success of cost reduction changes.

The example illustrated in Fig. A.7.1 shows how a change in cost is easily detected in terms of marginal cost performance but has much less immediate effect on the apparent cumulative average cost trend.

The sensitivity of marginal cost trends also aids in the evaluation of actual data for estimating purposes. Block data reflecting unusual nonrecurring effects of extraneous changes can be readily identified and eliminated from the data base where appropriate.

Improvement Curves with Different Slopes Should Be Treated Separately

Mathematically, the sum of two or more equations of the form $y = kx^{-n}$ with different values of n is not log-linear*, but plots as a slightly concave curve on logarithmic coordinates. In the case of improvement curve applications, this concavity or "flattening" of the aggregate curve with increasing quantity results from those constituents with the "steepest" slopes (largest absolute values of $-n$) becoming progressively less and less significant as a proportion of the total cost as quantity increases, and thus having less and less influence on the aggregate slope.

If two or more curves with the *same* n value are added, the total curve does plot as a straight line on logarithmic coordinates.†

In practice, the separate application of different improvement curves to appropriate groupings (e.g., segregating fabrication from assembly or composites from conventional metallic structure) will segregate trends with different "n factors" (see page 294), reducing to statistical insignificance any distortion that would result from the addition of dissimilar curves.

A Significant Change in a Trend May Indicate that Improvement Has "Started Over" due to a Change

The impact of a change in an ongoing production process‡ may be so significant that production costs after the change follow a new improvement curve trend. In more extreme cases, the plotting of the actual

---

*The sum is

$$\sum_{i=1}^{m} y_i = \sum_{i=1}^{m} k_i x^{-n} + k_2 x^{-n_2} + \cdots + k_m x^{-n_m}$$

and is not log-linear.

†The sum is

$$\sum_{i=1}^{m} y_q = \sum_{i=1}^{m} k_i x^{-n} \sum_{i=1}^{m} k_i = (k_1 + k_2 + \cdots + k_m)x^{-n}$$

and is log-linear.

‡Design changes to reduce production cost or a change to new tooling are examples which may be encountered.

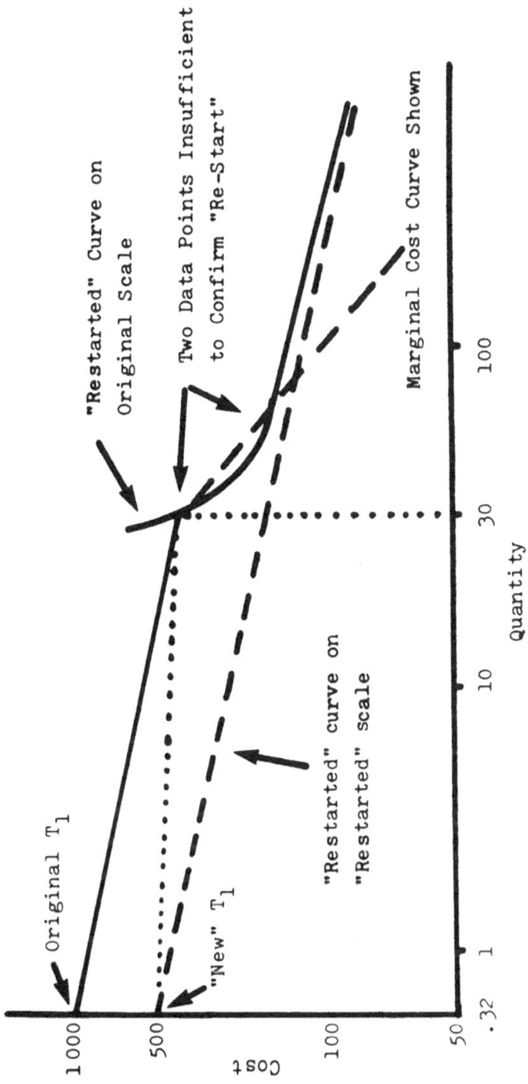

FIG. A.9.1 Rapidly decreasing cost after a major change may indicate a restarted trend.

data would show an interruption in the original log-linear trend followed by a concave curve after the occurrence of the change. This phenomenon is illustrated in Fig. A.9.1. If the prior units produced are ignored and the cost data for the new units are replotted on a new quantity scale starting at 1, the plot will appear as a straight line on log-log coordinates. This indicates that the improvement trend has "started over" and is often described as a "restart" of the improvement curve.

The example illustrated in Fig. A.9.1 involves a curve "restart" at a "new lower T_1" (and the same curve slope as before) after 30 units have been built due to a change in production at unit 31. The "new" marginal cost curve, when plotted on a "restarted" quantity scale starting at unit 1, is in this example parallel to the original curve, but when the same cost data are plotted on a scale starting at unit 31, the curve takes a concave shape which gradually "straightens out" again toward a log-linear trend as quantity increases.

The confirmation of this behavior obviously requires at least three data points. If only one or two actual data points are known, it *may* appear that the original trend has suddenly changed to a steeper slope. If a later change back to a shallower slope is also anticipated, no significant error will result as the resultant curve roughly approximates the restart curve shown.* Any erroneous assumption that the second steeper "leg" will continue for too long, however, may yield an underestimate of future production costs.* The "restart" curve is often a reliable mathematical tool for making extrapolations from the given data when a significant change occurs, as the exponential equation provides an "automatic" prediction of the apparent change back to a shallower curve slope.

The Wright Curve Is the Progenitor of These Mathematical Concepts

The earliest comprehensive learning curve analysis was published by T. P. Wright in the 1930s. The Wright Analysis was adopted for use in aircraft estimating and is still widely employed throughout the aerospace industry.

The Wright learning curve theory, like the McDonnell Douglas curve equations discussed above, is based on the concept of an exponential

---

*This steepening and then reflattening curve shape is often referred to colloquially as an "s-curve" or "dog-leg" effect.

†Of course, the rapid reduction in cost may well continue for some time. If conditions continue to change, as in the case of a continuing buildup in production rate, the *observed* improvement curve may continue on a steeply dropping trend until the changes are complete.

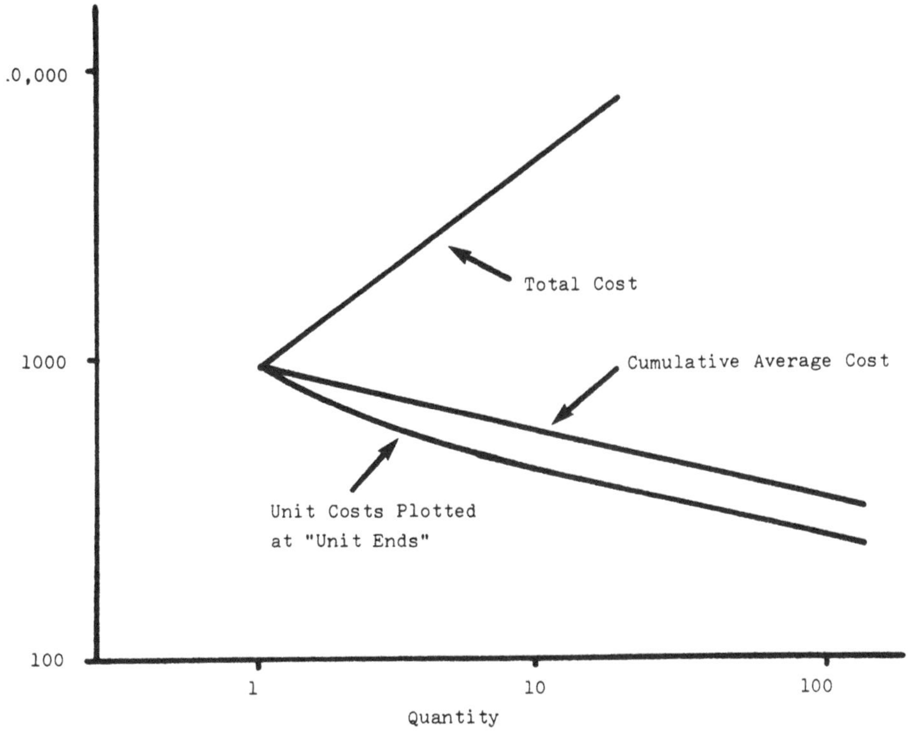

FIG. A.10.1 Original Wright curve differs from McDonnell curve only in plotting of unit costs.

or log-linear *cumulative average* cost trend. The Wright unit cost curve was illustrated graphically by plotting costs "at the ends of" the units, resulting in the curve shown in Fig. A.10.1, and when factor tables were developed for use in making computations, a separate set of factors was provided for unit costs.

These Wright unit factors and the Wright method of graphing unit costs may still be encountered and may at first cause some confusion. The McDonnell Douglas and Wright curve theories are identical, however, and the apparent differences in factors and unit cost curves involve differences only in presentation and terminology.

The Crawford Learning Curve Theory Is Based on Different Assumptions

The Crawford Theory Assumes a Log-Linear *Unit* Cost Trend

Another approach to learning curve analysis is embodied in the theory originally postualted by J. R. Crawford and still encountered in some quarters* that *unit* costs can be expected to follow a log-linear trend with increasing quantity.

If the unit cost trend is log-linear, then the cumulative average and cumulative total cost trends, as well as the first derivative of the cumulative total cost (marginal cost curve, discussed earlier) must not be log-linear. The Log-linear unit cost curve is defined by the equation

$$y_i = kx_i^{-n}$$

where y_i is the cost of unit x_i and k is the cost of the first unit.[†] The cumulative total cost is therefore

$$\sum_{i=1}^{x} y_i = k \sum_{i=1}^{x} x_i^{-n}$$

and the cumulative average is the total divided by quantity; clearly, neither is log-linear. Observations of actual cost behavior at McDonnell Douglas generally contradict the Crawford theory.

As shown in Fig. A.11.1, the cumulative average cost trend calculated using the Crawford curve does tend to approach a straight line asymptotically at large quantities, and any difference between the Crawford theory and the McDonnell/Wright observations becomes insignificant provided the comparison is confined to the large-quantity range. If the data to be analyzed involve smaller quantities, however, an entirely different situation exists.

The Extrapolation of Follow-On Production Costs from Small-Quantity Data Yields Lower Answers When the Crawford Theory is Used

Application of the two contradictory concepts to analyses of small-quantity data results in widely different conclusions. A projection of costs from two small-quantity lots using the Crawford theory indicates

---

*The Crawford curve is used at Boeing and is often called the Boeing curve.

†Note that despite the apparent similarity of this equation to the cumulative average cost equation discussed earlier, the variables represent different quantities and the equations describe entirely different trend predictions.

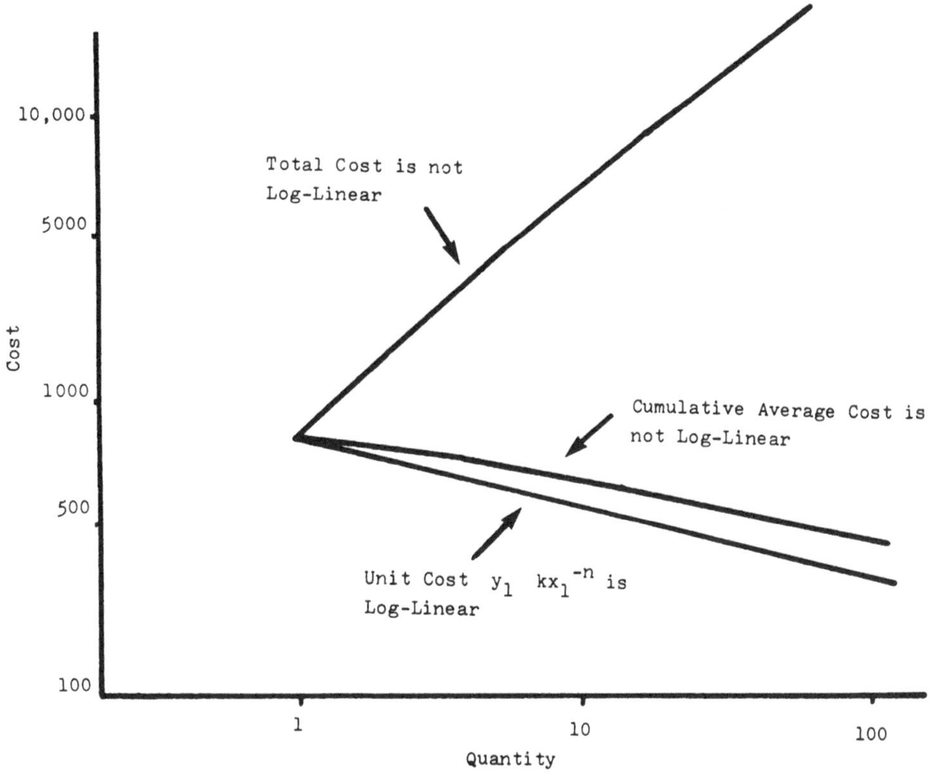

FIG. A.11.1 Crawford theory assumes log-linear unit cost trend.

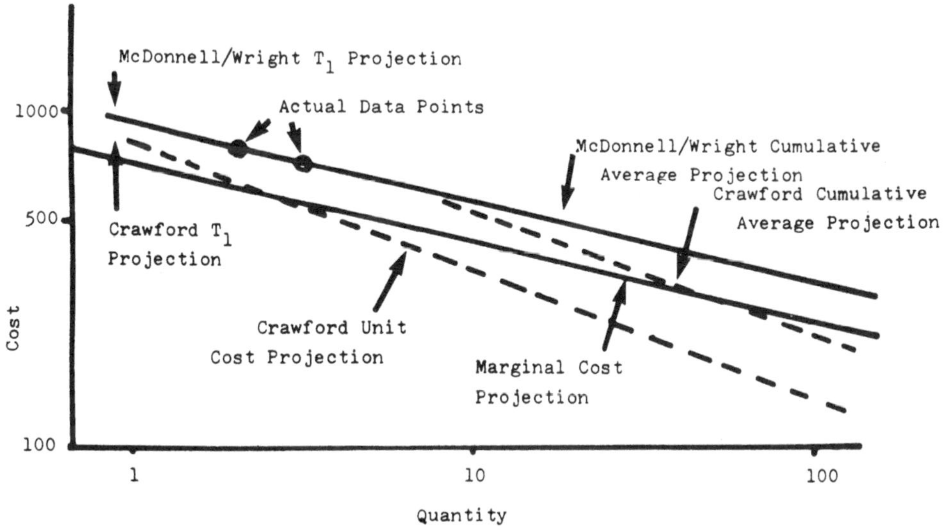

FIG. A.11.2 Crawford theory projects lower estimates for follow-on work.

a "steeper" curve, and therefore lower projections for follow-on production cost, than the MDC curve would predict. This phenomenon is illustrated in Fig. A.11.2.

In the example shown, the total cost incurred in the manufacture of each of the first two blocks (each of which comprises two units) is given, and it is desired to extrapolate these data to project the cost of 40 additional units, assuming a continuation of the improvement trend indicated by the actuals. Where McDonnell/Wright observations of log-linear cumulative average trend behavior indicate an 85% improvement curve (n - 0.23447) and a T_1 cost of 1000, application of the Crawford theory to the same data yields an assumed 76% curve slope. This measurement does not have the same mathematical meaning in the Crawford application, but the projection for the 40-unit follow-on block is lower, 301 hr per unit (average) versus 381 hr per unit indicated by the McDonnell/Wright approach—a difference of 21% in this case.

The Crawford Theory Also Indicates Lower Estimates for Small Quantities Based on Large-Quantity Data

Conversely, the calculation of an estimate for a small initial quantity using a cost estimating relationship (CER) developed from a trend established through several data points involving large quantities can be biased by the use of the Crawford theory. Because of the assumptions concerning the behavior of the trend in the small-quantity range, the T_1 and the early cost predictions for small quantities would be lower than the McDonnell/Wright concept would indicate, as shown in Fig.

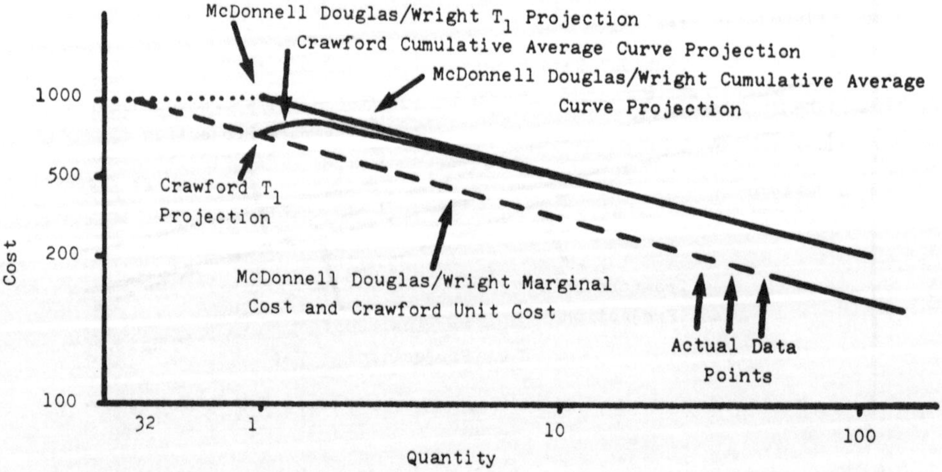

FIG A.11.3 Crawford theory yields lower estimates for small quantities.

A.11.3, even though the trends shown for larger quantities would be nearly identical for both concepts.

Figure A.11.3 illustrates the determination of a T_1 cost using a number of actual data points confined to quantities above 100 units. The example shown would yield a T_1 of 1000 and an 85% curve ($n = 0.23447$) using McDonnell/Wright methodology. The Crawford theory assumes nearly identical curve trends in the large-quantity range—the assumed curve slope is 85%—but the assumption that the *unit* cost trend is log-linear yields an assumed T_1 of 766.5. Note that this is the "intercept" of the McDonnell marginal cost curve through quantity $x = 1$; the Crawford theory yields a T_1 value which is lower and is $(1 - n)$ times the value indicated by McDonnell/Wright experience. As indicated on the graph, the Crawford theory would yield a significantly lower estimate for a small number of units if the source data were based on large quantities.

Appendix B
Manufacturing Engineering Terminology

Age Hardening: Hardening by aging, usually after rapid cooling or cold working.

Aging: In a metal or alloy, a change in properties that generally occurs slowly at room temperature and more rapidly at higher temperatures.

Algorithm: A method for deriving the solution to a problem by using the results from each cycle to refine the following cycle. The process is repeated until a satisfactory answer is obtained.

Allowance: A time increment included in the standard time for an operation to compensate the worker for production lost due to fatigue and normally expected interruptions, such as for personal and unavoidable delays. It is usually applied as a percentage of the normal or leveled time.

Aluminizing: Forming an aluminum or aluminum alloy coating on a metal by hot dipping, hot spraying, or diffusion.

Annealing: Heating to and holding at a suitable temperature and then cooling at a suitable rate, for such purposes as reducing hardness, improving machinability, facilitating cold working, producing a desired microstructure, or obtaining desired physical, mechanical, or other properties.

Anodizing: Forming a conversion coating on a metal surface by anodic oxidation. Most frequently applied to aluminum.

Arc Brazing: Brazing with an electric arc, usually with two nonconsumable electrodes.

Arc Welding: Welding with an electric arc-welding electrode.

Assemble: The basic element employed when one or more objects are put on or into another object so that they fit or contact each other in a predetermined relation to form a unit.

Assembly: Two or more parts or subassemblies joined together to form a complete machine, structure, or other article.

Assembly Line: The arrangement of machines, equipment, material, and workers which permits the work in process to progress sequentially from operation to operation until the product has been assembled.

Automatic Assembly: Assembly by machine consisting of a transfer system for moving the assemblies from station to station, automatic workheads to perform simple assembly operations, and inspection stations to check that the various operations have been completed successfully.

Automation: Substitution of machine labor for human labor, either manual or intellectual.

Avoidable Delay: Any time during an assigned work period which is within the control of the operator, and which he uses for idling, or for doing things unnecessary to the performance of the work. Such time does not include allowance for personal requirements, fatigue, and unavoidable delays.

Balanced Line: A series of progressive related operations with approximately equal standard times for each, arranged so that work flows at a desired steady rate from one operation to the next.

Bottom Drill: A flat-ended twist drill used to convert a cone at the bottom of a drilled hole into a cylinder.

Bottoming Tap: A tap with a chamfer of 1 to 1.5 threads in length.

Brazing: Joining metals by flowing a thin layer, capillary thickness, of non-ferrous filler metal into the space between them. Bonding results from the intimate contact produced by the dissolution of a small amount of base metal in the molten filler metal, without fusion of the base metal.

Basic Motion Time Study (BMT): A system of predetermined motion time standards. The essence of the system lies in the arbitrary definition of a basic motion as one that commences from rest and ends at rest. The system's purpose is to establish time standards for procedures that are composed of human motions controlled only by the individual performing them, and to do so without resorting to time study.

Bit: A binary digit, a quantum of information, a single pulse in a group of pulses.

Broach: A bar-shaped cutting tool provided with a series of cutting edges or teeth that increase in size or change in shape from the starting to the finishing end. The tool cuts in the axial direction when pushed or pulled, and is used to shape either holes or outside surfaces.

Byte: A sequence of adjacent binary digits operated upon as a unit, and usually shorter than a word.

Carbide: A compound or carbon with one or more metallic elements.

Carbide Tools: Cutting tools made of tungsten carbide, titanium carbide, tantalum carbide, or combinations of these, in a matrix of cobalt or nickel, having sufficient wear resistance and heat resistance to permit high machining speeds.

Case Hardening: Hardening a ferrous alloy so that the outer portion or case is made substantially harder than the inner portion or core. Typical processes used for case hardening are carburizing, cyaniding, carbonitriding, nitriding, induction hardening, and flame hardening.

Casting: An object at or near finished shape, obtained by solidification of a substance in a mold. The process of pouring molten metal into a mold to produce an object of desired shape.

Caustic Dip: A strongly alkaline solution in which metal is immersed for etching, neutralizing acid, or removing organic materials such as grease or paint.

Centerless Grinding: Grinding the outside or inside of a workpiece mounted on rollers rather than on centers. The workpiece may be in the form of a cylinder or the frustrum of a cone.

Chamfering: Making a sloping surface on the edge of a member.

Chromate Treatment: A treatment of metal in a solution of a hexavalent chromium compound to produce a conversion coating consisting of trivalent and hexavalent chromium compounds.

Cold Working: Deforming metal plasticially at a temperature lower than the recrystallization temperature.

Compound Die: Any die so designed that it performs more than one operation on a part with one stroke of the press, such as blanking and piercing where all functions are performed simultaneously within the confines of the particular blank size being worked.

Compression Molding: Pre-formed molding material is placed between two mold halves, which are heated and closed under pressure. The material then flows to fill the mold cavity, polymerizes from the heat and hardens. The mold then opens and the molded part is ejected or removed.

Constant Element: An element for which the leveled or normal time is always the same, regardless of the characteristics of the parts being worked upon, as long as the method and the working conditions are unchanged. An element for which, under a specified set of conditions, the standard time allowance should always be the same.

Continuous Timing: The procedure of timing used in making time studies whereby the watch is permitted to run continuously throughout the period of study while the observer notes and records the reading of the watch at the end of each element, delay, or any other type occurrence in the study, regardless of whether or not it has a direct bearing on the job. The elapsed times are obtained by subtracting the successive readings after the timing has been completed.

Contour Machining: Machining of irregular surfaces, such as those generated in tracer turning, tracer boring, and tracer milling.

Control Chart: A graphic comparison of process performance data to computed control limits drawn as limit lines on the chart.

Conversion Coating: A coating consisting of a compound of the surface metal, produced by chemical or electromechanical treatment of the metal. Examples include chromate coatings on zinc, cadmium, magnesium, and aluminum, plus oxides or phosphate coating on steel.

Cutting Speed: The relative velocity, usually expressed in feet per minute, between a cutting tool and the surface of the material from which it is removing stock.

Decimal Minute Stopwatch: A two-handed timing device whose movement may be started or stopped manually, and whose larger outer dial is divided into 100 spaces, each of which represents 0.01 min. The position of the large hand on the large dial indicates time in minutes to two decimal places, and the position of the small hand on a smaller, inner dial indicates time in whole minutes.

Delay Allowance: A time increment included in a time standard to allow for contingencies and minor delays beyond the control of the operator.

Diamond Tool: A diamond, shaped or formed to the contour of a single-pointed cutting tool, for use in the precision machining of nonferrous and nonmetallic materials.

Die: Various tools used to impart shape to material primarily because of the shape of the tool itself. Examples are blanking dies, cutting dies, drawing dies, forging dies, punching dies, and threading dies.

Die Casting: A casting process where molten metal is forced under high pressure into the cavity of a metal mold.

Dip Brazing: Brazing by immersion in a molten salt or metal bath. Where a metal bath is employed, it may provide the filler metal.

Downtime: A period of time that is usually equal to or greater than a specified minimum during which an operation is halted due to a lack of materials, a machinery breakdown, or the like.

Draft: The angle or taper on the surface of a punch or die, or the parts made with them, which facilitates the removal of the work. Taper put on the surfaces of a pattern so that it can be withdrawn successfully from the mold.

Drop Delivery: The method of introducing an object to the workpiece by gravity. A method whereby a chute or container is so placed that when work on a part is finished, it will fall or drop into a chute or container or onto a conveyer with little or no transport by the operator. The laying aside of a part by releasing it so that it falls or moves away from the work area either through the force of gravity or by mechanical or other means.

Economic Lot Size: That number of units of material or a manufactured item that can be purchased or produced within the lowest unit cost range. Its determination involves reconciling the decreasing trend in preparation unit costs and the increasing trend in unit costs of storage, interest, insurance, depreciation, and other costs incident to ownership as the size of the lot is increased.

Efficiency: The ratio of standard performance time to actual performance time, usually expressed as a percentage.

Effort Rating: That part of any performance rating technique concerned with evaluating the extent or degree to which the will to work is exhibited by an operator.

Elapsed Time: The actual time taken by an operator to complete a task, an operation, or an element of an operation.

Electroplating: Electrodepositing metal (may be an alloy) in an adherent form upon an object serving as a cathode.

Element: A subdivision of the work cycle, composed of a sequence of one of several fundamental motions and/or machine or process activities which is distinct, describable, and measurable.

Extrusion: Conversion of a billet into lengths of uniform cross section by forcing the plastic metal through a die orifice of the desired cross-sectional outline.

Fatigue Allowance: Time included in the production standard to allow for decreases or losses in production which might be attributed to fatigue. Usually applied as a percentage of the leveled time.

Fiber Optics: That branch of optics dealing with the transmission of light through small filamentary optical elements or fibers, which are transparent.

Fixture: A device which holds material in a desired position, but does not guide the machine or tools performing the necessary operations. Holds two or more parts in prescribed positions relative to each other while the parts are being assembled or joined together.

Flow Diagram: A graphic representation on a floor plan of the work area involved, the locations of workstations, and the paths of movement of men and materials.

Flow Process Chart: A graphic representation of the sequence of all operations, transportations, inspections, delays, and storages occurring during a process or procedure. It includes information considered desirable for analysis, such as time required and distance moved.

Foreign Element: An interruption which is not a regular occurrence in the work cycle, or operation, and one for which no provision was made in the normal sequence of elements of a time study.

Forging: Plastically deforming metal, usually hot, into desired shapes with compressive force, with or without dies.

Forming: Making a change, with the exception of shearing or blanking, in the shape or contour of a metal part without intentionally altering the thickness.

Foundry: A commercial establishment or building where metal castings are produced.

Free Machining: Pertains to the machining characteristics of an alloy to which an ingredient has been introduced to give small, broken chips, lower power consumption, better surface finish, and longer tool life. Among such additives are sulfur or lead to steel, lead to brass, lead or bismuth to aluminum, sulfur or selenium to stainless steel.

Frequency Distribution: A statistical tool for presenting data in a form which makes more clear the central tendency and the dispersion along the scale of measurement.

Fusion Welding: Welding, without pressure, in which a portion of the base metal is melted.

Gravity Feed: A method of supplying material to a machine or to a workstation by the force of gravity. Generally, a hopper or chute is used to store and guide the materials to the point of use.

Gripper: An appendage on the arm or manipulator of a robot used to grasp or pick up objects.

Heat Treatment: Heating and cooling a solid metal or alloy in such a way as to obtain desired conditions or properties.

Hydrogen Brazing: Brazing in a hydrogen atmosphere, usually in a furnace.

Idle Time: A time interval during which the operator, the equipment, or both do not perform useful work.

Injection Molding: Thermoplastic material is fed into a heated cylinder, then automatically ejected by plunger action, after it softens, into the mold cavities. The mold is cooled, causing the plasticized material to solidify.

Inspection: Examining an object for identification or checking it for verification of quality or quantity in any of its characteristics.

Investment Casting: Casting metal into a mold produced by investing an expendable pattern with a refractory slurry that sets at room temperature, after which the wax, plastic, or frozen mercury pattern is removed through the use of heat. Also known as the lost wax process.

Jig: A device which holds a piece of work in a desired position and guides the tool or tools which perform the necessary operations.

Jig Boring: Boring with a single-point tool where the work is set in position upon a table that can be located so as to bring any desired part of the work under the tool. Thus holes can be accurately spaced. This type of boring can be done on jig borers or milling machines.

Job Shop: A manufacturing enterprise devoted to producing special or custom-made parts or products, usually in small quantities for specific customers.

Knurling: Impressing a design into a metallic surface, usually by means of small, hard rollers that carry the corresponding design on their surfaces.

Laser: Light amplification by stimulated emission of radiation.

Layout: The arrangement of items within an area. The items may include roads, railroads, buildings, offices, departments, warehouses, equipment, machinery, furniture, facilities, aisles, and so on.

Learning Curve: A plot of productive output or unit work times of an individual or group as a function of time, or output per unit time.

Leveled Time: The average time, adjusted to account for differences in skill, effort, conditions, and consistency between operators and the factors surrounding an operation.

Line Production: A method of plant layout in which the machines and other equipment required, regardless of the operations they perform, are arranged in the order in which they are used in the process.

Liquid Honing: Polishing metal by bombardment with an air-ejected liquid containing fine, solid particles in suspension. If an impeller wheel is used to propel the suspension, the process is called wet blasting.

Loose Standard: An allowed or standard time greater than that required by a qualified operator performing his job with normal skill and effort and following the prescribed method.

Machine Element: A work cycle subdivision that is distinct, describable, and measurable, the time for which is entirely controlled by a machine and therefore is not influenced by the skill or effort of the operator.

Machining: Removing material, in the form of chips, from work, usually through the use of a machine.

Magnetic Disk: A flat, circular plate with a magnetic surface on which information may be placed in the form of magnetically polarized spots.

Magnetic Tapes: A tape or ribbon with a magnetic surface on which information may be placed in the form of magnetically polarized spots.

Manual Element: A distinct, describably, and measurable subdivision of a work cycle or operation performed by one or more human motions that are not controlled by process or machine.

Mass Production: A method of quantity production in which a high degree of planning, specialization of equipment and labor, and integrated utilization of all productive factors are the outstanding characteristics.

Material Flow: The progressive movement of material, parts, or products toward the completion of a production process between workstations, storage areas, machines, and the like.

Material Handling: The movement of materials, parts, subassemblies, or assemblies either manually or through the use of powered equipment.

Maximum Working Area: The area at the workplace which is bounded by an imaginary arc drawn by the operator's fingertips moving in the horizontal plane with the arm fully extended and moving about the shoulder as a pivot. The section where the maximum areas of the right and left hands overlap constitutes the maximum working area for the two hands.

Memomotion Study: A motion study technique that utilizes a motion picture camera operating at slower than normal speeds, such as one frame per second or one frame per one hundredth of a minute.

Memory: The capacity of a machine to store information subject to recall; the component of the computer system in which such information is stored.

Merit Rating: An organized and systematic evaluation of an employee's ability and job performance in terms of such factors as quality and quantity of work, knowledge, initiative, and dependability.

Method: The procedure or sequence of motions used by one or more operators to accomplish a given operation or work task. The sequence of operations or processes used to produce a given product or accomplish a given job. A specific combination of layout and working conditions, materials, equipment, and tools, as well as motion pattern involved in accomplishing a given operation or task.

Methods Study: The analysis of the sequence of motions used or proposed for use in performing an operation, and of the tools, equipment, and workstation layout used or proposed for use.

Methods Time Measurement (MTM): A system of predetermined motion time standards. It is a procedure which analyzes any operation into certain classifications of motions required to perform it, and assigns to each motion controlled only by the individual performing it a predetermined time standard determined by the nature of the motion and the conditions under which it is made.

Micromotion Study: The analysis of elements of motions too short or rapid for the eye to distinguish, by the use of motion pictures, sometimes in combination with an adequate time-indicating device.

Motion Study: The analysis of the manual and eye movements occurring in an operation or work cycle, for the purpose of eliminating wasted movements and establishing a better sequence and coordination of movements.

Motion Time Analysis (MTA): A system of predetermined motion time standards used for describing and recording an operation in terms of its motions. The value of each motion is predetermined as to both utility and time allowance.

Multiple Activity Process Chart: A synchronized graphic representation of operations performed simultaneously by two or more operators, two or more machines, or a combination of men and machines.

Normal Time: The time required by a qualified operator, working at a pace which is ordinarily used by workers when capably supervised, to complete an element, cycle, or operation when following the prescribed method.

Normalizing: Heating a ferrous alloy to a suitable temperature above the transformation range, and then cooling in air to a temperature substantially below the transformation range.

Numerical Control: Any system of control plus controlled equipment which accepts commands (data and instructions) in symbolic form as an input and converts this information into a physical output (into physical values such as dimensions or quantities).

Operation: The intentional changing of an object in any of its physical or chemical characteristics, including assembly or disassembly.

Operation Analysis: A study of the factors which affect the performance of an operation, such as purpose of the operation, other operation on the part, inspection requirements, materials used, material-handling methods, setup and tool equipment, existing working conditions, and methods employed.

Operation Process Chart: A graphic representation of the points at which materials are introduced into the process, and of the sequence of inspections and all operations except those involved in material handling.

Passivation: The changing of the chemically active surface of a metal to a much less reactive state.

Performance Rating: The act of comparing an actual performance by an operator against a defined concept of a normal performance.

Personal Allowance: Time included in the production standard to permit the operator to attend to personal necessities, such as getting a drink, trips to the restroom, and the like. Usually applied as a percentage of the leveled time.

Plating: Forming an adherent layer of metal upon an object.

Planning: The procedure for determining a course of action intended to accomplish a desired result.

Plant Layout: The physical arrangement, either existing or in plans, of industrial facilities.

Precipitation Hardening: Hardening caused by the precipitation of a constituent from a supersaturated solid solution.

Principles of Motion Economy: The rules and their corollaries that apply to human motions, which guide the development of the optimum way of accomplishing a given job.

Process: A planned series of actions or operations which advances a material or procedure from one stage of completion to another.

Process Capability: The minimum spread of a specific measurement variation which will include 99.7% of the measurements from the given process.

Process Chart: A graphic presentation of events occurring during a series of actions or operations, and of information pertaining to those events.

Processing: The act of prescribing the production process to produce a product as designed. This normally includes specifying the tools, equipment, fixtures, machines, and the like.

Production Control: The procedure of planning, routing, scheduling, dispatching, and expediting the flow of materials, parts, subassemblies, and assemblies within the plant from the raw state to the finished product in an orderly and efficient manner.

Production Study: A detailed record, often in the form of a time study or work sampling study, kept of an activity, operation, or group of activities or operations for a period of time to obtain reliable data concerning working time, idle time, downtime, personal time, machine breakdowns, quantity produced, and so on.

Productivity: The actual rate of output or production per unit of time worked.

Program: A sequence of instructions or commands that will cause a computer to perform a prescribed set of data processing functions.

Quality Control: The procedure of establishing acceptable limits of variation in size, weight, finish, and so forth for products or services, and of maintaining the resulting goods or services within these limits.

Quenching: Rapid cooling.

Reliability: The probability that a device will perform satisfactorily in its specified function for a specific period of time, under a given set of operating conditions.

Resistance Welding: Welding with resistance heating and pressure, the work being part of the electrical circuit.

Robot: A reprogrammable, multifunctional manipulator designed to move material, parts, tools, or specialized devices through variable, programmed motions for the performance of a variety of tasks.

Routing: A form listing, for the manufacture of a particular item, the sequence of operations, transportations, storages, and inspections to be used, and usually also the standard times applicable and the machines, equipment, tools, workstations, number of operators, material, parts, and the like that are to be used.

Setup: Making ready or preparing for the performance of a job or operation. Machine setup involves equipping a machine with the appropriate accessories, tools, and fixtures, setting the proper feed, speed, and depth of cut, and so forth. In manual work setup, it is the arrangement, prior to commencing work, of the tools, accessories, component parts, and details involved.

Snapback Method: The procedure of timing, used in making time studies, whereby the stopwatch is read and the watch hand is returned to zero at the termination of each element or work cycle.

Soldering: Similar to brazing, with the filler metal having a melting temperature range below an arbitrary value, generally 800°F. Soft solders usually are lead-tin alloys.

Spot Welding: Welding of lapped parts in which fusion is confined to a relatively small circular area. It is generally resistance welding, but may also be gas-shielded tungsten-art, gas-shielded metal-arc, or submerged arc welding.

Standard Time: The time which is determined to be necessary for a qualified operator, working at a normal pace and experiencing normal fatigue and delays, to perform a defined amount of work, of a specified quality, when following the prescribed method.

Stress Relieving: Heating to a suitable temperature, holding long enough to reduce residual stresses, and then cooling slowly enough to minimize the development of new residual stresses.

Subassembly: Two or more parts joined together to form a unit which is only a part of a complete machine, structure, or product.

Taper Tap: A tap with a chamfer seven to nine threads in length.

Template: A gauge or pattern used as a guide to the size and shape of a part to be made.

Time Formula: A collection of standard time data arranged in the form of an algebraic expression for determining the standard time for an operation.

Time Study: The procedure by which the actual elapsed time for performing an operation or subdivision of elements thereof is determined by the use of a suitable timing device and is recorded. The procedure usually includes adjustments of the actual time as the result of performance rating to derive the time which should be required to perform the task.

Tolerance: A permissible variation in a product or process characteristic, usually shown on a drawing or specification.

Tool Design: That division of mechanical design which specializes in the design of tools. The complete specification of materials, dimensions, tolerances, finishes, heat treatment, and identification for a given tool.

Tungsten-Arc Welding: Inert-gas-shielded arc welding using a tungsten electrode.

Unavoidable Delay: An occurrence that is essentially outside the operator's control or responsibility and that prevents him from doing productive work.

Vapor Degreasing: Degreasing work in vapor over a boiling liquid solvent, the vapor being considerably heavier than air. At least one constituent of the soil must be soluble in the solvent.

Variable Element: An element for the leveled time, under the same methods and working conditions, will change because of the varying characteristics of the parts being worked upon, such as size, weight-shape, density, hardness, viscosity, tolerance requirements, and finish.

Welding: Joining two or more pieces of material by applying heat, pressure, or both, with or without filler material, to produce localized union through fusion or recrystallization across the interface.

Weld Schedule: A record of all welding machine settings plus the identification of the machine for a given material, size, and finish.

Work Cycle: A pattern of motions or processes that is repeated with negligible variation each time an operation is performed. A succession of operations or processes that is repeated with little variation each time a unit or production is completed.

Work Factor: A system of predetermined motion time standards employing the work factor as an index of motion difficulty. The system is used for determining efficient methods and setting performance time standards.

Work Sampling Study: A statistical sampling technique employed to determine the proportion of delays or other activity present in the total work cycle.

Workstation: That section of a production center where the operator performs his assigned tasks, including space required for tools, a workbench, or a machine, with any stands, containers, conveyors, and the like for the material being worked on.

Appendix C
Manufacturing Engineering Data

1. Decimal Equivalents of Drills
2. Decimal Equivalents of Standard Gauges of Sheet Steel and Sheet Aluminum
3. Table of Metal Forming Bend Allowances
4. Tables of Sheet Metal Punching Pressures
5. Standard Machine Shop Formulas and Terms
6. Basic Welding Symbols
7. Number of Feet Per Pound of Lead Wire, Solid Wire Solder, and Tin Wire
8. NAS, AN and MS Screw and Bolt Selection Guide
9. The Statistics of Process Control

Table 1 Decimal Equivalents of Drills

Drill	Diam.	Drill	Diam.	Drill	Diam.	Drill	Diam.
80	.0135	35	.1100	15/64	.2344	Y	.4040
79	.0145	2.8 mm	.1102	6. mm	.2362	13/32	.4062
1/64	.0156	34	.1110	B	.2380	Z	.4130
78	.0160	33	.1130	6.1 mm	.2401	10.5 mm	.4134
77	.0180	2.9 mm	.1142	C	.2420	27/64	.4219
.5 mm	.0197	32	.1160	6.2 mm	.2441	11. mm	.4330
76	.0200	3. mm	.1181	D	.2460	7/16	.4375
75	.0210	31	.1200	6.3 mm	.2480	11.5 mm	.4528
74	.0225	3.1 mm	.1220	E	.2500	29/64	.4531
.6 mm	.0236	1/8	.1250	1/4	.2500	15/32	.4687
73	.0240	3.2 mm	.1260	6.4 mm	.2520	12. mm	.4724
72	.0250	30	.1285	6.5 mm	.2559	31/64	.4843
71	.0260	3.3 mm	.1299	F	.2570	12.5 mm	.4921
.7 mm	.0276	3.4 mm	.1339	6.6 mm	.2598	1/2	.5000
70	.0280	29	.1360	G	.2610	13. mm	.5118
69	.0292	3.5 mm	.1378	6.7 mm	.2638	33/64	.5156
68	.0310	28	.1405	17/64	.2656	17/32	.5312
1/32	.0313	9/64	.1406	H	.2660	13.5 mm	.5315
.8 mm	.0315	3 6 mm	.1417	6.8 mm	.2677	35/64	.5469
67	.0320	27	.1440	6.9 mm	.2716	14. mm	.5512
66	.0330	3.7 mm	.1457	I	.2720	9/16	.5625
65	.0350	26	.1470	7. mm	.2756	14.5 mm	.5709
.9 mm	.0354	25	.1495	J	.2770	37/64	.5781
64	.0360	3.8 mm	.1496	7.1 mm	.2795	15. mm	.5906
63	.0370	24	.1520	K	.2811	19/32	.5937
62	.0380	3.9 mm	.1535	9/32	.2812	39/64	.6094
61	.0390	23	.1540	7 2 mm	.2835	15.5 mm	.6102
1. mm	.0394	5/32	.1562	7.3 mm	.2874	5/8	.6250
60	.0400	22	.1570	L	2900	16. mm	.6299
59	.0410	4. mm	.1575	7.4 mm	.2913	41/64	.6406
58	.0420	21	.1590	M	.2950	16.5 mm	.6496
57	.0430	20	.1610	7.5 mm	.2953	21/32	.6562
1.1 mm	.0433	4.1 mm	.1614	19/64	.2968	17. mm	.6693
56	.0465	4.2 mm	.1654	7 6 mm	.2992	43/64	.6719
3/64	.0469	19	.1660	N	.3020	11/16	.6875
1.2 mm	.0472	4.3 mm	.1693	7.7 mm	.3031	17.5 mm	.6890
1.3 mm	.0512	18	.1695	7 8 mm	.3071	45/64	.7031
55	.0520	11/64	.1719	7.9 mm	.3110	18 mm	.7087
54	.0550	17	.1730	5/16	.3125	23/32	.7187
1.4 mm	.0551	4.4 mm	.1732	8. mm	.3150	18.5 mm	.7283
1.5 mm	.0591	16	.1770	O	.3160	47/64	.7344
53	.0595	4.5 mm	.1771	8.1 mm	.3189	19. mm	.7480
1/16	.0625	15	.1800	8.2 mm	.3228	3/4	.7500
1.6 mm	.0629	4.6 mm	.1811	P	.3230	49/64	.7656
52	.0635	14	.1820	8.3 mm	.3268	19.5 mm	.7677
1.7 mm	.0669	13	.1850	21/64	.3281	25/32	.7812
51	.0670	4.7 mm	.1850	8.4 mm	.3307	20. mm	.7874
50	.0700	3/16	.1875	Q	.3320	51/64	.7969
1.8 mm	.0709	4.8 mm	.1890	8 5 mm	.3346	20.5 mm	.8071
49	.0730	12	.1890	8.6 mm	.3386	13/16	.8125
1.9 mm	.0748	11	.1910	R	.3390	21. mm	.8268
48	.0760	4.9 mm	.1929	8.7 mm	.3425	53/64	.8281
5/64	.0781	10	.1935	11/32	.3437	27/32	.8437
47	.0785	9	.1960	8.8 mm	.3465	21.5 mm	.8465
2. mm	.0787	5. mm	.1968	S	.3480	55/64	.8594
46	.0810	8	.1990	8.9 mm	.3504	22. mm	.8661
45	.0820	5.1 mm	.2008	9. mm	.3543	7/8	.8750
2.1 mm	.0827	7	.2010	T	.3580	22.5 mm	.8858
44	.0860	13/64	.2031	9 1 mm	.3583	57/64	.8906
2.2 mm	.0866	6	.2040	23/64	.3594	23. mm	.9055
43	.0890	5.2 mm	.2047	9 2 mm	.3622	29/32	.9062
2.3 mm	.0905	5	.2055	9 3 mm	.3661	59/64	.9219
42	.0935	5.3 mm	.2087	U	3680	23.5 mm	.9252
3/32	.0937	4	.2090	9 4 mm	.3701	15/16	.9375
2.4 mm	.0945	5.4 mm	.2126	9.5 mm	.3740	24. mm	.9449
41	.0960	3	.2130	3/8	.3750	61/64	.9531
40	.0980	5 5 mm	.2165	V	.3770	24.5 mm	.9646
2.5 mm	.0984	7/32	2187	9.6 mm	.3780	31/32	.9687
39	0995	5 6 mm	.2205	9 7 mm	.3819	25. mm	.9843
38	.1015	2	.2210	9.8 mm	.3858	63/64	.9844
2.6 mm	.1024	5.7 mm	.2244	W	.3860		
37	.1040	1	.2280	9.9 mm	.3898		
2.7 mm	.1063	5.8 mm	.2283	25/64	.3906		
36	.1065	5.9 mm	.2323	10. mm	.3937		
7/64	.1093	A	.2340	X	.3970	1"	1.0000

Table 2 Decimal Equivalents of Standard Gauges of Sheet Steel and Sheet Aluminum

Number of Gauge	ALUMINUM (B & S)	SHEET STEEL (Mfg. Std.)	Number of Gauge	ALUMINUM (B & S)	SHEET STEEL (Mfg. Std.)
	(Thickness in decimal parts of an inch)			(Thickness in decimal parts of an inch)	
000000	.5800	......	23	.0226	.0269
00000	.5165	......	24	.0201	.0239
0000	.4600	......	25	.0179	.0209
000	.4096	......	26	.0159	.0179
00	.3648	......	27	.0142	.0164
0	.3249	......	28	.0126	.0149
1	.2893	......	29	.0113	.0135
2	.2576	......	30	.0100	.0120
3	.2294	.2391	31	.0089	.0105
4	.2043	.2242	32	.0080	.0097
5	.1819	.2092	33	.0071	.0090
6	.1620	.1943	34	.0063	.0082
7	.1443	.1793	35	.0056	.0075
8	.1285	.1644	36	.0050	.0067
9	.1144	.1495	37	.0045	.0064
10	.1019	.1345	38	.0040	.0060
11	.0907	.1196	39	.0035	....
12	.0808	.1046	40	.0031	....
13	.0720	.0897	41	.0028	....
14	.0641	.0747	42	.0025	....
15	.0571	.0673	43	.0022	....
16	.0508	.0598	44	.0020	....
17	.0453	.0538	45	.00176	....
18	.0403	.0478	46	.00157	....
19	.0359	.0418	47	.00140	....
20	.0320	.0359	48	.00124	....
21	.0285	.0329	49	.00099	....
22	.0253	.0299	50	.00088	....

Table 3 Table of Metal Forming Bend Allowances

WORKING FROM THE INSIDE, THE FIGURES SHOWN ARE THE CORRECT BEND ALLOWANCES.
SEE EXAMPLE FOR CLARITY

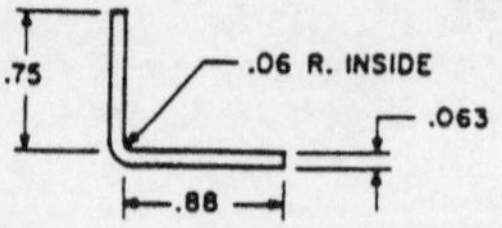

DEVELOPED LENGTH
.750 + .880 + .015 = 1.645

MATERIAL THICKNESS	INSIDE RADIUS OF BEND													
	.016	.030	.048	.063	.094	.125	.156	.188	.219	.250	.312	.375	.437	.500
.016	.004	.003	.010	.016	.029	.043	.056	.070	.083	.097	.124	.150	.177	.204
.020	.007	.000	.007	.014	.027	.041	.054	.067	.080	.094	.121	.148	.175	.201
.025	.010	.004	.004	.010	.023	.037	.051	.064	.077	.091	.118	.144	.171	.198
.030	.013	.007	.001	.007	.020	.034	.047	.061	.074	.088	.115	.141	.168	.195
.032	.014	.008	.000	.006	.019	.033	.046	.059	.073	.086	.113	.140	.167	.194
.036	.017	.010	.003	.003	.016	.030	.044	.057	.070	.084	.111	.137	.164	.191
.040	.020	.013	.006	.001	.014	.027	.040	.054	.067	.081	.108	.135	.161	.188
.045	.022	.015	.009	.002	.011	.025	.038	.052	.065	.078	.105	.132	.159	.186
.048	.025	.018	.011	.004	.008	.022	.035	.049	.062	.076	.103	.130	.156	.183
.050	.027	.020	.013	.007	.007	.020	.033	.047	.060	.074	.101	.127	.154	.181
.060	.032	.026	.019	.013	.001	.014	.028	.041	.054	.068	.095	.122	.148	.175
.063	.036	.029	.022	.015	.003	.011	.025	.038	.052	.065	.092	.119	.145	.172
.075	-	.036	.029	.022	.009	.005	.018	.031	.045	.058	.085	.112	.139	.165
.080	-	.039	.033	.026	.012	.000	.014	.028	.041	.055	.081	.108	.135	.162
.090	-	-	.040	.033	.020	.006	.007	.021	.034	.047	.074	.101	.128	.155
.125	-	-	-	.056	.042	.029	.015	.002	.011	.025	.051	.078	.105	.132
.160	-	-	-	.076	.063	.049	.036	.022	.009	.004	.031	.058	.085	.112
.190	-	-	-	-	.083	.070	.056	.043	.030	.016	.011	.038	.064	.091

NOTE: FIGURES TO THE LEFT OF THE HEAVY LINE ARE TO BE ADDED
FIGURES TO THE RIGHT OF THE HEAVY LINE ARE TO BE SUBTRACTED

Gauge	28	26	24	22	20	18	16	14	12	10	5/32	3/16	1/4
Material thickness	.0149	.0179	.0239	.0299	.0359	.0478	.0598	.0747	.1046	.1345	.1562	.1875	.250
Hole Diameter						PRESSURE IN TONS							
.125	.2	.2	.2	.3	.4	.5	.6	.7	1.0	1.3	1.5	1.8	2.5
.1875	.2	.3	.4	.4	.5	.7	.9	1.1	1.5	2.0	2.3	2.8	3.7
.250	.3	.4	.5	.6	.7	.9	1.2	1.5	2.1	2.6	3.1	3.7	4.9
.3125	.4	.4	.6	.7	.9	1.2	1.5	1.8	2.6	3.3	3.8	4.6	6.1
.375	.4	.5	.7	.9	1.1	1.4	1.8	2.2	3.1	4.0	4.6	5.5	7.4
.4375	.5	.6	.8	1.0	1.2	1.6	2.1	2.6	3.6	4.6	5.4	6.4	8.6
.500	.6	.7	.9	1.2	1.4	1.9	2.3	2.9	4.1	5.3	6.1	7.4	9.8
.5625	.7	.8	1.1	1.3	1.6	2.1	2.6	3.3	4.6	5.9	6.9	8.3	11.1
.625	.7	.9	1.2	1.5	1.8	2.4	2.9	3.7	5.1	6.6	7.7	9.2	12.3
.6875	.8	1.0	1.3	1.6	1.9	2.6	3.2	4.0	5.7	7.3	8.4	10.1	13.5
.750	.9	1.1	1.4	1.8	2.1	2.8	3.5	4.4	6.2	7.9	9.2	11.0	14.7
.8125	1.0	1.1	1.5	1.9	2.3	3.1	3.8	4.8	6.7	8.6	10.0	12.0	16.0
.875	1.1	1.2	1.6	2.1	2.5	3.3	4.1	5.1	7.2	9.2	10.7	12.9	17.2
.9375	1.2	1.3	1.8	2.2	2.6	3.5	4.4	5.5	7.7	9.9	11.5	13.8	18.4
1.0000	1.3	1.4	1.9	2.4	2.8	3.8	4.7	5.9	8.2	10.6	12.3	14.7	19.6

General Information:

Formula determining punching pressure for round holes in mild sheet steel: D × Thickness × 25 = Pressure in Tons.

Formula for determining blanking pressure in mild sheet steel: Shear Length × Thickness × 25 = Pressure in Tons.

Formula for stripping pressures: Shear Length × Thickness × 3500 = Pressure in Pounds.

Table 4b Approximate Pressures Required for Punching Round Holes in Aluminum 5052-H32 1/4 HD Sheet Metal

Gauge	28	26	24	22	20	18	16	14	12	10	5/32	3/16	1/4
Material thickness	.0149	.0179	.0239	.0299	.0359	.0478	.0598	.0747	.1046	.1345	.1562	.1875	.250
Hole Diameter	PRESSURE IN TONS												
.125	.1	.1	.1	.2	.2	.3	.4	.4	.6	.8	.9	1.1	1.5
.1875	.1	.2	.2	.2	.3	.4	.5	.7	.9	1.2	1.4	1.7	2.2
.250	.2	.2	.3	.4	.4	.5	.7	.9	1.3	1.6	1.9	2.2	2.9
.3125	.2	.2	.4	.4	.5	.7	.9	1.1	1.6	2.0	2.3	2.8	3.7
.375	.2	.3	.4	.5	.7	.8	1.1	1.3	1.9	2.4	2.8	3.3	4.4
.4375	.3	.4	.5	.6	.7	1.0	1.3	1.6	2.2	2.8	3.2	3.8	5.2
.500	.4	.4	.5	.7	.8	1.1	1.4	1.7	2.5	3.2	3.7	4.4	5.9
.5625	.4	.5	.7	.8	1.0	1.3	1.6	2.0	2.8	3.5	4.1	5.0	6.7
.625	.4	.5	.7	.9	1.1	1.4	1.7	2.2	3.1	4.0	4.6	5.5	7.4
.6875	.5	.6	.8	1.0	1.1	1.6	1.9	2.4	3.4	4.4	5.0	6.1	8.1
.750	.5	.7	.8	1.1	1.3	1.7	2.1	2.6	3.7	4.7	5.5	6.6	8.8
.8125	.6	.7	.9	1.1	1.4	1.9	2.3	2.9	4.0	5.2	6.0	7.2	9.6
.875	.7	.7	1.0	1.3	1.5	2.0	2.5	3.1	4.3	5.5	6.4	7.7	10.3
.9375	.7	.8	1.1	1.3	1.6	2.1	2.6	3.3	4.6	5.9	6.9	8.3	11.0
1.0000	.8	.8	1.1	1.4	1.7	2.3	2.8	3.5	4.9	6.4	7.4	8.8	11.8

Table 4c Approximate Pressures Required for Punching Round Holes in Stainless Steel 302 Sheet Metal

Gauge	28	26	24	22	20	18	16	14	12	10	5/32	3/16	1/4
Material thickness	.0149	.0179	.0239	.0299	.0359	.0478	.0598	.0747	.1046	.1345	.1562	.1875	.250
Hole Diameter	PRESSURE IN TONS												
.125	.3	.3	.3	.5	.6	.8	1.0	1.1	1.6	2.1	2.4	2.9	4.0
.1875	.3	.5	.6	.6	.8	1.1	1.4	1.8	2.4	3.2	3.7	4.5	5.9
.250	.5	.6	.8	1.0	1.1	1.4	1.9	2.4	3.4	4.2	5.0	5.9	7.8
.3125	.6	.6	1.0	1.1	1.4	1.8	2.4	2.9	4.2	5.3	6.1	7.4	9.9
.375	.6	.8	1.1	1.4	1.8	2.2	2.9	3.5	5.0	6.4	7.4	8.8	11.9
.4375	.8	1.0	1.3	1.6	1.9	2.6	3.4	4.2	5.8	7.4	8.6	10.2	13.9
.500	1.0	1.1	1.4	1.9	2.2	3.0	3.7	4.6	6.6	8.5	9.8	11.8	15.8
.5625	1.1	1.3	1.8	2.1	2.6	3.4	4.2	5.3	7.5	9.4	11.0	13.3	17.9
.625	1.1	1.4	1.9	2.4	2.9	3.8	4.6	5.9	8.3	10.6	12.3	14.7	19.9
.6875	1.3	1.6	2.1	2.6	3.0	4.2	5.1	6.4	9.2	11.7	13.4	16.2	21.7
.750	1.4	1.8	2.2	2.9	3.4	4.5	5.6	7.0	10.0	12.3	14.7	17.6	23.6
.8125	1.6	1.8	2.4	3.0	3.7	5.0	6.1	7.7	10.8	13.8	16.0	19.2	25.7
.875	1.8	1.9	2.6	3.4	4.0	5.3	6.6	8.2	11.6	14.7	17.1	20.6	27.6
.9375	1.9	2.1	2.9	3.5	4.2	5.6	7.0	8.8	12.4	15.9	18.4	22.1	29.5
1.0000	2.1	2.2	3.0	3.8	4.5	6.1	7.5	9.4	13.2	17.0	19.7	23.5	31.5

Table 5a Standard Machine Shop Formulas

Parameter	Turning	Milling	Drilling
1. Cutting speed (FMP)	$S_c = .262 \times D_t \times RPM$	$S_c = .262 \times D_m \times RPM$	$S_c = .262 \times D_d \times RPM$
2. Revolutions/Minute (RPM)	$RPM = 3.82 \times \frac{S_c}{D_t}$	$RPM = 3.82 \times \frac{S_c}{D_m}$	$RPM = 3.82 \times \frac{S_c}{D_d}$
3. Feed Rate (Inches/Min.)	$F_m = F_r \times RPM$	$F_m = Ft \times N \times RPM$	$F_m = F_r \times RPM$
4. Feed Per Tooth	—	$F_t = \frac{F_m}{N \times RPM}$	—
5. Cutting Time (Minutes)	$t = \frac{L}{F_m}$	$t = \frac{L}{F_m}$	$t = \frac{L}{F_m}$
6. Rate of Metal Removal (Inches³/Min.)	$R = 12 \times d \times F_r \times S_c$	$R = W \times d \times F_m$	$R = \frac{D^2 d}{4} \times F_m$
7. Horsepower Required at Spindle	$HP_s = R \times P$	$HP_s = R \times P$	$HP_s = R \times P$
8. Horsepower Required at Motor	$HP_m = \frac{R \times P}{E}$	$HP_m = \frac{R \times P}{E}$	$HP_m = \frac{R \times P}{E}$
9. Torque at Spindle (Inch-Pounds)	$T_s = \frac{63030\ HP_s}{RPM}$	$T_s = \frac{63030\ HP_s}{RPM}$	$T_s = \frac{63030\ HP_s}{RPM}$

Table 5b Machine Shop Formula Terms

d	Depth of cut, inches
D_t	Diameter of workpiece in thrning, inches
D_m	Diameter of milling cutter, inches
D_d	Diameter of drill, inches
E	Efficiency of spindle drive
F_m	Feed rate, inches per minute
F_r	Feed, inches per revolution
F_t	Feed, inches per tooth
HP_m	Horsepower at motor
HP_s	Horsepower at spindle
L	Length of cut, inches
N	Number of teeth in cutter
P	Unit power, horsepower per cubic inch per minute
R	Rate of metal removal, cubic inches per minute
RPM	Revolutions per minute of work or cutter
S_c	Cutting speed, feet per minute
t	Cutting time, minutes
T_s	Torque at spindle, inch-pounds
W	Width of workpiece, inches

Table 6 Basic Welding Symbols

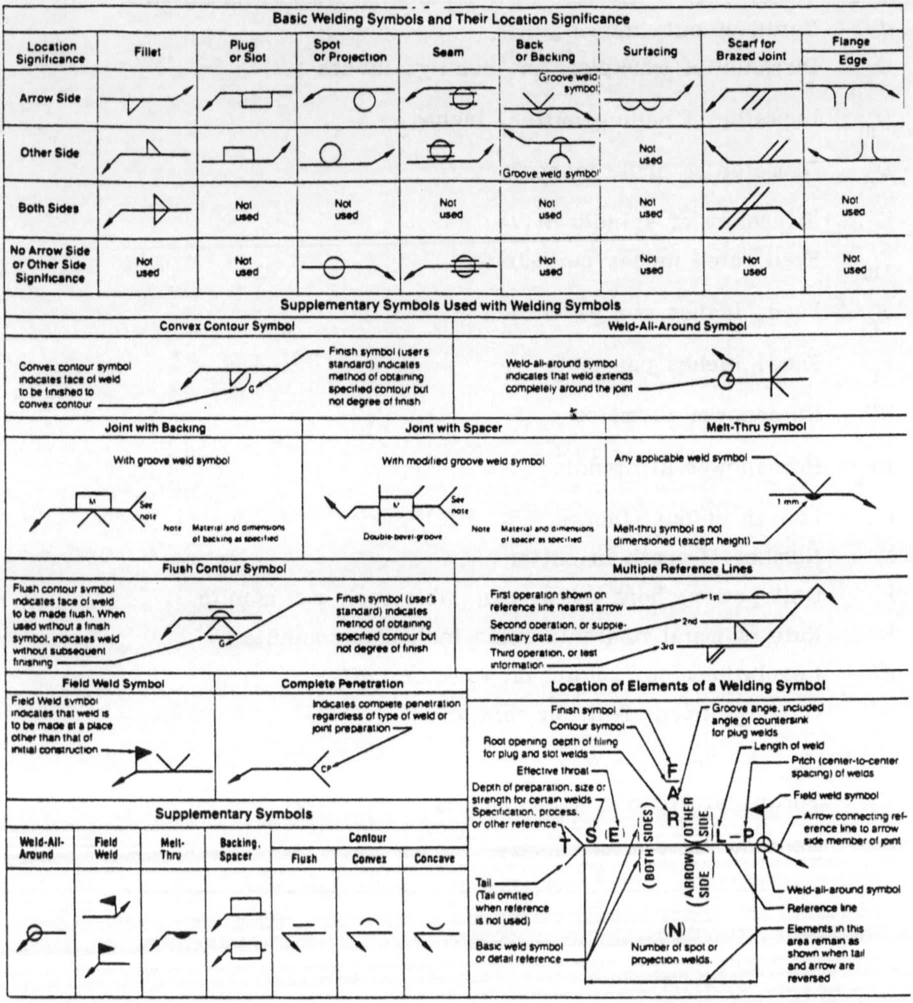

Table 7 Number of Feet per Pound of Lead Wire, Solid Wire Solder, and Tin Wire

Diameter Inch	Area Sq. Inch	Lead Wire Ft. per Lb.	30/70 Solder Ft. per Lb.	40/60 Solder Ft. per Lb.	50/50 Solder Ft. per Lb.	60/40 Solder Ft. per Lb.	Tin Wire Ft. per Lb.
.032	.00080	254	297	310	325	340	396
.036	.00113	180	211	220	230	241	281
.040	.00126	161	188	196	206	216	251
.045	.00159	128	150	156	164	172	200
.050	.00196	104	122	127	133	129	162
.056	.00246	82.6	96.6	101	106	111	129
.063	.00312	65.1	76.2	79.4	83.3	87.2	102
.071	.00396	51.3	60.0	62.6	65.7	68.7	80.0
.080	.00503	40.0	47.3	49.3	51.7	54.1	63.0
.090	.00636	32.0	37.4	39.0	41.0	42.9	49.9
.100	.00785	25.9	30.3	31.6	33.2	34.7	40.4
.112	.00985	20.6	24.1	25.1	26.4	27.6	32.1
.125	.01227	16.6	19.4	20.2	21.2	22.2	25.9
.140	.01539	13.2	15.4	16.1	16.9	17.7	20.6
.160	.02011	10.1	11.8	12.3	12.9	13.5	15.8
.180	.02545	8.0	9.4	9.8	10.2	10.7	12.5
.200	.03142	6.5	7.6	7.9	8.3	8.7	10.1

Note: The number of linear feet per lb. of flux-core wire solder will be somewhat greater than the above figures for solid wire. Where the amount of flux is 1.1% by weight, linear footage is increased by 9% over solid wire, with 3.2% the increase is 15%, and with 3.3% the increase is 27%. (Courtesy National Lead Co.)

Table 8 NAS, AN and MS Screw and Bolt Selection Guide

NAS 144 thru NAS 158
NAS 172, 174 and 176
Internal Wrenching Bolt
160,000 psi Min. T.S

NAS 220 thru NAS 224
Screw, Brazier Head
Phillips Recess, Alum., Bronze,
Alloy Steel, CRES

NAS 333 thru NAS 340
100° Flush Head Bolt
Phillips Recess
95,000 psi Min S.S.*

NAS 428
Crown Hex Head Bolt
125,000 psi Min. T.S.

NAS 464
Hex Head Bolt
95,000 psi Min. S.S.

NAS 501
Stainless Steel
Hex Head Bolt
75,000 psi Min. T.S.

NAS 514
Screw, Mach 100° Flat Head
Full Threaded, Alloy Steel
125,000 psi Min. T.S.

NAS 517
100° Flush Head Bolt
95,000 psi Min. S.S.

NAS 560
Screw—Hi Temp
100° Flush Head
321, A286 or Inconel "X"

NAS 563-572
Hex Head Bolt
160,000 psi Min. T.S

NAS 600 thru NAS 606
Screw, Mach Pan Head, Phillips,
Full Threaded, Alloy Steel
160,000 psi Min. T.S.

NAS 608-NAS 609
Std. Socket Head Cap Screw

NAS 610 thru NAS 616
Pan Head Screw
Reed & Prince Recess
160,000 psi Min. T.S.

NAS 623
Pan Head Screw
Phillips Recess
160,000 psi Min. T.S

Table 8 (Continued)

NAS 624 thru NAS 644
12 Point Bolt
180,000 psi Min T S

NAS 653 thru NAS 658
Hex Head Bolt, Titanium
95,000 psi Min S S

NAS 663 thru NAS 668
Flush Head Bolt, Titanium
95,000 psi Min. S.S

NAS 673 thru NAS 678
Hex Head Bolt, Titanium
95,000 psi Min. S.S

NAS 1003 thru NAS 1020
Hex Head Bolt
140,000 psi Min. T.S.

NAS 1083 thru NAS 1088
100° Flush Head Bolt, Titanium
Phillips Recess
95,000 psi Min. S.S

NAS 1096
Screw, Hex Head Recessed
Full Thread
125,000 psi Min. T.S

NAS 1103 thru NAS 1120
Hex Head Bolt
95,000 psi Min S S

NAS 1202 thru NAS 1210
Flush Head Bolt
95,000 psi Min. S.S.

NAS 1297
Hex Head Shoulder Bolt
125,000 psi Min. T.S.

NAS 1298
Braxier Head Shoulder Screw
Phillips Recess
125,000 psi Min. T.S.

NAS 1299
100° Flat Head Shoulder Screw
Phillips Recess
125,000 psi Min. T.S.

NAS 1303 thru NAS 1320
Hex Head Bolt
160,000 psi Min. T.S.

NAS 1402 thru NAS 1406
Pan Head Screw, Phillips
160,000 psi Min. T.S.

NAS 1603 thru NAS 1610
.0312 Oversize Shank
100° Flush Head, Phillips Recess
160,000 psi Min. T.S.

NAS 1703 thru NAS 1710
.0156 Oversize Shank,
100° Flush Head, Phillips Recess
160,000 psi Min T.S.

NAS 2903 thru NAS 2920
.0156 Oversize Shank
Hex Head Bolt
160,000 psi Min. T.S

NAS 3003 thru NAS 3020
.0312 Oversize Shank
Hex Head Bolt
160,000 psi Min T.S

AN

AN 3 thru AN 20
Hex Head Bolt
Steel & Stainless Steel
125,000 psi Min T S
Aluminum
62,000 psi Min T.S.

AN 73 thru AN 81
Hex Head Bolt
125,000 psi Min. T S

AN 173 thru AN 186
Hex Head Bolt, Close Tol Shank
Alloy & Stainless Steel & Aluminum

AN 500 thru AN 501
Screw, Mach Fillister Head, Slotted
Carbon Steel, CRES, Brass

AN 502 thru AN 503
Screw, Mach. Drilled Fillister, Slotted
Alloy Steel
125,000 psi Min T.S

AN 504
Screw, Mach Tapping, Thread Cutting,
Round Head, Carbon Steel and CRES;
Slotted and Recessed

AN 505
Screw, Mach Flat Head 82°
Carbon Steel, Brass, CRES Alum.l

AN 506
Screw, Mach. Tapping, Thread Cutting
Flat Head 82°
Carbon Steel, CRES; Slotted & Recessed

AN 507
Screw, Mach Flat Head 100°
Carbon Steel, CRES, Brass, Alum.,
Slotted & Recessed

AN 509
Screw Mach Flat Head 100°
Alloy Steel, Alum., CRES, Recessed

AN 510
Screw, Mach Flat Head 82°
Carbon Steel, CRES, Brass, Alum.,
Recessed

AN 515 and AN 520
Screw, Mach Round Head
Carbon Steel, Brass, CRES, Alum.,
Recessed

AN 525
Screw Mach. Washer Head
Steel, Alum.; Slotted & Recessed

AN 526
Screw, Mach Truss Head
Carbon Steel, CRES, Alum.,
Recessed

AN 530
Screw, Tapping Thread Forming or
Thread Cutting, Round Head, Spaced Thd.,
Recessed, Carbon or CRES Steel

Table 8 (Continued)

AN 531
Screw, Tapping Thread Forming or Thread Cutting 82° Flat Head, Spaced Thd. Recessed, Carbon or CRES Steel

AN 101001 thru AN 108200
Hex Head Bolt
Alloy & CRES Steel
125,000 psi Min T S

AN 148551 thru AN 149350
Internal Wrenching Bolt
140,000 psi Min T S

MS

MS 9033 thru MS 9039
MS 9060 thru MS 9066
12 Point Bolt-A286-1200°
130,000 psi Min. T.S.

MS 9088 thru MS 9094
12 Point Bolt-Steel
125,000 psi Min T.S.

MS 9122 thru MS 9123
Slotted Hex Head Mach. Screw
125,000 psi Min. T.S.

MS 9146 thru MS 9152
MS 9157 thru MS 9163
MS 9169 thru MS 9175
12 Point Bolt
125,000 psi Min. T.S.

MS 9177 thru MS 9178
12 Point Bolt, A286-1200°
130,000 psi Min. T.S.

MS 9183 thru MS 9186
MS 9189 thru MS 9192
12 Point Bolt-Steel
125,000 psi Min. T.S.

MS 9187 thru MS 9188
12 Point Bolt-A286-1200°
130,000 psi Min. T.S.

MS 9224
12 Point Bolt-A286-1200°
130,000 psi Min. T.S.

MS 9316 thru MS 9317
Slotted Hex Head Mach. Screw
140,000 psi Min. T.S.

MS 16219
Flat Countersunk Head, Slotted, Nonmagnetic, CRES Mach. Screw

MS 16200
Pan Head Slotted CRES Mach Screw

MS 16637 thru 16638
Screw Shoulder Socket Head, Hex Alloy Steel, uncoated, Cad. or Zinc

MS 20004 thru MS 20024
Internal Wrenching Bolt
160,000 psi Min. T.S.

MS 20033 thru MS 20046
Hex Head Bolt-1200°
110,000 psi Min T S

MS 20073 thru MS 20074
Hex Head Bolt
125,000 psi Min T S

MS 21250
12 Point Bolt
180,000 psi Min T S

MS 24583
Screw, Mach Flat Countersunk Cross Recessed, Carbon Steel, Cadmium

MS 24584
Screw Mach Pan Head Cross Recessed, Carbon Steel, Cadmium

MS24615 thru MS 24616
Screw, Tapping, Thread Forming, Type A, Flat Countersunk, Cross Recessed, Carbon Steel, Cad. Plated or CRES

MS 24617 thru MS 24618
Screw, Tapping, Thread Forming, Type A, Pan Head, Cross Recessed, Carbon Steel, Cad. Plated or CRES

MS 24619 thru MS 24620
Screw, Tapping, Thread Forming, Type B, Flat Countersunk, Cross Recessed, Carbon Steel, Cad. Plated or CRES

MS 24621 thru MS 24622
Pan Head, Self Tapping, Thread Forming, Cross Recess, Type B, Carbon Steel, Cad. Plated or CRES

MS 24623 thru MS 24624
Flat Head, Self Tapping, Thread Cutting, Cross Recess, Type BF, BG or BT; Carbon Steel, Cad. Plated or CRES

MS 24625 thru MS 24626
Pan Head, Self Tapping, Thread Cutting, Cross Recess, Type BF, BG or BT; Carbon Steel, Cad. Plated or CRES

MS 24627 thru MS 24628
Flat Head, Self Tapping, Thread Cutting, Cross Recess, Type D, F, G or T; Carbon Steel, Cad Plated or CRES

MS 24629 thru MS 24630
Pan Head, Self Tapping, Thread Cutting, Cross Recess, Type D, F, G or T; Carbon Steel, Cad, Plated or CRES

MS 24635 thru MS 24636
Flat Head, Self-Tapping, Thread Forming Slotted, Type A, Carbon Steel, Cad. Plated or CRES

MS 24637 thru MS 24638
Pan Head, Self Tapping, Thread Forming, Slotted, Type A, Carbon Steel, Cad. Plated or CRES

MS 24639 thru MS 24640
Flat Head, Self Tapping, Thread Forming, Slotted, Type B, Carbon Steel, Cad: Plated or CRES

Table 8 (Continued)

MS 24641 thru MS 24642
Pan Head, Self Tapping Thread Forming, Slotted Type B, Carbon Steel, Cad. Plated or CRES

MS 24643 thru MS 24644
Flat Head, Self Tapping, Thread Cutting, Spaced Threads, Slotted, Type BF, BG or BT, Carbon Steel, Cad. Plated or CRES

MS 24645 thru MS 24646
Pan Head, Self Tapping, Thread Cutting Spaced Threads, Slotted, Type BF, BG or BT, Carbon Steel, Cad. Plated or CRES

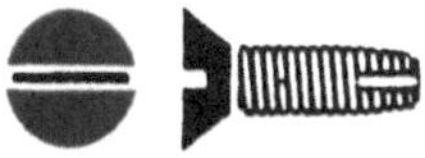

MS 24647 thru MS 24648
Flat Head, Self Tapping, Thread Cutting, Slotted, Type D, F, G, or T;
Carbon Steel, Cad Plated or CRES

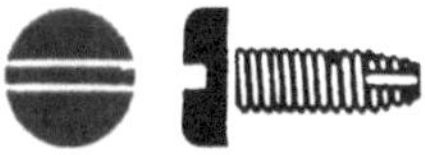

MS 24649 thru MS 24650
Pan Head, Self Tapping, Thread Cutting, Slotted, Type D, F, G or T, Carbon Steel, Cad. Plated or CRES

MS25087 A
Screw, Externally Relieved Body

MS 35188 thru MS 35203
Flat Head Machine Screw, Cross Recess: Steel, Brass, Alum., CRES: Plain, Cadmium or Zinc Plated: Phosphate, Black Oxide, Anodized or Passivated

MS 35204 thru MS 35219
Pan Head Machine Screw, Cross Recess: Steel, Brass, Alum., CRES: Plain, Cadmium or Zinc Plated, Phosphate, Black Oxide, Anodized or Passivated

MS 35221 thru MS 35236
Pan Head Machine Screw, Slotted: Steel, Brass, Alum., CRES: Plain, Cadmium or Zinc Plated: Phosphate, Black Oxide, Anodized or Passivated

MS 35237 thru MS 35251
and MS 35262
Flat Head Machine Screw, Slotted; Steel, Brass, Alum., CRES; Plain, Cadmium or Zinc Plated: Phosphate, Black Oxide, Anodized or Passivated

MS 35263 thru MS 35278
Fillister Head Machine Screw, Drilled, Slotted; Steel, Brass, Alum., CRES; Plain, Cadmium or Zinc Plated; Phosphate, Black Oxide, Anodized or Passivated

MS 35455 thru MS 35458 and
MS 35459 thru MS 35461
Socket Head Cap Screws, Alloy Steel and CRES, uncoated, Cadmium or Zinc Plated: Phosphate Treated or Passivated, etc.

Table 9 The Statistics of Process Control

CONTROL CHARTS FOR VARIABLES

Calculate the Average ($\overline{X}$) and Range (R) of each subgroup

$$\overline{X} = \frac{X_1 + X_2 + \ldots + X_n}{n}$$

$$R = X_{max} - X_{min}$$

Plot $\overline{X}$ and R

where: $\overline{X}$ = Average of measurements in each subgroup
R = Range of measurements in each subgroup
n = Number of measurements per subgroup

Calculate the Control Limits

$UCL_x = \overline{\overline{X}} + A_2\overline{R}$ $\quad UCL_R = D_4\overline{R}$

$LCL_x = \overline{\overline{X}} - A_2\overline{R}$ $\quad LCL_R = D_3\overline{R}$

where: $\overline{\overline{X}} = \dfrac{\overline{X}_1 + \overline{X}_2 + \ldots + \overline{X}_k}{k}$

$\overline{R} = \dfrac{R_1 + R_2 + \ldots + R_k}{k}$

$\overline{\overline{X}}$ is the Process Average and $\overline{R}$ is the Range Average for "k" subgroups

Factors

n	3	4	5	6	7	8
D_4	2.57	2.28	2.11	2.04	1.92	1.86
D_3	0	0	0	0	.076	.136
A_2	1.02	.73	.58	.48	.42	.37

← VARIABLES DATA

- FOR CONTROL OF INDIVIDUAL CHARACTERISTICS — USE $\overline{X}$ AND R CHARTS

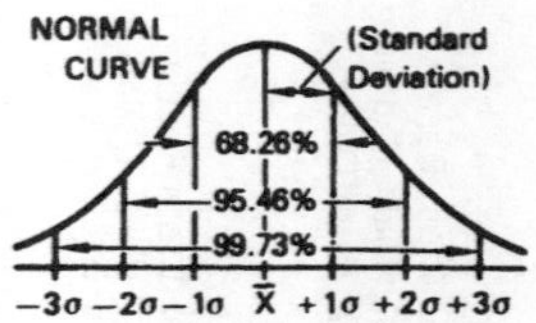

ATTRIBUTE DATA →

- FOR CONTROL OF OVERALL FRACTION DEFECTIVE — USE p CHARTS
- FOR CONTROL OF OVERALL NUMBER OF DEFECTS — USE c CHARTS
- FOR CONTROL OF OVERALL NUMBER OF DEFECTS PER UNIT — USE u CHARTS

CONTROL CHARTS FOR ATTRIBUTES

p Chart — % Defective

Plot p — number of defectives in each subgroup divided by number inspected in each subgroup (n)

Calculate the Control Limits

$$UCL_p = \overline{p} + \frac{3\sqrt{\overline{p}(1-\overline{p})}}{\sqrt{n}}$$

$$LCL_p = \overline{p} - \frac{3\sqrt{\overline{p}(1-\overline{p})}}{\sqrt{n}}$$

where: $\overline{p} = \dfrac{\text{total number of defectives}}{\text{total number inspected}}$

c Chart — Number of Defects

Plot c — Defect count in a subgroup

Calculate the Control Limits

$$UCL_c = \overline{c} + 3\sqrt{\overline{c}}$$

$$LCL_c = \overline{c} - 3\sqrt{\overline{c}}$$

where: $\overline{c} = \dfrac{\text{total number of defects}}{\text{total number of subgroups}}$

u Chart — Defects per unit

Plot u = Total defects in each subgroup divided by number units in subgroup (n)

Calculate the Control Limits

$UCL_u = \overline{u} + \dfrac{3\sqrt{\overline{u}}}{\sqrt{n}}$ $\quad LCL_u = \overline{u} - \dfrac{3\sqrt{\overline{u}}}{\sqrt{n}}$

where: $\overline{u} = \dfrac{\text{total number of defects}}{\text{total number inspected}}$

Index